Advances in Mathematical and Computational Modeling of Engineering Systems

The text covers a wide range of topics such as mathematical modeling of crop pest control management, water resources management, impact of anthropogenic activities on atmospheric carbon dioxide concentrations, impact of climate changes on melting of glaciers and polar bear populations, dynamics of slow–fast predator-prey system, and spread and control of HIV epidemic. It emphasizes the use of mathematical modeling to investigate the fluid flow problems including the breaking of viscoelastic jet, instability arising in nanofiber, flow in an annulus channel, and thermal instability in nano-fluids in a comprehensive manner. This book will be a readily accessible source of information for students, researchers, and policymakers interested in the application of mathematical and computational modeling techniques to investigate various biological and engineering phenomena.

Features

- Focuses on the current modeling and computational trends to investigate various ecological, epidemiological, and engineering systems.
- Presents the mathematical modeling of a wide range of ecological and environmental issues including crop pest control management, water resources management, the effect of anthropogenic activities on atmospheric carbon dioxide concentrations, and impact of climate changes on melting of glaciers and polar bear population.
- Covers a wide range of topics including the breaking of viscoelastic jet, instability arising in nanofiber, flow in an annulus channel, and thermal instability in nano-fluids.
- Examines evolutionary models i.e., models of time-varying processes. Highlights the recent developments in the analytical methods to investigate the nonlinear dynamical systems.
- Showcases diversified applications of computational techniques to solve practical biological and engineering problems.

The book focuses on the recent research developments in the mathematical modeling and scientific computing of biological and engineering systems. It will serve as an ideal reference text for senior undergraduate, graduate students, and researchers in diverse fields including ecological engineering, environmental engineering, computer engineering, mechanical engineering, mathematics, and fluid dynamics.

Smart Technologies for Engineers and Scientists

Series Editor:
Mangey Ram

The aim of this new book series is to publish the research studies and articles that bring up the latest development and research applied to mathematics and its applications in the manufacturing and management sciences areas. Mathematical tool and techniques are the strength of engineering sciences. They form the common foundation of all novel disciplines as engineering evolves and develops. The series will include a comprehensive range of applied mathematics and its application in engineering areas such as optimization techniques, mathematical modeling and simulation, stochastic processes and systems engineering, safety-critical system performance, system safety, system security, high assurance software architecture and design, mathematical modeling in environmental safety sciences, finite element methods, differential equations, reliability engineering, etc.

For more information about this series, please visit: www.routledge.com/Mathematical-Engineering-Manufacturing-and-Management-Sciences/book-series/CRCMEMMS

Advances in Mathematical and Computational Modeling of Engineering Systems

Edited by
Mukesh Kumar Awasthi, Maitri Verma
and Mangey Ram

CRC Press
Taylor & Francis Group
Boca Raton London New York

CRC Press is an imprint of the
Taylor & Francis Group, an **informa** business

First edition published 2023
by CRC Press
6000 Broken Sound Parkway NW, Suite 300, Boca Raton, FL 33487-2742

and by CRC Press
4 Park Square, Milton Park, Abingdon, Oxon, OX14 4RN

CRC Press is an imprint of Taylor & Francis Group, LLC

Library of Congress Cataloging-in-Publication Data
Names: Awasthi, Mukesh Kumar, editor. | Verma, Maitri, editor. |
Ram, Mangey, editor.
Title: Advances in mathematical and computational modeling of engineering systems / edited by Mukesh Kumar Awasthi, Maitri Verma, and Mangey Ram.
Description: First edition. | Boca Raton : CRC Press, [2023] |
Series: Mathematical engineering, manufacturing, and management sciences |
Includes bibliographica l references and index.
Identifiers: LCCN 2022040818 (print) | LCCN 2022040819 (ebook) |
ISBN 9781032392912 (hbk) | ISBN 9781032434599 (pbk) | ISBN 9781003367420 (ebk)
Subjects: LCSH: Systems engineering–Mathematics. | Engineering models. |
Biological systems–Mathematical models.
Classification: LCC TA168 .A288 2023 (print) | LCC TA168 (ebook) |
DDC 620.001/1–dc23/eng/20221110
LC record available at https://lccn.loc.gov/2022040818
LC ebook record available at https://lccn.loc.gov/2022040819

ISBN: 9781032392912 (hbk)
ISBN: 9781032434599 (pbk)
ISBN: 9781003367420 (ebk)

DOI: 10.1201/9781003367420

Typeset in Sabon
by codeMantra

Contents

Preface

The contemporary world is moving fast into the knowledge and innovation of science and technology. Mathematical models play a very crucial role in understanding various complex processes arising in natural and engineering sciences. The last few decades witnessed a rapid increase in the use of mathematical models in the comprehension of engineering systems. The researchers working in various areas of science and technology have successfully used mathematical and computational modeling techniques to address various crucial phenomena facing mankind, such as the emergence and re-emergence of infectious diseases, adverse impacts of climate changes on the ecosystem, water resources planning and management, crop pest control management, to name a few.

The analytical and computational techniques used to simulate and study complex mathematical models have evolved significantly in recent years, which has widened the applicability of mathematical models in the investigation of real-world problems. Although mathematical models find application in nearly every area of engineering science, they are most extensively applied in basic scientific research in the areas of mathematical ecology, epidemiology, environmental systems, and the study of fluid flow problems. One of the basic aims of the book is to cover the recent developments in mathematical modeling in these areas of research. The articles presented in the book are aimed to provide the readers a comprehensive insight into the recent developments in various aspects of modeling and simulation of engineering systems from diverse areas of engineering, such as mechanical engineering, computer science engineering, biological engineering, environmental engineering, and so on.

This book will serve as a reference for researchers, academicians, engineering professionals, policymakers, and graduate students interested in the modeling and computation of applied engineering problems. This book, for sure, equips the readers with the tools and techniques used for the formulation of a mathematical framework that bridges the gap between the core theory and experimentation in different areas of engineering sciences. A special emphasis of the book is on the mathematical modeling of the spread and control of some of the most burning issues of the current time, such as the control of the HIV pandemic, crop-pest management, rainfall-runoff modeling, and emission and impacts of global warming gases. This book is also intended to make the readers aware of the diversified applications of computational techniques to solve practical engineering problems, including analysis of Richtmyer–Meshkov Instability in shock-refrigerant bubble interaction, numerical investigation of wave pattern evolution in the Gray-Scott Model, investigation of heat and mass transfer in the convective flow of nanofluid, assessment of Blasius boundary layer flow for a parallel duct, etc.

We wish that this book aids the readers to develop a scientific understanding to model the complex systems in various areas of engineering and also make them aware of the recent advancements in the computation techniques used to analyze the mathematical models.

Dr. Mukesh Kumar Awasthi
Dr. Maitri Verma
Dr. Mangey Ram

Editors

Dr. **Mukesh Kumar Awasthi** received his PhD on the topic "Viscous Correction for the Potential Flow Analysis of Capillary and Kelvin-Helmholtz instability". He is working as an Assistant Professor in the Department of Mathematics at Babasaheb Bhimrao Ambedkar University, Lucknow. Dr. Awasthi is specialized in the mathematical modeling of flow problems. He has taught courses of Fluid Mechanics, Discrete Mathematics, Partial differential equations, Abstract Algebra, Mathematical Methods, and Measure theory to postgraduate students. He has acquired excellent knowledge in the mathematical modeling of flow problems, and he can solve these problems analytically as well as numerically. He has a good grasp of the subjects such as viscous potential flow, electro-hydrodynamics, magneto-hydrodynamics, heat, and mass transfer. He has excellent communication skills and leadership qualities. He is self-motivated and responds to suggestions in a more convincing manner.

Dr. **Awasthi** was qualified for National Eligibility Test (NET) conducted on all India level in 2008 by the Council of Scientific and Industrial Research (CSIR) and received Junior Research Fellowship (JRF) and Senior Research Fellowship (SRF) for doing research. He has published 115 plus research publications (journal articles/books/book chapters/conference articles) in Elsevier, Taylor & Francis, Springer, Emerald, World Scientific, and many other national and international journals and conferences. Also, he has published seven books. He has attended many symposia, workshops, and conferences in mathematics as well as fluid mechanics. He received the "Research Awards" consecutively four times from 2013 to 2016 by the University of Petroleum and Energy Studies, Dehradun, India. He also received the start-up research fund for his project "Nonlinear study of the interface in multilayer fluid system" from UGC, New Delhi.

Dr. Maitri Verma received her PhD in Mathematics from Banaras Hindu University, Varanasi, India on the topic "Mathematical modeling and analysis of greenhouse gases and their control". She is working as an Assistant Professor in the Department of Mathematics at Babasaheb Bhimrao Ambedkar University, Lucknow. Dr. Verma is specialized in mathematical modeling of ecological, environmental, and engineering systems. She is actively working in the modeling areas of mathematical modeling and analysis of emission and control of greenhouse gases, modeling of the dynamics of predator-prey systems, dynamics of computer virus in a network, etc. She has 14 research papers to her credit in the journals of international repute. She has taught courses of Dynamical Systems, Mathematical modeling, Ordinary differential equations, Real analysis, Linear Algebra, Functional analysis, and Probability and probability distributions to postgraduate students. Dr. Verma is a life member of the Indian Mathematical Society, Indian Science Congress, and Biomathematical Society of India.

Dr. Verma was qualified for National Eligibility Test (NET) conducted by the Council of Scientific and Industrial Research (CSIR), India, in years 2009 and 2010 and availed Junior Research Fellowship (JRF) and Senior Research Fellowship (SRF) during the PhD program. She has been awarded a postdoctoral fellowship from the National Board of Higher Mathematics, Department of Atomic Energy, Mumbai, India in 2015. She has attended many workshops and conferences and won the best paper presentation award two times. She is involved as a reviewer in many international journals. She has received the start-up research grant for her project entitled "Causes and impacts of global warming: A mathematical study" from the University Grants Commission, New Delhi, India.

Prof. Mangey Ram received the PhD degree major in Mathematics and minor in Computer Science from G. B. Pant University of Agriculture and Technology, Pantnagar, India. He has been a Faculty Member for around 12 years and has taught several core courses in pure and applied mathematics at undergraduate, postgraduate, and doctorate levels. He is currently the Research Professor at Graphic Era (Deemed to be University), Dehradun, India. Before joining the Graphic Era, he was a Deputy Manager (Probationary Officer) with Syndicate Bank for a short period. He is Editor-in-Chief of International Journal of Mathematical, Engineering and Management Sciences, Journal of Reliability and Statistical Studies, Editor-in-Chief of six Book Series with Elsevier, CRC Press-A Taylor and Frances Group, Walter De Gruyter Publisher Germany, River Publisher and the Guest Editor & Member of the editorial board of various journals. He has published 225 plus research publications (journal articles/books/book chapters/conference articles) in IEEE, Taylor & Francis, Springer, Elsevier, Emerald, World Scientific, and many other national and international journals and conferences. Also, he has published more than 50 books (authored/edited) with international publishers like Elsevier, Springer Nature, CRC Press, Taylor & Francis Group, Walter De Gruyter Publisher Germany, River Publisher. His fields of research are reliability theory and applied mathematics. Dr. Ram is a Senior Member of the IEEE, Senior Life Member of Operational Research Society of India, Society for Reliability Engineering, Quality and Operations Management in India, Indian Society of Industrial and Applied Mathematics. He has been a member of the organizing committee of several international and national conferences, seminars, and workshops. He has been conferred with the "Young Scientist Award" by the Uttarakhand State Council for Science and Technology, Dehradun, in 2009. He has been awarded the "Best Faculty Award" in 2011; the "Research Excellence Award" in 2015; and recently the "Outstanding Researcher Award" in 2018 for his significant contribution in academics and research at Graphic Era Deemed to be University, Dehradun, India.

Contributors

Sagar Adhurya
Systems Ecology and Ecological
    Modeling Laboratory,
    Department of Zoology
Visva-Bharati University
Santiniketan, India

Fahad Al Basir
Department of Mathematics
Asansol Girls' College
Asansol, India

Mukesh Kumar Awasthi
Department of Mathematics
Babasaheb Bhimrao Ambedkar
    University
Lucknow, India

Rakesh C. Bhadula
Graphic Era Hill University
Dehradun, India

Dinesh C. S. Bisht
Department of Mathematics
Jaypee Institute of Information
    Technology
Noida, India

Alok Dhaundiyal
Centre for Energy Research
Budapest, Hungary

Nitesh Dutt
Mechanical Engineering
    Department
College of Engineering
Roorkee, India

Ananya Dwivedi
Department of Mechanical
    Engineering
Galgotias College of Engineering
    and Technology
Greater Noida, India

Jamila Elalami
LASTIMI, Higher School of
    Technology in Salé
University Mohammed V
Rabat, Morocco

Azeddine Elmajidi
LASTIMI, Higher School of
    Technology in Salé
University Mohammed V
Rabat, Morocco

Elhousseine Elmazoudi
CISIEV, FSJES
University Cadi Ayyad
Marrakech, Morocco

Mudavath Gopal Naik
Department of Civil Engineering,
University College of
Engineering
Osmania University
Hyderabad, India

Harish Gupta
Department of Civil Engineering,
University College of
Engineering
Osmania University
Hyderabad, India

Reshu Gupta
Applied Science Cluster
(Mathematics)
University of Petroleum and Energy
Studies
Dehradun, India

Syed Aftab Haider
Department of Mathematics
Shia P. G. College
Lucknow, India

Sandeep Hamsa
Water Resources
Water and Land Management
Training and Research Institute
Hyderabad, India

Akmal Husain
Department of Mathematics,
Applied Science Cluster, SoE
University of Petroleum and Energy
Studies
Dehradun, India

Vijaya Nand Kala
GBPEC
Pauri Grahwal, India

Kunwer Singh Mathur
Department of Mathematics and
Statistics
Dr. Harisingh Gour
Vishwavidyalaya
Sagar, India

Manne Mohan Raju
Nagarjunasagar Project, Irrigation
& CAD (PW) Department
Government of TS
Hyderabad, India

Prakash Narayan
Department of Mathematics
Pratap University
Jaipur, India

Aadhi Naresh
Department of Civil Engineering,
University College of
Engineering
Osmania University
Hyderabad, India

Ram Naresh
Department of Mathematics,
School of Basic & Applied
Sciences
Harcourt Butler Technical
University
Kanpur, India

Pallav Jyoti Pal
Department of Mathematics
Krishna Chandra College
Hetampur, India

Alok Kumar Pandey
Department of Mathematics
Graphic Era Deemed to be
University
Dehradun, India

Santanu Ray
Systems Ecology and Ecological
    Modeling Laboratory,
    Department of Zoology
Visva-Bharati University
Santiniketan, India

Tapan Saha
Department of Mathematics
Presidency University
Kolkata, India

Ankit R. Singh
Mechanical Engineering
    Department
Indian Institute of Technology
Bombay, India

Jaipal Singh
DBS (P. G.) College
Dehradun, India

Satyvir Singh
School of Physical and
    Mathematical Sciences
Nanyang Technological University
Singapore

Vijai Krishna Singh
Department of Applied Sciences
Institute of Engineering &
    Technology
Lucknow, India

Anand Kumar Solanki
Mechanical Engineering
    Department
Gayatri Vidya Parishad College of
    Engineering
Visakhapatnam, India

Laszlo Toth
Institute of Technology
Hungarian University of
    Agriculture and Life Sciences,
Godollo, Hungary

Agraj Tripathi
Department of Basic Science and
    Humanities
Pranveer Singh Institute of
    Technology
Kanpur, India

Himanshu Upreti
Department of Allied Sciences
Graphic Era Hill University
Haldwani, India

Alok Kumar Verma
Department of Mathematics,
    School of Physical and Decision
    Science
Babasaheb Bhimrao Ambedkar
    University
Lucknow, India

Maitri Verma
Department of Mathematics,
    School of Physical and Decision
    Science
Babasaheb Bhimrao Ambedkar
    University
Lucknow, India

Chandra Bhushan Vishwakarma
Department of Mechanical
    Engineering
Galgotias College of Engineering
    and Technology
Greater Noida, India

Devendra Yadav
Department of Mechanical
    Engineering
Galgotias College of Engineering
    and Technology
Greater Noida, India

Chapter 1

Parameter identification and fuzzy T-S robust static output stabilization for a carbon dioxide model

Moroccan context

Azeddine Elmajidi and Jamila Elalami
University Mohammed V

Elhousseine Elmazoudi
University Cadi Ayyad

CONTENTS

DOI: 10.1201/9781003367420-1

1

NOMENCLATURE

T-S	Takagi Sugeno Fuzzy models
PDC	Parallel Distributed Compensation
SOFC	Static Output Feedback Control
LMI	Linear Matrix Inequalities
BMI	Bilinear Matrix Inequalities

1.1 INTRODUCTION: BACKGROUND AND DRIVING FORCES

The level of atmospheric carbon dioxide (CO_2) has continuously increased since the advent of the industrial era in the beginning of the 20th century, which risks distorting human and ecosystem life quality by causing global warming. The greenhouse effect is mainly due to the transparency of CO_2 to solar radiation and the reflection of thermal radiation from the earth. In addition, the forest, the main actor in absorbing and minimizing the CO_2 effects, has decreased drastically by deforestation actions.

On the one hand, there are natural CO_2 emissions that are related to volcanic activity, fires, and other natural causes, and on the other hand, we have anthropogenic emissions, which imply human activities. The anthropogenic part is constantly growing, which yields environmental impacts such as pollution, ocean acidification, and the greenhouse effect. Those emissions must be mitigated to preserve the environment for the next generations. Previously, mitigation attempts were limited to awareness-raising and training actions on such concentration effects, especially when the CO_2 levels were acceptable. However, with the phenomenal concentration levels and to understand the kinetics of such factors to control their growth over time, it is interesting to switch to more suitable and complex methods.

Several works have been devoted to the environmental systems study and its interaction with other components, to elucidate the factors and the climate change consequences. Most of them use mathematical models, including differential equations, which are useful to explore the impact of various factors on the dynamics of atmospheric greenhouse gases. These models are based on different approaches, from modeling to adopting several tools to perform system analysis.

This being said, most research works mention deforestation as a general cause of high CO_2 concentrations. It is clear that forests are natural regulators and sinks of greenhouse gas emissions, and any reduction will increase these emissions. For more details, one can refer to the works (Agarwal et al., 2010; Agarwal & Pathak, 2015; Bjornlund, 2009; Bremner et al., 2010; Chaudhary et al., 2015; Ciesla & FAO, 1997; Devi & Gupta, 2020; Devi & Mishra, 2020; Dignon, 1995; Dubey et al., 2009; Goreau, 1992; D. F. Karnosky et al., 2009; D. Karnosky & IUFRO Task Force on Environmental Change, 2001; Madhu & Hatfeld, 2013; Misra et al., 2014, 2015; Misra & Verma, 2013; N A S Colloqium, 1997; Shukla et al., 1989; Shukla & Dubey, 1997; Sundar, 2015; Tennakone, 1990; Verma & Misra, 2018; Woodwell et al., 1983). From another perspective, other researchers have claimed overpopulation as another cause aggravating CO_2 emissions (Agarwal et al., 2010; Bremner et al., 2010; Dubey et al., 2009; Misra et al., 2014, 2015; Misra & Verma, 2013; Shukla et al., 1989; Sundar, 2015). In addition, another theme studying the impact of industrial pressure on the environmental problem resurfaces through the works of (Agarwal et al., 2010; Chaudhary et al., 2015; Dubey et al., 2009). However, the problem was first discussed in the 90s by Goreau (1992); Shukla et al. (1989); Tennakone (1990). Finally, other factors, such as the poverty and behavior impacts, are highlighted by other works, or even the impact of pollution type on the reduction of forest resources (Bremner et al., 2010; Chaudhary et al., 2015), where it is shown that the equilibrium point of the biomass decreases against any growth in the human population and pollution. This decay can lead to extinction if the growth of other parameters becomes uncontrolled (Shukla et al., 1989).

The research works carried out lately to resolve the environmental problem, propose solutions which alternate between classic and innovative, mainly about the limitation of the deforestation as studied (Misra & Verma, 2013); the protection and the preservation of a part of the forest biomass, by dividing it into two parts, the first one being unreserved forest biomass which is available for the human uses, and the second one concerns the reserved forest biomass that has no human direct connection and remains preserved by a strict authority (Devi & Mishra, 2020); the adoption of planting plans by raising the importance of the time gap between the afforestation needs and the actual planting action (Devi & Gupta, 2020; Misra et al., 2015; Shukla et al., 1989).

On the other hand, other works such as those of (Agarwal & Pathak, 2015; Verma & Misra, 2018) raise solutions, rather, innovative by using alternative resources or scientific methods leading to a reduction and lesser exploitation of forest resources and fossil fuel combustion with energy-saving technologies. Moreover, noteworthy that the low-carbon fuels switch leads to CO_2 emissions reduction.

During the past few decades, the Takagi–Sugeno Fuzzy modeling technique, a method based on fuzzy logic, developed by Lotfi Zadeh, has emerged (Zadeh, 1965). It was used later to model systems based on human reasoning (Lilly, 2010a, b, e; Mamdani & Assilian, 1975; Takagi & Sugeno, 1985). This new alternative technique (fuzzy T-S modeling) (Takagi & Sugeno, 1985, 1992) makes a significant contribution to the study of complex nonlinear dynamical systems by allowing the modeling through extracted data, giving an equivalent representation to nonlinear models, and proposing a command pulled out directly from the original fuzzy model. In addition, when combined with the nonlinearity sector approach (Tanaka & Wang, 2003f), the T-S modeling offers a good balance between complexity and accuracy. It allows the rewriting of a nonlinear system as a sum of several weighted linear subsystems by nonlinear activation functions that satisfies the convexity property (Chadli et al., 2001; Nagy-Kiss, 2010; Tanaka & Sugeno, 1992). In addition, due to its equivalency (Tanaka & Wang, 2003e), it's more convenient to study a given nonlinear system with its corresponding Takagi-Sugeno model. Furthermore, nonlinear system transformations to T-S models are not unique and may be configurable following the designer strategy, which may help in both controller and observer design (Sugeno et al., 1999). Consequently, the stability conditions, through T-S modeling, give a balance between computational complexity and conservativeness. And even if the T-S conditions may sometimes have a huge numerical calculation, it is possible to decrease the numerical resolution significantly with good premise variables choice. Moreover, according to several studies (Chadli, 2002; Chadli et al., 2001; Tanaka & Sugeno, 1992; Tanaka & Wang, 2003b, f), the distributed parallel compensation (PDC) control technique is the most appropriate method for T-S fuzzy systems, given the direct connection between the control law elaboration and the original T-S model.

Finally, uncertainty finds always its place to model parameters through inaccurate estimation or disturbance presence. Hopefully, the robust theory is itemized as a convenient way to extract stability and stabilization conditions, assuming that certain parameters will be unknown but bounded (Manai & Benrejeb, 2012; Nagy-Kiss, 2010; Scherer, 2001; Tanaka & Wang, 2003d).

Afterward, by using T-S Fuzzy modeling, this chapter deals with the chosen model's (Misra & Verma, 2013) parameter identification in the Moroccan context, deforestation impact, and stability conditions. Also, it explores the stabilization performance for the retransformed (forced) model obtained by assuming deforestation as an input variable, according to the considered output matrix C shape, LMI conditions, and parameters uncertainty. Finally, the study can be extended to other regions adopting the same proof line.

1.2 MODEL PRESENTATION

In this section, we will focus on nonlinear delay-free systems study through an environmental model (1.1) presented by (Misra & Verma, 2013). The model is given as follows[1]:

---

[1] $C(t)$: Atmospheric Carbon Dioxide level in (ppm)

$N(t)$: Human population in (person)

$F(t)$: Forest biomass in (ton)

Q_0: Natural atmospheric Carbon Dioxide elevation rate in (ppm/year)

λ: Anthropogenic atmospheric Carbon Dioxide elevation rate coefficient in (ppm.[person. year] $^{-1}$)

α: Natural atmospheric Carbon Dioxide depletion rate coefficient in (year^{-1})

λ_1: Atmospheric Carbon Dioxide depletion rate coefficient due to forest biomass in ([year.ton] $^{-1}$)

s: Intrinsic Human population growth rate in (year^{-1})

L: Human population carrying capacity in (person)

Θ: Human population depletion rate coefficient due to Carbon Dioxide in ([ppm.year] $^{-1}$)

π: Human population growth ratio due to forest biomass in (person.ton^{-1})

ϕ: Deforestation rate coefficient in ([person.year]$^{-1}$)

μ: Intrinsic Forest biomass growth rate in (year^{-1})

M: Forest biomass carrying capacity in (ton)

π_1: Forest biomass growth ratio due to CO_2 in (ton/ppm)

$$\dot{X} = \begin{pmatrix} Q_0 + \lambda N - \alpha C - \lambda_1 CF \\ sN(1 - \dfrac{N}{L}) - \theta CN + \pi \phi NF \\ \mu F(1 - \dfrac{F}{M}) - \phi NF + \pi_1 \lambda_1 CF \end{pmatrix} \qquad (1.1)$$

where

$$X = \begin{pmatrix} C \\ N \\ F \end{pmatrix} \qquad (1.2)$$

1.3 CARBON DIOXIDE MODEL T-S CONSTRUCTION

1.3.1 Model value space

From a theoretical point of view, the variables C, N and F from the model (1.1) have a wide operating space ($\mathbb{R}_+^3$); however, it can be shown that they are limited by maximum values. Therefore, smaller operating areas must be chosen to reflect real functional space, defined by the set Ω. To find this operating space, several methods are available; one should refer to (Freedman & So, 1985; Hahn, 1967) that describe a simplified method for obtaining maximum values for systems whose shape is close to that of food chains (Prey–Predator models). In our present study, the interest in the operating area is mainly linked to Takagi-Sugeno fuzzy model transformation.

$$\Omega = (C, N, F)/C_m \leqslant C \leqslant C_M, \ N_m \leqslant N \leqslant N_M, \ F_m \leqslant F \leqslant F_M \qquad (1.3)$$

1.3.1.1 Variable C operating subspace

Let's take the system's (1.1) first element, and then the critical points likely to be an extremum are the C values satisfying the condition $\dot{C} = 0$. Using (1.1), this is equivalent to:

$$Q_0 + \lambda N - \alpha C - \lambda_1 CF = 0 \qquad (1.4)$$

which leads to:

$$C = \frac{Q_0 + \lambda N}{\alpha + \lambda_1 F} \qquad (1.5)$$

Finally:

$$
\begin{cases}
C_M = \dfrac{Q_0 + \lambda N_M}{\alpha + \lambda_1 F_m} \\[3mm]
C_m = \dfrac{Q_0 + \lambda N_m}{\alpha + \lambda_1 F_M}
\end{cases}
\tag{1.6}
$$

1.3.1.2 Variable N operating subspace

In the same way $\dot{N} = 0$ means that:

$$
sN\left(1 - \frac{N}{L}\right) - \theta CN + \pi\phi NF = 0
\tag{1.7}
$$

Then:

$$
N = 0 \quad \text{or} \quad N = L(1 - \frac{\theta}{s}C + \frac{\pi\phi}{s}F)
\tag{1.8}
$$

Finally:

$$
\begin{cases}
N_M = L(1 - \dfrac{\theta}{s}C_m + \dfrac{\pi\phi}{s}F_M) \\[3mm]
N_m = 0
\end{cases}
\tag{1.9}
$$

1.3.1.3 Variable F operating subspace

In the same way $\dot{F} = 0$ means:

$$
\mu F(1 - \frac{F}{M}) - \phi NF + \pi_1 \lambda_1 CF = 0
\tag{1.10}
$$

This leads to:

$$
F = 0 \quad \text{or} \quad F = M(1 - \frac{\phi}{\mu}N + \frac{\pi_1 \lambda_1}{\mu}C)
\tag{1.11}
$$

Finally to:

$$
\begin{cases}
F_M = M(1 - \dfrac{\phi}{\mu}N_m + \dfrac{\pi_1 \lambda_1}{\mu}C_M) \\[3mm]
F_m = 0
\end{cases}
\tag{1.12}
$$

1.3.1.4 Combined value space

Based on Equations (1.6), (1.9), and (1.12),[2] it can be noticed that the system's upper and down bounds are manifested in a cyclic form. To extract the values, we will have to rewrite these equations. To simplify the calculation, we will consider the upper bound of the human population N by eliminating the element $-\dfrac{\theta}{s}C_m$ from Equation (1.9). Then, by injecting (1.6) and then (1.9) into (1.12), we can come to a definition of F_M according to the parameters of the (1.1), and subsequently N_m, C_M and C_m, where:

$$C_M = \frac{Q_0 + \lambda N_M}{\alpha} \qquad C_m = \frac{Q_0}{\alpha + \lambda_1 F_M} \tag{1.13}$$

$$N_M = L(1 + \frac{\pi\phi}{s} F_M) \qquad N_m = 0 \tag{1.14}$$

$$F_M = sM\frac{[\mu\alpha + \pi_1\lambda_1(Q_0 + \lambda L)]}{(\mu s\alpha - \pi\pi_1\lambda_1\lambda\phi ML)} \qquad F_m = 0 \tag{1.15}$$

1.3.2 Equilibrium points

1.3.2.1 Definition

The system's equilibrium points are the solutions of the system equations below:

$$\begin{cases} Q_0 + \lambda N - \alpha C - \lambda_1 CF & = 0 \\[2mm] sN(1 - \dfrac{N}{L}) - \theta CN + \pi\phi NF & = 0 \\[2mm] \mu F(1 - \dfrac{F}{M}) - \phi NF + \pi_1\lambda_1 CF & = 0 \end{cases} \tag{1.16}$$

which means:

$$\begin{cases} C & = \dfrac{Q_0 + \lambda N}{\alpha + \lambda_1 F} \\[2mm] s(1 - \dfrac{N}{L}) - \theta C + \pi\phi F & = 0 \text{ or } N = 0 \\[2mm] \mu(1 - \dfrac{F}{M}) - \phi N + \pi_1\lambda_1 C & = 0 \text{ or } F = 0 \end{cases} \tag{1.17}$$

---

[2] The form of Equations (1.6), (1.9) and (1.12) conclude on the minimum values down bounding the operating space. Moreover, they show that $C > 0$, $N \geq 0$ and $F \geq 0$.

By combining all possibilities, we can note the omnipresence of carbon in all four equilibrium points (Misra & Verma, 2013):

$$E_1 = \left(\frac{Q_0}{\alpha}, 0, 0\right), \quad E_2 = (C_2, N_2, 0),$$

$$E_3 = (C_3, 0, F_3), \quad E_4 = (C^*, N^*, F^*) \tag{1.18}$$

I.3.2.2 E_4 equilibrium point

In the rest of this chapter, we will focus on the interior equilibrium point E_4, since it ensures both humanity's survival by the presence of the human population and forest biomass existence that represents both the natural sink of CO_2 emissions CO_2 and food source for humans. To obtain the inner equilibrium point E_4, definition (1.17) is rewritten as ensuring the triplet $\{C, N, F\}$ non-nullity, which means:

$$
\begin{cases}
C = \dfrac{Q_0 + \lambda N}{\alpha + \lambda_1 F} \\[2mm]
s(1 - \dfrac{N}{L}) - \theta C + \pi \phi F = 0 \\[2mm]
\mu(1 - \dfrac{F}{M}) - \phi N + \pi_1 \lambda_1 C = 0
\end{cases}
\tag{1.19}
$$

However, its existence depends on finding a triplet $\{C, N, F\}$ that solves Equation (1.19) and the solutions are strictly positive. These conditions have a physical meaning related to system context (Misra & Verma, 2013). By injecting (1.19).1 into (1.19).2 and (1.19).3, we come across a second-degree equation system with two variables N and F. Therefore, based on the isoclines[3] of these two equations, we can conclude the following conditions of existence (Misra et al., 2015; Misra & Verma, 2013).

$$
\begin{cases}
s > \dfrac{\theta Q_0}{\alpha} \\[2mm]
\alpha \phi > \pi_1 \lambda_1 \lambda \\[2mm]
(\mu \alpha + \pi_1 \lambda_1 Q_0)(s\alpha + \theta \lambda L) > L(s\alpha - \theta Q_0)(\alpha \phi - \pi_1 \lambda_1 \lambda)
\end{cases}
\tag{1.20}
$$

---

[3] A curve of solutions of a differential equation (i.e., $\dot{x} = fx(t)$) that accept the same rate of change or slope (i.e., $f(x(t))$) $= cste$ (Bailey et al., 1977; Edwards & Penney, 2013; Zill & Wright, 2012).

1.3.3 T-S model

The purpose of this part is to transform the nonlinear model (1.1) into a set of time-invariant linear sub-models, called T-S multi-models, using the extremum values of the premise variables, which are directly related to system nonlinearity. The overall system results from the sum of the weighted sub-models with functions called firing force.

1.3.3.1 Change of coordinate

To begin with, a change of variable around the equilibrium point is necessary to end up with an equilibrium point at the origin. Let's take:

$$x = X - E_4^T \tag{1.21}$$

where

$$X = \begin{pmatrix} C \\ N \\ F \end{pmatrix} \quad \text{and} \quad x = \begin{pmatrix} x_1 \\ x_2 \\ x_3 \end{pmatrix}$$

The model (1.1) is then equivalent to the model (1.22) given as follows:

$$\dot{x} = \begin{pmatrix} Q_0 + \lambda(x_2 + N^*) - \alpha(x_1 + C^*) - \lambda_1(x_1 + C^*)(x_3 + F^*) \\ (x_2 + N^*)\left[s(1 - \dfrac{(x_2 + N^*)}{L}) - \theta(x_1 + C^*) + \pi\phi(x_3 + F^*) \right] \\ (x_3 + F^*)\left[\mu(1 - \dfrac{(x_3 + F^*)}{M}) - \phi(x_2 + N^*) + \pi_1\lambda_1(x_1 + C^*) \right] \end{pmatrix} \tag{1.22}$$

Using (1.17) and more precisely the fact that:

$$\begin{cases} Q_0 + \lambda N^* - \alpha C^* + \lambda_1 C^* F^* &= 0 \\ s(1 - \dfrac{N^*}{L}) - \theta C^* + \pi\phi F^* &= 0 \\ \mu(1 - \dfrac{F^*}{M}) - \phi N^* + \pi_1\lambda_1 C^* &= 0 \end{cases} \tag{1.23}$$

Equation (1.22) can be reduced to:

$$\dot{x} = \begin{pmatrix} \lambda x_2 - \alpha x_1 - \lambda_1 (x_1 F^* + C^* x_3 + x_1 x_3) \\ (x_2 + N^*)\left[-s\dfrac{x_2}{L} - \theta x_1 + \pi \phi x_3 \right] \\ (x_3 + F^*)\left[-\mu\dfrac{x_3}{M} - \phi x_2 + \pi_1 \lambda_1 x_1 \right] \end{pmatrix} \tag{1.24}$$

Thus, the behavior of the model (1.24) can be divided into two parts as in Equation (1.25), one part with linear behavior and another defining the nonlinear behavior of the system away from the equilibrium point $x = 0$.

$$\dot{x} = J^* x + g(x) \tag{1.25}$$

where

$$J^* = \begin{pmatrix} -(\alpha + \lambda_1 F^*) & \lambda & -\lambda_1 C^* \\ -\theta N^* & -s\dfrac{N^*}{L} & \pi \phi N^* \\ \pi_1 \lambda_1 F^* & -\phi F^* & -\mu\dfrac{F^*}{M} \end{pmatrix},$$

$$g(x) = \begin{pmatrix} -\lambda_1 x_1 x_3 \\ x_2\left[-s\dfrac{x_2}{L} - \theta x_1 + \pi \phi x_3 \right] \\ x_3\left[-\mu\dfrac{x_3}{M} - \phi x_2 + \pi_1 \lambda_1 x_1 \right] \end{pmatrix}$$

1.3.3.2 T-S model transformation

Before using the nonlinear sector method presented before (Lilly, 2010d; Tanaka & Wang, 2003f), one must first write the vector function $g(x)$ in the matrix form $A_x x$. However, this rewriting is not unique and one may end up with several representations, especially since there is no systematic way to obtain the optimal configuration which will depend on the objectives of the modeling (Lilly, 2010e; Sugeno et al., 1999; Tanaka & Sugeno, 1992). From this point of view, it would be wise to find a compromise between the number of nonlinearities in the chosen representation, the obtained model complexity, and finally the process of obtaining and setting up the controllers and observers. For example, we can quote three representations in our case for the matrix A_x, which seem distinct:

$$
{}^{1}A_x = \begin{pmatrix} -\lambda_1 x_3 & 0 & 0 \\ -\theta x_2 & -s\dfrac{x_2}{L} & \pi\phi x_2 \\ \pi_1\lambda_1 x_3 & -\phi x_3 & -\mu\dfrac{x_3}{M} \end{pmatrix} \tag{1.26}
$$

$$
{}^{2}A_x = \begin{pmatrix} 0 & 0 & -\lambda_1 x_1 \\ 0 & -\theta x_1 - s\dfrac{x_2}{L} + \pi\phi x_3 & 0 \\ 0 & 0 & \pi_1\lambda_1 x_1 - \phi x_2 - \mu\dfrac{x_3}{M} \end{pmatrix} \tag{1.27}
$$

$$
{}^{3}A_x = \begin{pmatrix} -\lambda_1 x_3 & 0 & 0 \\ 0 & -\theta x_1 - s\dfrac{x_2}{L} + \pi\phi x_3 & 0 \\ 0 & 0 & \pi_1\lambda_1 x_1 - \phi x_2 - \mu\dfrac{x_3}{M} \end{pmatrix} \tag{1.28}
$$

The premise variable choice is different following the configuration choice. For instance, ${}^{1}A_x$ contains 2 nonlinear terms against 3 in ${}^{2}A_x$ and ${}^{3}A_x$. In the following, we will use the ${}^{1}A_x$ configuration. Thus, the system (1.24) will become:

$$
\dot{x} = Ax \tag{1.29}
$$

where

$$
A = J^* + {}^{1}A_x
$$

Therefore, the premise vector z is defined as follows:

$$
z_1 = x_2 \tag{1.30}
$$

$$
z_2 = x_3 \tag{1.31}
$$

Using the nonlinearity sector and the model variables' extreme values, we can write the premise vector as[4]:

$$z_1 = (N_M - N^*)M_1^1 + (\epsilon_N - N^*)M_2^1 \tag{1.32}$$

$$z_2 = (F_M - F^*)M_1^2 + (\epsilon_F - F^*)M_2^2 \tag{1.33}$$

Knowing that the membership functions respect:

$$\begin{cases} M_1^1 + M_2^1 &= 1 \\ M_1^2 + M_2^2 &= 1 \end{cases} \tag{1.34}$$

Thus, the functions M_j^i can be defined as:

$$M_1^1 = \frac{z_1 + N^* - \epsilon_N}{N_M - \epsilon_N} \qquad M_2^1 = \frac{N_M - N^* - z_1}{N_M - \epsilon_N} \tag{1.35}$$

$$M_1^2 = \frac{z_2 + F^* - \epsilon_F}{F_M - \epsilon_F} \qquad M_2^2 = \frac{F_M - F^* - z_2}{F_M - \epsilon_F} \tag{1.36}$$

where ϵ_N and ϵ_F present the chosen shift to avoid the other equilibrium points attraction domains. In addition, if we use a linguistic abbreviation to designate these fuzzy sets, for example *Large, Small* for the functions of z_1 and *Positive, Negative* for those of z_2, we will have $2^2 = 4$ possible combinations which will designate the rules of the nonlinear (1.29) system according to fuzzy T-S models (Tanaka & Sugeno, 1992; Tanaka & Wang, 2003b).

Rule 01: If z_1 is Large and z_2 is Positive, Then $\dot{x} = A_1x$ \hfill (1.37)

Rule 02: If z_1 is Large and z_2 is Negative, Then $\dot{x} = A_2x$ \hfill (1.38)

Rule 03: If z_1 is Small and z_2 is Positive, Then $\dot{x} = A_3x$ \hfill (1.39)

Rule 04: If z_1 is Small and z_2 is Negative,Then $\dot{x} = A_4x$ \hfill (1.40)

---

[4] In the following, we choose to build a T-S model by bounds presenting the whole operating space except for the minimal operating area of the equilibrium points E_1, E_2 and E_3 which are not of particular interest for the present study. Also, we note that no rule objects to the T-S model with a narrower operating area.

where

$$
A_1 = \begin{pmatrix} -(\alpha + \lambda_1 F_M) & \lambda & -\lambda_1 C^* \\ -\theta N_M & -s\dfrac{N_M}{L} & \pi\phi N_M \\ \pi_1 \lambda_1 F_M & -\phi F_M & -\mu\dfrac{F_M}{M} \end{pmatrix} \quad A_2 = \begin{pmatrix} -\alpha - \lambda_1 \epsilon_F & \lambda & -\lambda_1 C^* \\ -\theta N_M & -s\dfrac{N_M}{L} & \pi\phi N_M \\ \pi_1 \lambda_1 \epsilon_F & -\phi\epsilon_F & -\mu\dfrac{\epsilon_F}{M} \end{pmatrix}
$$

$$
A_3 = \begin{pmatrix} -(\alpha + \lambda_1 F_M) & \lambda & -\lambda_1 C^* \\ -\theta\epsilon_N & -s\dfrac{\epsilon_N}{L} & \pi\phi\epsilon_N \\ \pi_1 \lambda_1 F_M & -\phi F_M & -\mu\dfrac{F_M}{M} \end{pmatrix} \quad A_4 = \begin{pmatrix} -\alpha - \lambda_1 \epsilon_F & \lambda & -\lambda_1 C^* \\ -\theta\epsilon_N & -s\dfrac{\epsilon_N}{L} & \pi\phi\epsilon_N \\ \pi_1 \lambda_1 \epsilon_F & -\phi\epsilon_F & -\mu\dfrac{\epsilon_F}{M} \end{pmatrix}
$$

$$(1.41)$$

Thereafter, the activation functions w_i and the normalized firing probability will be used to obtain the global behavior of the system:

$$
w_i = \begin{cases} M_1^1 M_1^2 & \text{For the fuzzy set\{Large and Positive\}} \\ M_1^1 M_2^2 & \text{For the fuzzy set\{Large and Negative\}} \\ M_2^1 M_1^2 & \text{For the fuzzy set\{Small and Positive\}} \\ M_2^1 M_2^2 & \text{For the fuzzy set\{Small and Negative\}} \end{cases}, \quad h_i = \dfrac{w_i}{\displaystyle\sum_{i=1}^{4} w_i}
$$

$$(1.42)$$

Unlike switched systems, systems described by T-S multi-models can simultaneously belong to all fuzzy set values as long as the activation function w_i measuring the degree of membership is different from 0 ($w_i \neq 0$) (Lilly, 2010e). By blending all the rules set through the firing probability h_i, the system (1.29) will then be presented by:

$$
\dot{x} = \sum_{i=1}^{4} h_i A_i x \tag{1.43}
$$

Remark 1.1

The use of $^3 A_x$ will lead to an unforced nonlinear system representation in eight sub-models.

1.4 STABILITY ANALYSIS

1.4.1 Local stability

A given nonlinear system local stability around an equilibrium point, can be done with its Jacobian Matrix, which must have eigenvalues with negative real parts.

1.4.2 Global stability

The T-S model global stability refers to (Tanaka & Sugeno, 1992; Tanaka & Wang, 2003b, f) work, which can be described in the model (1.43) by finding a symmetric positive definite matrix P that verifies (1.44):

$$\left\{ \begin{array}{l} A_1^T.P + P.A_1 < 0 \\ A_2^T.P + P.A_2 < 0 \\ A_3^T.P + P.A_3 < 0 \\ A_4^T.P + P.A_4 < 0 \\ \quad\ P \end{array} \right. \tag{1.44}$$

1.4.3 Attraction domain

The attraction domain notion is meaningless in the case of global stability since it is the entire operating space (Amato et al., 2006). In the opposite case, several studies have focused on estimating the region in which the system can re-converge to the chosen equilibrium point (Amato et al., 2006; Burchardt & Ratschan, 2007; Khalil, 2002; Vannelli & Vidyasagar, 1985). In the following, we choose to use a method proposed by (Khalil, 2002) to estimate the attraction region. The objective of the method is to estimate an attraction region using sub-levels of the Lyapunov function $(V < c)$ where $c > 0$ a value to be defined and verifies $\dot{V} < 0$, since this kind of subspace is proved invariant (Gu et al., 2003; Hahn, 1967; Khalil, 2002). Moreover, it should be noted that the value of c is strongly related to the region geometry.

Therefore, as long as the matrix J^* is a Hurwitz matrix, then there exist symmetric and positive matrices P, Q such that:

$$J^{*T}P + PJ^* = -Q \tag{1.45}$$

And let V be a quadratic Lyapunov function associated with (1.43):

$$V = x^T P x \tag{1.46}$$

Knowing that the matrices P and Q are symmetric, then they are diagonalizable and their transition matrices are orthogonal, which can be translated as:

$$P = M_P^T D_P M_P \tag{1.47}$$

$$Q = M_Q^T D_Q M_Q \tag{1.48}$$

where $M_P^T M_P = \mathcal{I}_n$ and $M_Q^T M_Q = \mathcal{I}_n$

This leads to (Khalil, 2002):

$$\lambda_{\min P} \parallel x \parallel^2 \leqslant x^T P x \leqslant \lambda_{\max P} \parallel x \parallel^2 \tag{1.49}$$

$$\lambda_{\min Q} \parallel x \parallel^2 \leqslant x^T Q x \leqslant \lambda_{\max Q} \parallel x \parallel^2 \tag{1.50}$$

Thus:

$$V < c \Rightarrow \lambda_{\min P} \parallel x \parallel^2 < c \tag{1.51}$$

By injecting (1.25) and (1.45) into the derivative of (46), we find:

$$\dot{V} = -x^T Q x + 2 x^T P g(x) \tag{1.52}$$

Using the matrix norm sub-multiplication property and (1.50) into (1.52) implies:

$$\dot{V} \leqslant -\lambda_{\min Q} \parallel x \parallel^2 + 2 \parallel x \parallel \parallel P g(x) \parallel \tag{1.53}$$

Knowing that for any matrix M and vector v of appropriate dimensions, we have:

$$\parallel Mv \parallel = \left\| \begin{pmatrix} m_{11} & \cdots & m_{1j} & \cdots & m_{1n} \\ \vdots & \vdots & \vdots & \vdots & \vdots \\ m_{i1} & \cdots & m_{ij} & \cdots & m_{in} \\ \vdots & \vdots & \vdots & \vdots & \vdots \\ m_{n1} & \cdots & m_{nj} & \cdots & m_{nn} \end{pmatrix} \begin{pmatrix} v_1 \\ \vdots \\ v_i \\ \vdots \\ v_n \end{pmatrix} \right\| \tag{1.54}$$

$$= \left\| \sum_{j=1}^{n} \begin{pmatrix} m_{1j} \\ \vdots \\ m_{ij} \\ \vdots \\ m_{nj} \end{pmatrix} v_j \right\| \leqslant \max_j \parallel M_j \parallel \left(\sum_{j=1}^{n} \mid v_j \mid \right) \tag{1.55}$$

As a result, by injecting (1.55) into (1.53), we get:

$$\dot{V} \leqslant -\lambda_{\min Q} \parallel x \parallel^2 + 2 \parallel x \parallel \max_j \parallel P_j \parallel \left(\sum_{j=1}^{3} |g_j(x)| \right) \qquad (1.56)$$

$$\leqslant -\lambda_{\min Q} \parallel x \parallel^2 + 2 \parallel x \parallel^3 \max_{1 \leqslant j \leqslant 3} \parallel P_j \parallel \left(\lambda_1 + \frac{s}{L} + \theta + \pi\phi + \frac{\mu}{M} + \phi + \pi_1 \lambda_1 \right) \quad (1.57)$$

$$\leqslant - \parallel x \parallel^2 \left[\lambda_{\min Q} - 2 \parallel x \parallel \max_{1 \leqslant j \leqslant 3} \parallel P_j \parallel \left(\lambda_1 + \frac{s}{L} + \theta + \pi\phi + \frac{\mu}{M} + \phi + \pi_1 \lambda_1 \right) \right] (1.58)$$

Finally, to ensure $\dot{V} < 0$, it is sufficient to take a subspace verifying:

$$\parallel x \parallel < \frac{\lambda_{\min Q}}{2 \max\limits_{1 \leqslant j \leqslant 3} \parallel P_j \parallel \left(\lambda_1 + \dfrac{s}{L} + \theta + \pi\phi + \dfrac{\mu}{M} + \phi + \pi_1 \lambda_1 \right)} = r \qquad (1.59)$$

where

$$c = \lambda_{\min P} r^2 \qquad (1.60)$$

1.5 CO_2 T-S MODEL STATIC OUTPUT CONTROLLER

The model (1.25) used so far characterizes a nonlinear autonomous unforced system. To control it, we choose to define a parameter that can be considered as an input. In the following, the deforestation rate ϕ is taken as a parameter control.

1.5.1 T-S representation

1.5.1.1 Forced system

Let's take:

$$u = \phi - \phi^* \qquad (1.61)$$

where ϕ^* is the actual deforestation rate coefficient and ϕ is the controlled deforestation rate coefficient.

Then the system will be described as:

$$\begin{cases} \dot{x} = Ax + Bu \\ y = Cx \end{cases} \qquad (1.62)$$

where

$$A = \begin{pmatrix} -(\alpha + \lambda_1(F^* + x_3)) & \lambda & -\lambda_1 C^* \\ -\theta(N^* + x_2) & -s\dfrac{(N^* + x_2)}{L} & \pi\phi^*(N^* + x_2) \\ \pi_1\lambda_1(F^* + x_3) & -\phi^*(F^* + x_3) & -\mu\dfrac{(F^* + x_3)}{M} \end{pmatrix} \text{ and }$$

$$B = \begin{pmatrix} 0 \\ \pi(x_2 x_3 + N^* x_3) \\ -(x_2 x_3 + F^* x_2) \end{pmatrix}$$

1.5.1.2 Premise variables

Knowing the non-existence of a systematic method for obtaining the premise variables of a given system, a first choice would be to take the vector z as:

$$z_1 = N^* + x_2 \tag{1.63}$$

$$z_2 = F^* + x_3 \tag{1.64}$$

$$z_3 = \pi(x_2 x_3 + N^* x_3) \tag{1.65}$$

$$z_4 = -(x_2 x_3 + F^* x_2) \tag{1.66}$$

But it will result in a T-S Model with 16 different rules. Still, Sugeno et al. (1999); Takagi and Sugeno (1985), (1992) propose guidelines based on partitioning the value space to obtain a minimalist z premise vector. This being said, we note that z_3 and z_4 can be expressed by a linear combination of z_1 and z_2 added to a new variable $z_3 = x_2 x_3$. Therefore, the adopted premise vector is such that:

$$z_1 = N^* + x_2 \tag{1.67}$$

$$z_2 = F^* + x_3 \tag{1.68}$$

$$z_3 = (N^* + x_2)(F^* + x_3) \tag{1.69}$$

which means:

$$
A = \begin{pmatrix}
-(\alpha + \lambda_1 z_2) & \lambda & -\lambda_1 C^* \\
-\theta z_1 & -s\dfrac{z_1}{L} & \pi \phi^* z_1 \\
\pi_1 \lambda_1 z_2 & -\phi^* z_2 & -\mu\dfrac{z_2}{M}
\end{pmatrix}
\tag{1.70}
$$

$$
B = \begin{pmatrix}
0 \\
\pi(z_3 - F^* z_1) \\
-(z_3 - N^* z_2)
\end{pmatrix}
\tag{1.71}
$$

1.5.1.3 T-S rule construction

Assuming that the modified value of the parameter ϕ will have an impact on the model's upper bounds in (1.62) and to avoid a possible repercussion on the model's control through the null elements of the system matrix, we redefine these values as:

$$
C_M = \frac{Q_0 + \lambda N_M}{\alpha}, \quad C_m = \frac{Q_0}{\alpha + \lambda_1 F_M}
\tag{1.72}
$$

$$
N_M = L(1 + \frac{\pi \phi_{max}}{s} F_M), \quad N_m = \epsilon_N
\tag{1.73}
$$

$$
F_M = sM \frac{[\mu\alpha + \pi_1 \lambda_1 (Q_0 + \lambda L)]}{(\mu s\alpha - \pi \pi_1 \lambda_1 \lambda \phi_{max} ML)}, \quad F_m = \epsilon_F
\tag{1.74}
$$

where ϵ_N and ϵ_F represent the respective lower bounds of human population and forest biomass and ϕ_{max} represents the maximum deforestation rate.

Using the nonlinearity sector, the premise vector can be expressed as:

$$
z_1 = (N_M) M_1^1 + (\epsilon_N) M_2^1
\tag{1.75}
$$

$$
z_2 = (F_M) M_1^2 + (\epsilon_F) M_2^2
\tag{1.76}
$$

$$
z_3 = (N_M F_M) M_1^3 + (\epsilon_N \epsilon_F) M_2^3
\tag{1.77}
$$

where

$$M_1^1 = \frac{z_1 - \epsilon_N}{N_M - \epsilon_N}, \quad M_2^1 = \frac{N_M - z_1}{N_M - \epsilon_N} \tag{1.78}$$

$$M_1^2 = \frac{z_2 - \epsilon_F}{F_M - \epsilon_F}, \quad M_2^2 = \frac{F_M - z_2}{F_M - \epsilon_F} \tag{1.79}$$

$$M_1^3 = \frac{z_3 - \epsilon_N \epsilon_F}{N_M F_M - \epsilon_N \epsilon_F}, \quad M_2^3 = \frac{N_M F_M - z_3}{N_M F_M - \epsilon_N \epsilon_F} \tag{1.80}$$

The system (1.62) will then be presented by:

$$\dot{x} = \sum_{i=1}^{8} h_i (A_i x + B_i u) \tag{1.81}$$

where

$$A_1 = A_2 = \begin{pmatrix} -(\alpha + \lambda_1 F_M) & \lambda & -\lambda_1 C^* \\ -\theta N_M & -s\dfrac{N_M}{L} & \pi\phi^* N_M \\ \pi_1 \lambda_1 F_M & -\phi^* F_M & -\mu\dfrac{F_M}{M} \end{pmatrix},$$

$$A_3 = A_4 = \begin{pmatrix} -(\alpha + \lambda_1 \partial_F) & \lambda & -\lambda_1 C^* \\ -\theta N_M & -s\dfrac{N_M}{L} & \pi\phi^* N_M \\ \pi_1 \lambda_1 \epsilon_F & -\phi^* \epsilon_F & -\mu\dfrac{\epsilon_F}{M} \end{pmatrix}$$

$$A_5 = A_6 = \begin{pmatrix} -(\alpha + \lambda_1 F_M) & \lambda & -\lambda_1 C^* \\ -\theta \epsilon_N & -s\dfrac{\epsilon_N}{L} & \pi\phi^* \epsilon_N \\ \pi_1 \lambda_1 F_M & -\phi^* F_M & -\mu\dfrac{F_M}{M} \end{pmatrix},$$

$$A_7 = A_8 = \begin{pmatrix} -(\alpha + \lambda_1 \epsilon_F) & \lambda & -\lambda_1 C^* \\ -\theta \epsilon_N & -s\dfrac{\epsilon_N}{L} & \pi \phi^* \epsilon_N \\ \pi_1 \lambda_1 \epsilon_F & -\phi^* \epsilon_F & -\mu \dfrac{\epsilon_F}{M} \end{pmatrix}$$

$$B_1 = \begin{pmatrix} 0 \\ \pi(N_M F_M - F^* N_M) \\ -(N_M F_M - N^* F_M) \end{pmatrix}, \quad B_2 = \begin{pmatrix} 0 \\ \pi(\epsilon_N \epsilon_F - F^* N_M) \\ -(\epsilon_N \epsilon_F - N^* F_M) \end{pmatrix}$$

$$B_3 = \begin{pmatrix} 0 \\ \pi(N_M F_M - F^* N_M) \\ -(N_M F_M - N^* \epsilon_F) \end{pmatrix}, \quad B_4 = \begin{pmatrix} 0 \\ \pi(\epsilon_N \epsilon_F - F^* N_M) \\ -(\epsilon_N \epsilon_F - N^* \epsilon_F) \end{pmatrix}$$

$$B_5 = \begin{pmatrix} 0 \\ \pi(N_M F_M - F^* \epsilon_N) \\ -(N_M F_M - N^* F_M) \end{pmatrix}, \quad B_6 = \begin{pmatrix} 0 \\ \pi(\epsilon_N \epsilon_F - F^* \epsilon_N) \\ -(\epsilon_N \epsilon_F - N^* F_M) \end{pmatrix}$$

$$B_7 = \begin{pmatrix} 0 \\ \pi(N_M F_M - F^* \epsilon_N) \\ -(N_M F_M - N^* \epsilon_F) \end{pmatrix}, \quad B_8 = \begin{pmatrix} 0 \\ \pi(\epsilon_N \epsilon_F - F^* \epsilon_N) \\ -(\epsilon_N \epsilon_F - N^* \epsilon_F) \end{pmatrix}$$

And the firing probability h_i can be obtained using:

$$w = \begin{pmatrix} M_1^1 M_1^2 M_1^3 \\ M_1^1 M_1^2 M_2^3 \\ M_1^1 M_2^2 M_1^3 \\ M_1^1 M_2^2 M_2^3 \\ M_2^1 M_1^2 M_1^3 \\ M_2^1 M_1^2 M_2^3 \\ M_2^1 M_2^2 M_1^3 \\ M_2^1 M_2^2 M_2^3 \end{pmatrix}, \quad h_i = \dfrac{w_i}{\displaystyle\sum_{i=1}^{8} w_i} \tag{1.82}$$

1.5.2 Static output feedback controller

The literature represents several techniques to control fuzzy models (Ait Kaddour et al., 2015; Benzaouia & El Hajjaji, 2014; Chadli, 2002; Chadli et al., 2001, 2002; Chadli & El Hajjaji, 2005; Lilly, 2010c; Manai & Benrejeb, 2012; Nagy-Kiss, 2010; Oudghiri et al., 2007; Su et al., 2013; Tanaka & Sugeno, 1992; Tanaka & Wang, 2003a, c, f). Most of them use state feedback as common starting point. However, in real plants, its availability or estimation is not always possible and may require a lot of effort (Nguyen et al., 2017). In addition, it is sometimes preferable to proceed with output feedback control despite the increased LMI number to solve (Ait Kaddour et al., 2015; Chadli, 2002; Elmajidi et al., 2019; Nguyen et al., 2017).

In the following, a method known by Output-PDC (Chadli, 2002, Chapter 4) will be used. In this method, the controller is fuzzy type where each control rule is obtained from its equivalent rule in the T-S fuzzy model. Finally, the controller of the whole system is the sum of all local controllers weighted by their corresponding firing force.

$$\begin{cases} \dot{x}(t) = \sum_{i=1}^{8}\sum_{j=1}^{8} h_i h_j A_{ij} x(t) \\ y(t) = Cx(t) \end{cases} \qquad (1.83)$$

where

$$A_{ij} = A_i + B_i.F_j.C$$

The application of Lyapunov theorem on the Equation (1.83) gives rise to BMI (Bilinear Matrix Inequalities) impossible to solve directly, which leads to the division of the equation into two parts, a dominant part where $(i = j)$ and a crossed part where $(i \neq j)$.

1.5.3 LMI formulation

We intend to use different LMI conditions (Theorem 4.1 (MSC), Theorem 4.2 (MSij) and Corollary 4.1 (MDR)) stated in (Chadli, 2002, Chapter 4).

1.6 PARAMETER IDENTIFICATION FOR MOROCCAN CASE

To accomplish this goal, it's necessary to have a history of the CO_2 rate, human population, and forest biomass. Despite the lack of national data on this subject, referring to studies carried out on a global scale or in regions close to Moroccan territories, show that the country is subject to the same international constraints (IPCC, 2007; MTEDD, 2020; N A S Colloqium, 1997; Pittock, 2019) that can be considered as a starting point for estimating the designated parameters.

1.6.1 Variables background

1.6.1.1 CO$_2$ level

The CO$_2$ rate in the atmosphere is calculated through the ratio between the CO$_2$ molecules and the atmospheric air while considering the density of both substances. Moreover, it is to be specified that the theoretical world average concentration of CO$_2$ uses the weight of the world's atmospheric air which amounts to $P_{AT} = 5.148 * 10^{15}$ tons. To estimate the CO$_2$ levels in the atmosphere, we propose two methods based on two different assumptions:

1.6.1.1.1 Worldwide Balanced and Uniform CO$_2$ Level

Despite that CO$_2$ emissions are different from one region to another, it can be assumed that it tends toward an equilibrium by considering the migrations of the emissions between the regions (Friedlingstein et al., 2020a, b). In this case, we limit ourselves to the CO$_2$ values from the National Oceanic and Atmospheric Administration (*NOAA*) portal (Dlugokencky Ed, 2016; Ritchie et al., 2020).To express the CO$_2$ emission values, the following conversion equations are often used (Friedlingstein et al., 2020b):

$$1Gt\, C \Leftrightarrow \frac{44.01}{12.011} Gt\, CO_2 \tag{1.84}$$

$$1\, \text{ppm} \Leftrightarrow 2.124 Gt\, C \tag{1.85}$$

1.6.1.1.2 Region Balanced and Uniform CO$_2$ Level

Given that the Moroccan land is about $710,850$ km^2, which represents 0.4773% of the world area and assuming that the air is uniformly distributed throughout the world and referring to the formula below (Teesing, n.d.):

$$C\, (\text{ppm}) = 24.45 \times \frac{C\left(\frac{mg}{m^3}\right)}{P_M\left(44.01\frac{g}{mol}\right)} \tag{1.86}$$

We can estimate the CO$_2$ concentration in Morocco from the world concentration by using the following rule:

$$C_{\text{Mor}} = C_W \frac{E_{\text{Mor}}}{q E_W} \tag{1.87}$$

where

C_W: Global CO_2 concentration in year N (ppm).

C_{Mor}: Concentration CO_2 in Morocco in year N (ppm).

E_W: World CO_2 emissions in year N (ton).

E_{Mor}: Emissions CO_2 in Morocco in year N (ton).

q: Represents the volumeproportion ($V_{Mor} = q.V_W$).

P_M: CO_2 Molecular weight.

Based on the inventories of greenhouse gas emissions in Morocco, we note a clear increase in CO_2 emissions over the years. We introduce all the data in the following Table 1.1 (Dlugokencky Ed, 2016; Friedlingstein et al., 2020a, b; MTEDD, 2020; Ritchie et al., 2020). However, it should be noted that there is an inconsistency between the national data established in (MTEDD, 2020) and other data established in international databases, which can lead to false estimates.

1.6.1.2 Human population

Referring to the Moroccan "Haut Commissariat au Plan" (HCP) data, we can identify the national data for this variable (HCP-Population, 2014). The data from 2014 onward are only estimates based on the 2014 national census.

1.6.1.3 Forest biomass

By referring to the FAO or the World Bank data, we established the forest history in Morocco, nevertheless, these data, although appearing correct, are incoherent compared to the data of the Moroccan HCP commission (HCP, 2014). Also, we could not find data in tons to be able to incorporate them directly into the studied model, except for year 2018 in (Hajji, 2018) which is estimated approximately to 11 Million tons.

Table 1.1 Global versus Moroccan Carbon Dioxide Emissions

Year	2004	2010	2018	2020
Moroccan CO_2 emissions in Mt CO_2 eq	58.7	75	90.9	95.5
Worldwide CO_2 emissions in Gt CO_2 eq	34	38.47	42.1	39.94
Worldwide CO_2 level in ppm	378	390	409	414
Estimated CO_2 level for Morocco in ppm	137	159	185	208

A way to quantify the emissions of various gases on an equal footing (same measurement scale) to take into consideration their effect on the climate. It describes, for a given mixture and quantity of greenhouse gases, the quantity of CO_2 that would have the same global warming power.
The CO_2 emissions in Morocco in 2020 are estimated by data extrapolation between 2004 and 2018. The international data also include CO_2 emissions due to land use change, which are continuously decreasing, but these emissions are between 4 and 5 Gt CO_2 equivalent.

1.6.2 Parameters identification

1.6.2.1 Parameter Q_0

Due to the distinction vagueness between natural and anthropogenic CO_2 emissions and the lack of national data concerning these emissions, we choose to use studies carried out in regions nearby Morocco to get an idea about Q_0 parameter. Using the main natural sources of CO_2 emissions and their degree of involvement in global warming, namely, volcanic eruptions (Cooper et al., 2018; Gerlach, 2011), human respiration, and forest fires (GEO, n.d.), the impact of the above natural sources through the parameter Q_0 is estimated to be 10% [5,6,7] of the average differences in CO_2 concentrations between two successive years.

1.6.2.2 Parameter λ

The parameter can be computed by calculating the CO_2 annual anthropogenic increase rate emissions and dividing it by the human population.

1.6.2.3 Parameter α

This coefficient has a close relationship with carbon's lifetime in nature (Misra et al., 2015; Misra & Verma, 2013).

1.6.2.4 Parameter λ_l

In absence of national data relating to CO_2 emissions' absorption by forests and according to (Harris et al., 2021), forests absorb $7.6 * 10^9$ tonne/year, which represents twice what they emit CO_2 per year, which induces a CO_2 reduction of $3.8 * 10^9$ tonne/year. In addition, on a global scale, we got:

$$7.7826 \, Gt \, CO_2 eq \Leftrightarrow 1 \, ppm \tag{1.88}$$

Then the rate of reduction CO_2 due to the forest is identified as:

$$T_{Red_{An}} = 0.4883 \, ppm/year \tag{1.89}$$

Therefore, referring to national data regarding variable values, we get:

$$\lambda_1 = \frac{T_{Red_{An}}}{CF} \tag{1.90}$$

---

[5] Volcanic activities are sources of CO_2 emissions up to 0.5 GtCO$_2$eq/year approximating 1.25% of the world emissions.

[6] Each person releases 300 kt/year of CO_2 by breathing process, which represents up to 6% of global CO_2 emissions.

[7] The 2020 summer fires caused nearly 1.3 GtCO$_2$ eq of emissions, equivalent to 3.25% of emissions.

1.6.2.5 Parameter s

According to Farkas (2001), the human population's intrinsic growth rate can be computed by:

$$s = \frac{ln(\frac{N(t_2)}{N(t_1)})}{(t_2 - t_1)} \tag{1.91}$$

1.6.2.6 Parameter L

According to Farkas (2001), the parameter L can be estimated graphically. From a practical point of view, one can use the Verhulst model (Magal & Malthus, n.d.; Mujib et al., 2019) which states that a population can be presented in the form:

$$N(t) = \frac{L}{A.e^{-s.t} + 1} \tag{1.92}$$

where

$$A = \frac{L - N_0}{N_0}$$

As a result (Mujib et al., 2019):

$$L = \frac{(N_2 * (N_1 * N_2 - 2 * N_1 * N_3 + N_2 * N_3))}{(N_2^2 - N_1 * N_3)} \tag{1.93}$$

where N_i represent the populations in three different years.

1.6.2.7 Parameter θ

In the paper (Bressler, 2021) a new indicator called *"Human Cost of Carbon"* is calculated, which establishes the additional deaths generated by the emission of one ton CO_2. According to the study, 4,434 tons of CO_2 are enough to kill one person in the world in the next 80 years. For example, using the CO_2 emissions in Morocco in 2018 (see Table 1.1), this amount (leads to 256 additional deaths per year from 2018 and due to 2018 emissions. However, if we take into consideration the cumulative 80 years preceding the selected date, the amount will increase drastically to:

$$D_{CO_2} = 20,500 \text{ pers/year} \tag{1.94}$$

Which begets:

$$\theta = \frac{D_{CO_2}}{CN} \tag{1.95}$$

1.6.2.8 Parameter ϕ

Due to the lack of data on this coefficient, we will try to estimate it indirectly. Based on the data found, the legally extracted wood volume from Moroccan forests is comprised between 600,000 and 800,000 m³/year, while the amount extracted illegally has reached "three times the legally extracted wood" (MTEDD, 2020). Moreover, the Moroccan forest density woods (ELLATIFI, 2012) has an average of 600 kg/m³. This leads to forest biomass anthropogenic reduction around:

$$\text{Foret}_D = 1 \text{ Mt/year} \tag{1.96}$$

Consequently:

$$\phi = \frac{\text{Foret}_D}{\text{NF}} \tag{1.97}$$

1.6.2.9 Parameter μ

From (Farkas, 2001) the forest biomass intrinsic growth rate is calculated by:

$$\mu = \frac{ln(\frac{F(t_2)}{F(t_1)})}{(t_2 - t_1)} \tag{1.98}$$

Remark 1.2

In the absence of forests data in tons, we assume henceforth, that each square kilometer (km²) will generate the same amount of biomass in tons.

1.6.2.10 Parameter M

The parameter M is estimated by referring to (Magal & Malthus, n.d.; Mujib et al., 2019).

1.6.2.11 Proportion π

To calculate this parameter, it's necessary to possess data on the positive, direct, and limited-time effect, that the human population will have as a result of deforestation.[8] For lack of data, we choose to validate this parameter by aligning with the work of (Misra et al., 2015; Misra & Verma, 2013).

---

[8] Although the relationship between deforestation and human population growth is complex, we choose to differentiate the notion linked to the π parameter from that which evokes deforestation due to human population growth, while specifying that the first notion is local and limited in time while the latter is global and continuous in time.

1.6.2.12 Proportion π_1

It represents the fertilizing effect of CO_2 on forest biomass. Although many studies confirm this effect, there is a conflict about the sustainability of this positive action. Due to lack of data, we choose to validate this parameter by aligning with the work of (Misra et al., 2015; Misra & Verma, 2013).

1.7 CO_2 T-S MODEL ROBUST CONTROLLER SYNTHESIS

Robust control is a common way to counteract the effects of uncertainty in a given dynamical system. Uncertainty is mainly due to the model's intentional simplicity, unknown driving forces' existence, or purely due to random events such as stochastic perturbations. In this perspective and given the uncertain character of the parameters estimated in the previous section, we choose to propose a PDC robust control law for the environmental model (1.1). In the following, we choose to use a method illustrated in (Benzaouia & El Hajjaji, 2014, Section 1.5.2), which takes up the work of (Chadli et al., 2002; Chadli & El Hajjaji, 2005).

1.7.1 Forced uncertain model

The T-S uncertain forced system can be presented as follows:

$$
\begin{cases}
\dot{x}(t) = \sum_{i=1}^{8}\sum_{j=1}^{8} h_i h_j (A_{ij} + \Delta_{A_{ij}}) x(t) \\
y(t) = Cx(t)
\end{cases}
\tag{1.99}
$$

Where $A_{ij} = A_i + B_i.F_j.C$ and $\Delta_{A_{ij}} = \Delta_{A_i} + \Delta_{B_i}.F_j.C$

Assuming that the uncertainties are structured according to:

$$
\begin{aligned}
\Delta_{A_i} &= D_{A_i}\Delta_i(t)E_{A_i} \\
\Delta_{B_i} &= D_{B_i}\Delta_i(t)E_{B_i}
\end{aligned}
\tag{1.100}
$$

where D_{A_i}, D_{B_i}, E_{A_i} and E_{B_i} are known matrices with constant parameters and $\Delta_i(t)$ an uncertainty verifying the condition below:

$$
\Delta_i^T(t)\Delta_i(t) < \mathbb{I}, \ \forall \ 1 \leqslant i \leqslant r
\tag{1.101}
$$

where $\mathbb{I}$ is an identity matrix of appropriate dimension.

1.7.2 LMI formulation

The application of the Theorem 1.15 from (Benzaouia & El Hajjaji, 2014, Section 1.5.2) on the model (1.99) can be summarized by finding matrices N_i, W, S_{ij} and Q and scalars ϵ_{ij}, δ_{ij} such that the following inequalities are verified, $\forall (i,j) \in 1...8 / i < j$:

$$Q > 0$$

$$\begin{pmatrix} T_{ii} + S_{ii} & * & * \\ E_{A_i}Q & -\epsilon_{ii}I & * \\ E_{B_i}N_iC & 0 & -\delta_{ii}I \end{pmatrix} < 0$$

$$\begin{pmatrix} T_{ij} + T_{ji} + S_{ij} + S_{ij}^T & * & * \\ \begin{bmatrix} E_{A_i} \\ E_{A_j} \end{bmatrix}Q & -\begin{bmatrix} \epsilon_{ij}I & 0 \\ 0 & \epsilon_{ji}I \end{bmatrix} & * \\ \begin{bmatrix} E_{B_i}N_i \\ E_{B_j}N_j \end{bmatrix}C & 0 & -\begin{bmatrix} \delta_{ij}I & 0 \\ 0 & \delta_{ji}I \end{bmatrix} \end{pmatrix} < 0$$

$$\begin{pmatrix} S_{11} & S_{12} & ... & S_{18} \\ S_{12}^T & S_{22} & ... & S_{28} \\ \vdots & \vdots & \ddots & \vdots \\ S_{18}^T & S_{28}^T & ... & S_{88} \end{pmatrix} > 0$$

$$CQ = WC$$

$$(1.102)$$

where

$$T_{ij} = QA_i^T + A_iQ + C^TN_j^TB_i^T + B_iN_jC + \epsilon_{ij}D_{A_i}D_{A_j}^T + \delta_{ij}D_{B_i}D_{B_j}^T \qquad (1.103)$$

Then the uncertain system (1.99) with a common matrix C is asymptotically stable and the closed-loop gain by output feedback is given by $F_i = N_iCC^T(CQC^T)^{-1} = N_iW^{-1}$.

1.8 RESULTS AND DISCUSSION

1.8.1 Parameter identification summary for Morocco

Table 1.2 contains the system parameters estimates. However, it's practically impossible to validate those values due to lack and ambiguity in national data.

In addition, the incorporation of calculated parameters (Table 1.2) in model (1.1) allowed us to find the model's maximum values. However, the equilibrium point E_4 (1.18) cannot exist due to the non-respect of an existence's condition (1.20)-3. With qualitative analysis and by taking into consideration the different parameters' magnitude, this last condition can be converted into a simpler sufficient condition:

$$\mu - L\phi > 0 \tag{1.104}$$

Indeed, this last condition is revealed to be invalid with the available parameters. This invalidity can be explained by a very low forest biomass intrinsic growth rate compared to the human population carrying capacity and the much more important deforestation rate double impact. In addition, a computational inconsistency can also be considered due to the current form of the Moroccan forest biomass evolution far from an assumed logistic function, and consequently, the estimation would be imperfect. This being said, we conclude that a Moroccan ecosystem governed by the (1.1) model under Table 1.2's constraints, can only lead to equilibrium points where the extinction of the human population and the forest biomass reigns. Hereafter, we

Table 1.2 Model (1.1) Parameter Estimates Summary in Moroccan Context

Parameter	Estimated Value	Unit
Q_0	3.45×10^{-1}	ppm/year
λ	1.06×10^{-7}	ppm/(year.person)
α	1.16×10^{-2}	year^{-1}
λ_0	2.33×10^{-10}	(year.ton)$^{-1}$
s	1.01×10^{-2}	year^{-1}
L	5.27×10^7	person
θ	3.14×10^{-6}	(ppm.year)$^{-1}$
π	10^{-2}	person/ton
ϕ	5×10^{-9}	(person.year)$^{-1}$
μ	2×10^{-3}	year^{-1}
M	6.05×10^6	ton
π_1	10^{-2}	ton/ppm

choose to take other values for the parameters μ and ϕ^9 to guarantee the existence of the equilibrium point E_4.

$$\phi = 10^{-10} \tag{1.105}$$

$$\mu = 5 \times 10^{-2} \tag{1.106}$$

1.8.2 T-S model computation

To show the proposed method's feasibility and reliability, we choose to use the numerical values from Table 1.2 by using units "Million Tons" and "Million People" to adjust the problem's conditioning. Therefore, the equilibrium point E_4 and the maximum values for CO_2, human population, and forest biomass can be found. In addition, to compare the original (1.1) and T-S (1.43) models, we choose to use an identical initial value (1.108) that corresponds to the current situation in Morocco (Figure 1.1).

$$E_4 = \begin{pmatrix} 294.2341 \\ 47.85 \\ 5.4682 \end{pmatrix}, \quad X_M = \begin{pmatrix} 370.5915 \\ 52.66 \\ 6.047 \end{pmatrix} \tag{1.107}$$

$$X_0 = \begin{pmatrix} 208 \\ 37 \\ 5.97 \end{pmatrix}, \quad x_0 = \begin{pmatrix} -86.3857 \\ -10.879 \\ 0.5021 \end{pmatrix} \tag{1.108}$$

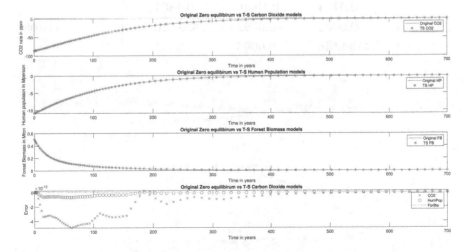

Figure 1.1 Carbon dioxide original versus T-S models.

---

[9] Following a discussion with a Moroccan Forest Department Responsible, we have been assured that $\dfrac{Foret_D}{F} = 0.3\%$.

1.8.3 T-S model behavior

By carrying out several simulations for the model under different conditions, we notice that the system always converges towards the equilibrium point (Figure 1.2). Regarding the CO_2 x_1 behavior, we can emphasize two graphical manifestations. On the one hand, for each initial condition doublet $[x_1, x_2]$, the x_1 behavior remains practically the same disregarding state x_3 initial condition. On the other hand, a small value x_2 leads the variable x_1 to decrease before starting a growth towards the equilibrium point E_4. In addition, the variable x_2 initial condition is the one that dominates the behavior of the human population. Finally, the behavior of the forest biomass is generally governed by its initial condition x_3, with a slight variation according to x_2 initial condition.

We can also note the impact of the parameters λ and ϕ on the behavior of the system. On the one hand, a growing λ will lead to a decrease of the equilibrium population against a growth of the equilibrium forest biomass. On the other hand, the growth of ϕ will impact the existence of the equilibrium point E_4 and transform it into a chaotic system (Misra & Verma, 2013).

1.8.4 Stability analysis

The study of the stability of the unforced nonlinear system (1.43) will be divided into two parts. First, the local stability is discussed according to the evaluation of the Jacobian matrix J^* at the equilibrium point E_4 :

$$J^* = \begin{pmatrix} -0.0184 & 0.1060 & -0.1301 \\ -0.0002 & -0.0092 & 4.7853*10^{-11} \\ 2.4184*10^{-11} & -0.0005 & -0.0452 \end{pmatrix} \tag{1.109}$$

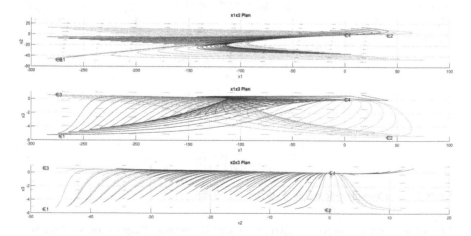

Figure 1.2 Carbon dioxide T-S model states evolution for different initial conditions.

which turns out to have eigenvalues all with negative real part, and therefore the system is locally asymptotically stable, and below the eigenvalues:

$$\Lambda_{vpj^*} = \begin{pmatrix} -0.0160 \\ -0.0116 \\ -0.0452 \end{pmatrix} \tag{1.110}$$

Afterward, the study of global stability is governed by the exploration of the matrix in equality (1.44). From a graphical point of view, we can notice in Figure 1.3 that the equilibrium point E_4 enjoys a large region of attraction, even for trajectories that pass near the other equilibrium points. Further analysis of 2D projection in Figure 1.3, gives an idea on the equilibrium point's E_4 nature by the trajectories shape and direction (Figure 1.4). To summarize, the equilibrium point E_4 has a shape that combines the classical "Stable focus node" and "Stable singular node" shapes on the x_1x_2 level, on the x_2x_3 level, it is more of the "Stable node", and finally on the x_1x_3 plan it is a "Stable singular node" type. Thereafter, the implementation of the matrix inequality by means of Yalmip (Löfberg, 2004) and Mosek (ApS, 2021) in Matlab leads to a feasible problem, and the matrix P allowing it, and consequently the system is globally asymptotically stable.

$$P = 10^4 * \begin{pmatrix} 0.0416 & -0.2124 & 0.1017 \\ -0.2124 & 8.4571 & -1.1666 \\ 0.1017 & -1.1666 & 6.1622 \end{pmatrix} \tag{1.111}$$

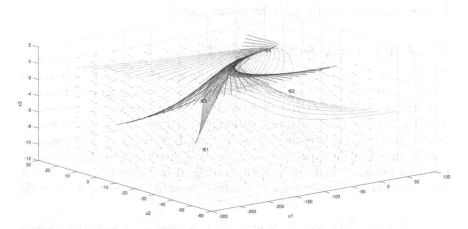

Figure 1.3 T-S model 3D gradient and states trajectories.

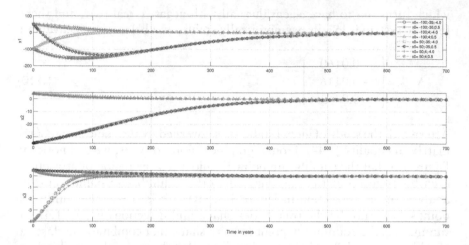

Figure 1.4 T-S 2D gradient and states trajectories.

Remark 1.3

The T-S model construction using $^{3}A_x$ matrix on the whole operating space leads to an unfeasible LMI, which further proves the premise variables' choice importance. It can be concluded that using three premise variables has increased the LMI conservatism since the solver tries to find a common P-matrix for all subsystems.

1.8.5 Forced system control

In this part, we will focus on the stabilization of the forced system assuming the deforestation rate ϕ as a control. Afterward, the results will be extended for the uncertain case through the robust control.

1.8.5.1 Static output feedback control

The forced system (1.62) stabilization has been performed taking into consideration different aspects: T-S structure choice; T-S structure bounding; stabilization methods; matrix C shape. It has been agreed by experimentation that the $^{1}A_x$ structure is the most stable structure[10] in terms of results, conservatism, and computational complexity. If the choice of the construction bounds heavily impacts the $^{2}A_x$ and $^{3}A_x$ structures by dissolving the optimization problem for the large operating areas, it does it also for the $^{1}A_x$ structure, but in a different manner. Indeed, we can always find a controller

---

[10]The word stable used here means the optimization problem-solving aptitude disregarding the T-S construction bounds, without forgetting that all $^{i}A_x$ structures are equivalent.

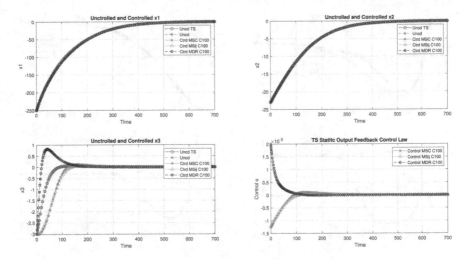

Figure 1.5 Control and states for $x_0 = [-250; -35; -4]$.

that will stabilize the system (1.62) in the ^1A_x structure, by solving one of the LMIs cited in Section 1.5.3, but the performance decreases drastically by choosing the whole operating space. This can be explained by taking into consideration all the possible positions that the system can take during the T-S transformation and consequently during the controller[11] K look-up.

1.8.5.1.1 Stabilization Techniques

By adopting the ^1A_x structure for the transformation to the T-S model, we choose to use output matrix $C = (1\ 0\ 0)$ to perform simulation for different initial conditions x_0 for each available method. Figures 1.5–1.7 show the three methods' ability to stabilize the forced system (1.62) for different initial conditions, but with contrasting performances that differ according to the initial condition, hereafter the controllers K according to the optimization method.

$$K_{1.5.3.\text{MSC}} = 10^{-4} \times [0.0031 \quad 0.0111 \quad -0.0432 \quad 0.0014 \quad -0.1030$$

$$0.0695 \quad -0.0542 \quad 0.0532] \tag{1.112}$$

$$K_{1.5.3.\text{MS}_{ij}} = 10^{-5} \times [0.0020 \quad 0.1070 \quad -0.4346 \quad 0.0672 \quad -0.7166$$

$$0.7205 \quad -0.6847 \quad 0.5300] \tag{1.113}$$

---

[11] Finding a K-controller means satisfying linear constraints that depend directly on the chosen A_x-matrix and the C-matrix.

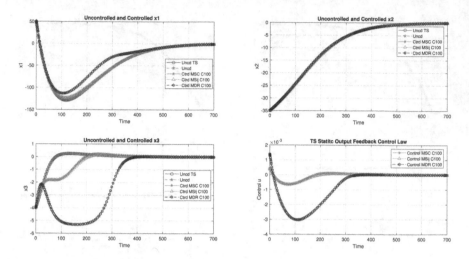

Figure 1.6 Control and states for $x_0 = [50; -23; -3]$.

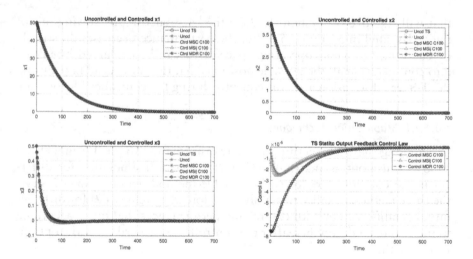

Figure 1.7 Control and states for $x_0 = [50; 4; 0.5]$.

$$K_{1.5.3.MDR} = 10^{-4} \times [-0.0126 \quad 0.0217 \quad -0.0797 \quad 0.1118 \quad -0.0990$$

$$0.1955 \quad -0.0905 \quad 0.1994] \tag{1.114}$$

In general, the performances of the controllers differ according to the initial values, but we can always verify that the decay rate method (MDR) ((Chadli et al., 2001).Cor.4.1) allows a faster convergence than the other methods, especially for the states x_1 and x_3. Also, we can notice that constructed controllers may affect slightly the state x_2 performance, which can be explained by the low value of π, which limits the effect of the controller K on the second equation of the model.

In terms of computational complexity, it is estimated that MSC method ((Chadli et al., 2001).TH.4.1) requires additional computation for almost identical results compared to MS_{ij} method ((Chadli et al., 2001).TH.4.2). However, the latter method's LMIs construction is more complex than the former. Moreover, during the computation of the K controllers, no reference was made to the performance of K through the objective function, except the decay rate maximization for the MDR method, to obtain "optimal" minimalist controllers. From this point of view, the solver lookup for optimal solutions which will simply ensure the system's stabilization.

1.8.5.1.2 Maximized Norm Controllers

To adjust the already used methods' performance and to promote the fast convergence to the equilibrium point, we also tested an additional requirement "maximize the controller norm K" as an "objective" function during the problem optimization. This requirement (Figures 1.8 and 1.9) often leads to an overrun during the convergence, however, it represents an additional cost that exceeds the system's natural limits, hereafter as an indication of the value of K using (Chadli et al., 2001).TH.4.2 (MS_{ij} method).

$$K_{\text{MaxNorm.1.5.3.TH.4.2}} = 10^{-5} \times [0.7294 \quad -0.7294 \quad 0.7294 \quad -0.7294 \quad 0.7294$$

$$-0.7294 \quad 0.7294 \quad -0.7294]$$

$$(1.115)$$

Remark 1.4

During the performed simulations we noted, sometimes, the need for a controller $u = f(\phi)$ that exceeds the parameter's natural limit. The most captivating situation is to get a negative $u = f(\phi)$ to ensure stability. This need can be explained by the fact that in these cases minimizing or even stopping deforestation will not correct the system and that it is necessary to reverse the deforestation pattern by putting up a reforestation plan.

1.8.5.2 Robust stabilization

In this part, we will develop a robust controller to counteract the possible computational hazards of the T-S model already presented in the previous sections. The error estimation in the model can be presented in two ways, where the use of one or the other will depend on the designers' needs:

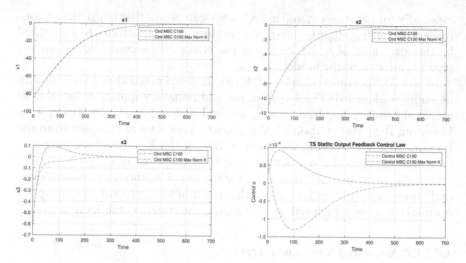

Figure 1.8 Control and states: Maximized Norm Controller for Method 1.5.3.Th4.1.

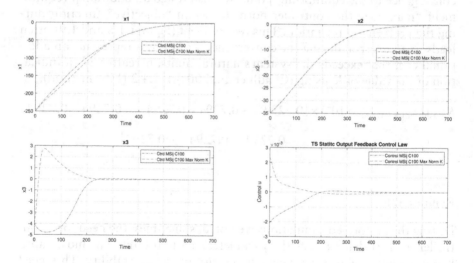

Figure 1.9 Control and states: Maximized Norm Controller for Method 1.5.3.Th4.2.

- Estimating the computational errors of each calculated parameter and deducing the maximum deviation at model level DA_i and DB_i, and it will have the following form:

$$DA_i = \begin{pmatrix} \delta * \alpha_{11} * a_{i11} & \cdots & \delta * \alpha_{i1r} * a_{i1r} \\ \vdots & \ddots & \vdots \\ \delta * \alpha_{r1} * a_{ir1} & \cdots & \delta * \alpha_{irr} * a_{rr} \end{pmatrix} \tag{1.116}$$

$$DB_i = \begin{pmatrix} \delta * \beta_1 * b_{i1} \\ \vdots \\ \delta * \beta_r * b_{ir} \end{pmatrix} \tag{1.117}$$

- Directly require the maximum deviation at the model level on which we intend to build the robust controller.

$$DA_i = \Delta * A_i \tag{1.118}$$

$$DB_i = \Delta * B_i \tag{1.119}$$

The idea of this section is to test the feasibility of the Theorem (Section 1.7.2) while identifying the maximum possible value of δ. In the general case, the incorporation of the scalar δ in the developed conditions will lead to a BMI. However, the use of Yalmip's "bisection" function will allow us to compute the operating range of the scalar Δ or δ. Consequently, the uncertainty margin allowing us to ensure the stabilization of the model, as well as the corresponding controller. Furthermore, there are several ways to decompose a matrix, however, few of them can ensure the expected results. This being said, we use the following decomposition to evaluate their alignment with the objectives set, namely, to cover the matrix space generated by the maximum deviations of the nominal model:

- AI decomposition which consists in taking $D_{A_i} = A_i$ and $D_{B_i} = B_i$, and consequently $E_{A_i} = I_3$ and $E_{B_i} = I_3$;
- IA decomposition which consists in taking $D_{A_i} = I_3$ and $D_{B_i} = I_3$, and therefore $E_{A_i} = A_i$ and $E_{B_i} = B_i$.

In the following Table 1.3, we present the results obtained for the case of model uncertainties. Although the decompositions above characterize the same matrix, we notice that the solutions are largely different according to the chosen decomposition. Also, the choice of the matrix $C = (001)$ allows us to ensure the stabilization for a maximum error margin of around 15%. Also, the matrix $C = (100)$ is the one that delivers the least satisfactory performances in terms of robustness. Finally, the adoption of the uncertainties at the level of the parameters has also been evaluated, where we notice that the use of this method with the choice of the matrices α and β as in (1.120) hereafter, leads to maximum margins either lower or equivalent to model uncertainties, but with different controllers.

$$\alpha = \begin{pmatrix} 2 & 1 & 2 \\ 2 & 3 & 3 \\ 3 & 2 & 3 \end{pmatrix}, \quad \beta = \begin{pmatrix} 0 \\ 3 \\ 3 \end{pmatrix} \tag{1.120}$$

Table 1.3 Maximal Uncertainty According to C Matrix and Decomposition

C Matrix	Decomposition	
	AI	IA
100	[−10.05,0]%	[−4.16,4.16]%
010	[−13.93,0]%	[−4.34,4.34]%
001	[−9.77,0]%	[−15.39,15.39]%
011	[−8.96,0]%	[−10.93,10.93]%

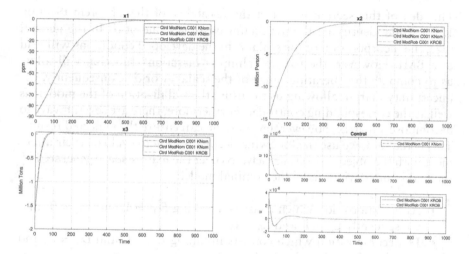

Figure 1.10 Uncertain and nominal System: States and Control for C = (0 0 1).

We choose to illustrate the efficiency of the controller obtained on an uncertain model in comparison with the nominal model in Figure 1.10, where it is obvious that the robust controller ensures the stabilization of the uncertain model, despite the variable uncertainty injected into it.

1.9 CONCLUSION

Referring to some previous results, this chapter deals with both nominal and robust stability and stabilization conditions of continuous nonlinear systems in the Moroccan context. First, an unforced mathematical model linking CO_2 level, human population, and forest biomass is used. Thereafter, the model's parameters for the Moroccan context are computed based on national and international data sources. At that point, using T-S Modeling, the unforced system's stability analysis according to the Lyapunov theory and the linear matrix inequalities (LMI) shows that the equilibrium point E_4 enjoys a large region of attraction. To perform stabilization, the deforestation parameter was considered as control input. Subsequently, the fuzzy model is transformed into forced model. In addition, appropriate LMIs are used to discuss the stabilization performance by reviewing different aspectsas T-S structure choice and bounding, C matrix's shape, the stabilization's technique, and control input. The simulations carried out exposed that the structure ^1A_x is more suitable for the chosen model. Also, it is noticed that the more the T-S bounding space is narrower, well the controller results are. Moreover, the decay rate technique (MDR) gives satisfying results even if the smaller πvalue affects the controller effect. At last, it's noteworthy the need for a controller that exceeds the parameter's natural limit to ensure stability, which proves that in some circumstances stopping deforestation alone cannot solve the problem of limiting the CO_2 levels and other measures must be put in place. Finally, to extract the maximum uncertainty at which the system would be robustly stabilized, an optimization problem is extended using existing LMIs. Though the AI and IA decompositions represent the same matrix, we notice that IA gives the best results far beyond AI. Also, the choice of the matrix $C = (001)$ allows us to ensure the stabilization for a maximum robustness margin. As a perspective it would be interesting to inspect the reforestation added value. The goal would be to combine the efforts of limiting deforestation with that of reforestation to ensure system's stability.

REFERENCES

Agarwal, M., Fatima, T., & Freedman, H. I. (2010). Depletion of forestry resource biomass due to industrialization pressure: A ratio-dependent mathematical model. *Journal of Biological Dynamics*, 4(4), 381–396. https://doi.org/10.1080/17513750903326639

Agarwal, M., & Pathak, R. (2015). Conservation of forestry biomass with the use of alternative resource. *Open Journal of Ecology*, 5(4), 87–109. https://doi.org/10.4236/oje.2015.54009

Ait Kaddour, A., El Mazoudi, E. H., Benjelloun, K., & Elalami, N. (2015). Static output-feedback controller design for a fish population system. *Applied Soft Computing*, 29, 280–287. https://doi.org/10.1016/j.asoc.2014.12.033

Amato, F., Cosentino, C., & Merola, A. (2006). On the region of asymptotic stability of nonlinear quadratic systems. In *14th Mediterranean Conference on Control and Automation, MED'06* (pp. 1–5). https://doi.org/10.1109/MED.2006.328727

ApS, M. (2021). *MOSEK Modeling Cookbook – Release 3.2.3. November* (pp. 1–127). https://docs.mosek.com/modeling-cookbook/index.html

Bailey, D. F., Boyce, W. E., DiPrima, R. C., & Braun, M. (1977). Elementary differential equations and boundary value problems. *The American Mathematical Monthly*, 84(8), 664–665. https://doi.org/10.2307/2321040

Benzaouia, A., & El Hajjaji, A. (2014). *Advanced Takagi–Sugeno Fuzzy Systems* (Vol. 8). Springer International Publishing. https://doi.org/10.1007/978-3-319-05639-5

Bjornlund, L. (2009). *Deforestation*. ReferencePoint Press. https://books.google. co.ma/books?id=M8VaPgAACAAJ

Bremner, J., Carr, D. L., Suter, L., & Davis, J. (2010). Population, poverty, environment, and climate dynamics in the developing world. *Interdisciplinary Environmental Review, 11*(2/3), 112. https://doi.org/10.1504/ier.2010.037902

Bressler, R. D. (2021). The mortality cost of carbon. *Nature Communications, 12*(1). https://doi.org/10.1038/s41467-021-24487-w

Burchardt, H., & Ratschan, S. (2007). Estimating the region of attraction of ordinary differential equations by quantified constraint solving. In *Proceedings of the 3rd WSEAS/IASME International Conference on Dynamical Systems and Control* (pp. 1–12).

Chadli, M. (2002). *Stabilité et commande de systèmes décrits par des multimodèles*. https://tel.archives-ouvertes.fr

Chadli, M., & El Hajjaji, A. (2005). Output robust stabilisation of uncertain Takagi-Sugeno model. In *Proceedings of the 44th IEEE Conference on Decision and Control, and the European Control Conference, CDC-ECC '05, 2005* (pp. 3393–3398). https://doi.org/10.1109/CDC.2005.1582686

Chadli, M., Maquin, D., & Ragot, J. (2001). Stability and stabilisability of continuous Takagi-Sugeno systems. *Journées Doctorales d'Automatique, 1*, 1–6. https://hal.archives-ouvertes.fr/hal-00980988

Chadli, M., Maquin, D., & Ragot, J. (2002). Static output feedback for Takagi-Sugeno systems: An LMI approach. In *10th Mediterranean Conference on Control and Automation MED2002*, CDROM. https://hal.archives-ouvertes.fr/hal-00278220

Chaudhary, M., Dhar, J., & Misra, O. P. (2015). A mathematical model for the conservation of forestry biomass with an alternative resource for industrialization: A modified Leslie Gower interaction. *Modeling Earth Systems and Environment, 1*(4). https://doi.org/10.1007/s40808-015-0056-8

Ciesla, W. M., & FAO. (1997). *Le Changement Climatique, les Forêts et L'aménagement Forestier: Aspects Généraux*. Food and Agriculture Organization of the United Nations. http://digitallibrary.un.org/record/259369

Cooper, C. L., Swindles, G. T., Savov, I. P., Schmidt, A., & Bacon, K. L. (2018). Evaluating the relationship between climate change and volcanism. In *Earth-Science Reviews* (Vol. 177, pp. 238–247). https://doi.org/10.1016/j.earscirev.2017.11.009

Devi, S., & Gupta, N. (2020). Comparative study of the effects of different growths of vegetation biomass on CO_2 in crisp and fuzzy environments. *Natural Resource Modeling, 33*(2), e12263. https://doi.org/10.1111/nrm.12263

Devi, S., & Mishra, R. P. (2020). Preservation of the forestry biomass and control of increasing atmospheric CO_2 using concept of reserved forestry biomass. *International Journal of Applied and Computational Mathematics, 6*(1). https://doi.org/10.1007/s40819-019-0767-z

Dignon, J. (1995). Impact of biomass burning on the atmosphere. In *Ice Core Studies of Global Biogeochemical Cycles* (Vol. 30, pp. 299–311). Springer. https://doi.org/10.1007/978-3-642-51172-1_16

Dlugokencky Ed, T. P. (2016). *Trends in Atmospheric Carbon Dioxide*. NOAA/ESRL. https://gml.noaa.gov/ccgg/trends/global.html

Dubey, B., Sharma, S., Sinha, P., & Shukla, J. B. (2009). Modelling the depletion of forestry resources by population and population pressure augmented
industrialization. *Applied Mathematical Modelling*, 33(7), 3002–3014.
https://doi.org/10.1016/j.apm.2008.10.028

Edwards, C. H., & Penney, D. E. (2013). *Elementary Differential Equations with
Boundary Value Problems Elementary*. Prentice Hall, Hoboken, NJ.

Ellatifi, M. (2012). L'économie de la forêt et des produits forestiers au Maroc: bilan
et perspectives, Thèse de Doctorat en Sciences Economiques.

Elmajidi, A., El Mazoudi, E. H., Elalami, J., & Elalami, N. (2019). A fuzzy logic
control of a polynomial carbon dioxide model. *Ecology, Environment and
Conservation*, 25(2), 876–887.

Farkas, M. (2001). Population dynamics in continuous time. In *Dynamical Models
in Biology* (pp. 17–61). https://doi.org/10.1016/b978-012249103-0/50002-0

Freedman, H. I., & So, J. W. H. (1985). Global stability and persistence of simple
food chains. *Mathematical Biosciences*, 76(1), 69–86. https://doi.org/10.
1016/0025-5564(85)90047-1

Friedlingstein, P., O'Sullivan, M., Jones, M. W., Andrew, R. M., Hauck, J., Olsen,
A., Peters, G. P., Peters, W., Pongratz, J., Sitch, S., Le Quéré, C., Canadell,
J. G., Ciais, P., Jackson, R. B., Alin, S., Aragão, L. E. O. C., Arneth, A.,
Arora, V., Bates, N. R., ... & Zaehle, S. (2020a). *Global Carbon Budget:
Supplemental data of Global Carbon Budget 2020 (Version 1.0) [Data set]*.
https://doi.org/10.18160/gcp-2020

Friedlingstein, P., O'Sullivan, M., Jones, M. W., Andrew, R. M., Hauck, J., Olsen,
A., Peters, G. P., Peters, W., Pongratz, J., Sitch, S., Le Quéré, C., Canadell,
J. G., Ciais, P., Jackson, R. B., Alin, S., Aragão, L. E. O. C., Arneth, A.,
Arora, V., Bates, N. R., ... & Zaehle, S. (2020b). Global carbon budget 2020.
Earth System Science Data, 12(4), 3269–3340. https://doi.org/10.5194/
essd-12-3269-2020

GEO. (n.d.). *Les feux de l'été ont causé des émissions record de CO$_2$*.
https://www.geo.fr/environnement/les-feux-de-lete-ont-cause-des-
emissions-record-de-co2-206360

Gerlach, T. (2011). Volcanic versus anthropogenic carbon dioxide. *Eos,
Transactions American Geophysical Union*, 92(24), 201–202. https://doi.org/
10.1029/2011EO240001

Goreau, T. J. (1992). Control of atmospheric carbon dioxide. *Global Environmental
Change*, 2(1), 5–11. https://doi.org/10.1016/0959-3780(92)90031-2

Gu, K., Chen, J., & Kharitonov, V. (2003). *Stability of Time-Delay Systems*.
Birkhauser, Boston.

Hahn, W. (1967). Stability of motion. In *Stability of Motion*. Springer, Berlin,
Heidelberg. https://doi.org/10.1007/978-3-642-50085-5

Hajji, A. (2018). *Introduction À la Conversion Énergétique De La Biomasse Au
Maroc*. Masen Talents Campus Universiapolis University, Morocco.

Harris, N. L., Gibbs, D. A., Baccini, A., Birdsey, R. A., de Bruin, S., Farina, M.,
Fatoyinbo, L., Hansen, M. C., Herold, M., Houghton, R. A., Potapov, P. V.,
Suarez, D. R., Roman-Cuesta, R. M., Saatchi, S. S., Slay, C. M., Turubanova,
S. A., & Tyukavina, A. (2021). Global maps of twenty-first century forest car-
bon fluxes. *Nature Climate Change*, 11(3), 234–240. https://doi.org/10.1038/
s41558-020-00976-6

HCP. (2014). https://www.hcp.ma

HCP-Population. (2014). https://www.hcp.ma/Population-du-Maroc-par-annee-civile-en-milliers-et-au-milieu-de-l-annee-par-milieu-de-residence-1960-2050_a677.html

IPCC. (2007). IPCC AR4 synthesis report. In *Change* (Vol.46). Cambridge University Press. www.ipcc.ch

Karnosky, D. F., Oksanen, E., Dickson, R. E., & Isebrands, J. G. (2009). Impacts of interacting greenhouse gases on forest ecosystems. In *The Impact of Carbon Dioxide and Other Greenhouse Gases on Forest Ecosystems. Report No.3 of the IUFRO Task Force on Environmental Change* (pp. 253–267). https://doi.org/10.1079/9780851995519.0253

Karnosky, D., & IUFRO Task Force on Environmental Change. (2001). The impact of carbon dioxide and other greenhouse gases on forest ecosystems. Report No.3 of the IUFRO task force on environmental change. In *The Impact of Carbon Dioxide and Other Greenhouse Gases on Forest Ecosystems. Report No.3 of the IUFRO Task Force on Environmental Change*. CABI Pub. https://doi.org/10.1079/9780851995519.0000

Khalil, H. (2002). *Nonlinear Systems;* 3rd ed. Prentice-Hall. https://cds.cern.ch/record/1173048

Lilly, J. H. (2010a). Basic concepts of fuzzy sets. In *Fuzzy Control and Identification* (pp. 11–26). John Wiley & Sons, Ltd. https://doi.org/10.1002/9780470874240.ch2

Lilly, J. H. (2010b). Mamdani fuzzy systems. In *Fuzzy Control and Identification* (pp. 27–45). John Wiley & Sons, Ltd. https://doi.org/10.1002/9780470874240.ch3

Lilly, J. H. (2010c). Modeling and control methods useful for fuzzy control. In *Fuzzy Control and Identification* (pp. 71–87). John Wiley & Sons, Ltd. https://doi.org/10.1002/9780470874240.ch5

Lilly, J. H. (2010d). Parallel distributed control with Takagi-Sugeno fuzzy systems. In *Fuzzy Control and Identification* (pp. 106–120). John Wiley & Sons, Ltd. https://doi.org/10.1002/9780470874240.ch7

Lilly, J. H. (2010e). Takagi-Sugeno fuzzy systems. In *Fuzzy Control and Identification* (pp. 88–105). John Wiley & Sons, Ltd. https://doi.org/10.1002/9780470874240.ch6

Löfberg, J. (2004). YALMIP: A toolbox for modeling and optimization in MATLAB. In *Proceedings of the IEEE International Symposium on Computer-Aided Control System Design* (pp. 284–289). https://doi.org/10.1109/cacsd.2004.1393890

Madhu, M., & Hatfeld, J. L. (2013). Dynamics of plant root growth under increased atmospheric carbon dioxide. *Agronomy Journal, 105*(3), 657–669. https://doi.org/10.2134/agronj2013.0018

Magal, C., & Malthus, A. M. (n.d.). Mathématiques en dynamique des populations (pp. 437–448). https://www.apmep.fr/IMG/pdf/AAA05047.pdf

Mamdani, E. H., & Assilian, S. (1975). An experiment in linguistic synthesis with a fuzzy logic controller. *International Journal of Man-Machine Studies, 7*(1), 1–13. https://doi.org/10.1016/S0020-7373(75)80002-2

Manai, Y., & Benrejeb, M. (2012). Robust stabilization for uncertain Takagi-Sugeno fuzzy continuous model with time-delay based on razumikhin theorem. In *Fuzzy Controllers- Recent Advances in Theory and Applications* (p. 22). https://doi.org/10.5772/48422

Misra, A. K., Lata, K., & Shukla, J. B. (2014). Effects of population and population pressure on forest resources and their conservation: A modeling study. *Environment, Development and Sustainability*, 16(2), 361–374. https://doi.org/10.1007/s10668-013-9481-x

Misra, A. K., & Verma, M. (2013). A mathematical model to study the dynamics of carbon dioxide gas in the atmosphere. *Applied Mathematics and Computation*, 219(16), 8595–8609. https://doi.org/10.1016/j.amc.2013.02.058

Misra, A. K., Verma, M., & Venturino, E. (2015). Modeling the control of atmospheric carbon dioxide through reforestation: Effect of time delay. *Modeling Earth Systems and Environment*, 1(3), 1–24. https://doi.org/10.1007/s40808-015-0028-z

MTEDD. (2020). *4ème Rapport sur l'État de l'Environnement du Maroc- version intégrale Ministère de la Transition Enérgetique et du developpement durable Royaume du Maroc.* https://www.environnement.gov.ma

Mujib, M., Mardiyah, M., Suherman, Rakhmawati, R., Andriani, S., Mardiyah, M., Suyitno, H., Sukestiyarno, S., & Junaidi, I. (2019). The application of differential equation of verhulst population model on estimation of bandar lampung population. *Journal of Physics: Conference Series*, 1155(1), 12017. https://doi.org/10.1088/1742-6596/1155/1/012017

N A S Colloqium. (1997). *(NAS Colloquium) Carbon Dioxide and Climate Change.* National Academies Press, Washington, DC. https://doi.org/10.17226/6238

Nagy-Kiss, A. M. (2010). *Analyse et synthèse de multimodèles pour le diagnostic. Application à une station d'épuration.* https://tel.archives-ouvertes.fr

Nguyen, A.-T., Tanaka, K., Dequidt, A., & Dambrine, M. (2017). Static output feedback design for a class of constrained Takagi–Sugeno fuzzy systems. *Journal of the Franklin Institute*, 354(7), 2856–2870. https://doi.org/10.1016/j.jfranklin.2017.02.017

Oudghiri, M., Chadli, M., & El Hajjaji, A. (2007). One-step procedure for robust output H∞ fuzzy control. In *2007 Mediterranean Conference on Control and Automation, MED* (pp. 1–6). https://doi.org/10.1109/MED.2007.4433964

Pittock, A. B. (2019). *Climate Change: The Science, Impacts and Solutions.* https://doi.org/10.1071/9780643098381

Ritchie, H., Roser, M., & Rosado, P. (2020). *CO2 and Greenhouse Gas Emissions - by Sector.* Our World in Data. https://ourworldindata.org/co2-and-other-greenhouse-gas-emissions

Scherer, C. (2001). *Theory of Robust Control* (pp. 1–160). Delft University of Technology. https://www.imng.uni-stuttgart.de/mst/files/RC.pdf

Shukla, J. B., & Dubey, B. (1997). Modelling the depletion and conservation of forestry resources: Effects of population and pollution. *Journal of Mathematical Biology*, 36(1), 71–94. https://doi.org/10.1007/s002850050091

Shukla, J. B., Freedman, H. I., Pal, V. M., Misra, O. P., Agarwal, M., & Shukla, A. (1989). Degradation and subsequent regeneration of a forestry resource: A mathematical model. *Ecological Modelling*, 44(3–4), 219–229. https://doi.org/10.1016/0304-3800(89)90031-8

Su, X., Shi, P., Wu, L., & Song, Y. D. (2013). A novel control design on discrete-time takagi-sugeno fuzzy systems with time-varying delays. *IEEE Transactions on Fuzzy Systems*, 21(4), 655–671. https://doi.org/10.1109/TFUZZ.2012.2226941

Sugeno, M., Nguyen, H. T., & Prasad, N. R. (1999). *Fuzzy Modeling and Control : Selected Works of M. Sugeno*. CRC Press, Boca Raton, FL.

Sundar, S. (2015). Effect of elevated carbon dioxide concentration on plant growth: A mathematical model. *American Journal of Applied Mathematics and Statistics*, 3(2), 59–67. https://doi.org/10.12691/ajams-3-2-3

Takagi, T., & Sugeno, M. (1985). Fuzzy identification of systems and its applications to modeling and control. *IEEE Transactions on Systems, Man and Cybernetics*, 15(1), 116–132. https://doi.org/10.1109/TSMC.1985.6313399

Tanaka, K., & Sugeno, M. (1992). Stability analysis and design of fuzzy control systems. *Fuzzy Sets and Systems*, 45(2), 135–156. https://doi.org/10.1016/0165-0114(92)90113-I

Tanaka, K., & Wang, H. O. (2003a). Fuzzy observer design. In *Fuzzy Control Systems Design and Analysis* (pp. 83–96). John Wiley & Sons, Ltd. https://doi.org/10.1002/0471224596.ch4

Tanaka, K., & Wang, H. O. (2003b). LMI control performance conditions and designs. In *Fuzzy Control Systems Design and Analysis* (pp. 49–82). John Wiley & Sons, Ltd. https://doi.org/10.1002/0471224596.ch3

Tanaka, K., & Wang, H. O. (2003c). New stability conditions and dynamic feedback designs. In *Fuzzy Control Systems Design and Analysis* (pp. 229–257). John Wiley & Sons, Ltd. https://doi.org/10.1002/0471224596.ch12

Tanaka, K., & Wang, H. O. (2003d). Robust fuzzy control. In *Fuzzy Control Systems Design and Analysis* (pp. 97–108). John Wiley & Sons, Ltd. https://doi.org/10.1002/0471224596.ch5

Tanaka, K., & Wang, H. O. (2003e). T-S fuzzy model as universal approximator. In *Fuzzy Control Systems Design and Analysis* (pp. 277–289). John Wiley & Sons, Ltd. https://doi.org/10.1002/0471224596.ch14

Tanaka, K., & Wang, H. O. (2003f). Takagi-Sugeno fuzzy model and parallel distributed compensation. In *Fuzzy Control Systems Design and Analysis* (pp. 5–48). John Wiley & Sons, Ltd. https://doi.org/10.1002/0471224596.ch2

Teesing. (n.d.). *mgm3 to ppm converter*. https://teesing.com

Tennakone, K. (1990). Stability of the biomass-carbon dioxide equilibrium in the atmosphere: Mathematical model. *Applied Mathematics and Computation*, 35(2), 125–130. https://doi.org/10.1016/0096-3003(90)90113-H

Vannelli, A., & Vidyasagar, M. (1985). Maximal lyapunov functions and domains of attraction for autonomous nonlinear systems. *Automatica*, 21(1), 69–80. https://doi.org/10.1016/0005-1098(85)90099-8

Verma, M., & Misra, A. K. (2018). Optimal control of anthropogenic carbon dioxide emissions through technological options: A modeling study. *Computational and Applied Mathematics*, 37(1), 605–626. https://doi.org/10.1007/s40314-016-0364-2

Woodwell, G. M., Hobbie, J. E., Houghton, R. A., Melillo, J. M., Moore, B., Peterson, B. J., & Shaver, G. R. (1983). Global deforestation: Contribution to atmospheric carbon dioxide. *Science*, 222(4628), 1081–1086. https://doi.org/10.1126/science.222.4628.1081

Zadeh, L. A. (1965). Fuzzy sets. *Information and Control*, 8(3), 338–353. https://doi.org/10.1016/S0019-9958(65)90241-X

Zill, D. G., & Wright, W. S. (2012). *Differential Equations with Boundary-Value Problems*. https://books.google.co.ma/books?id=0UX8e0xdOr0C

Chapter 2

Numerical investigation of wave pattern evolution in Gray–Scott model using discontinuous Galerkin finite element method

Satyvir Singh
Nanyang Technological University

CONTENTS

2.1 INTRODUCTION

In development biology, the evolution of pattern formation is a central but still unresolved challenging research topic. Many biological and biomedical concerns include evolution and form changes, which are the consequences of many nonlinear interactions (Harrison, 1993; Meinhardt, 1982; Murray, 1989). Therefore, mathematical modeling and numerical computing play a critical role in comprehending and forecasting the outcomes of such complicated interactions. The reaction and diffusion process of biochemical components generate a wide range of models that can be observed in nature; hence, the coupled reaction-diffusion system emerges.

According to Turing (1952), under some circumstances, a chemical reaction with diffusion may generate complex spatial pattern formation of chemical concentration. There are many prominent examples of coupled reaction-diffusion system, including the Gray–Scott model, which plays an important role in developed biology. This model was proposed by Gray and Scott (1983) as a substitution for the autocatalytic model of glycolysis initiated by Sel'kov (1984). Pearson (1993) made a significant contribution to the study of spot pattern creation in the two-dimensional Gray–Scott model by including the importance of space by removing the restriction of

DOI: 10.1201/9781003367420-2

47

a well-stirred tank. In his study, several complex spot-type patterns were illustrated for a Gray–Scott model via numerical experiments. Mazin et al. (1996) numerically investigated a class of non-equilibrium phenomena in the bistable Gray–Scott model, including stable localized structures, mixed Turing–Hopf modes, interacting fronts, global Turing structures, and spatiotemporal chaos. Reynolds et al. (1994) investigated the one-dimensional time-dependent wave replication, which is filled in the domain with a periodic array of spots. Doelman et al. (1997) studied numerically the one-dimensional Gray–Scott model and observed various wave patterns, including self-replicating, stationary, and traveling waves. Tok-Onarcan et al. (2019) investigated the wave pattern formation of the Gray–Scott reaction-diffusion model in one-dimensional space using trigonometric quadratic B-spline functions.

In this work, our aim is to explore numerically the wave pattern evolution in the one-dimensional Gray–Scott reaction-diffusion model using a high-order discontinuous Galerkin finite element method (DG-FEM). This method is considered a hybrid approach of Finite Element and Finite Volume methods. The DG-FEM method is rapidly being used in the last decades as a computational tool for solving nonlinear partial differential equations that arise in a broad spectrum of scientific and engineering problems (Singh, 2018, 2021a, b, c, 2022; Singh & Battiato, 2021; Singh et al., 2022).

This work provides a one-dimensional mixed modal DG-FEM scheme for solving the Gray–Scott model, which differs from earlier studies. The third-order scaled Legendre basis functions are adopted for DG spatial discretization, while a third-order TVD Runge–Kutta scheme is employed as a temporal discretization. The remaining of this chapter is planned as follows. The mathematical model for Gray–Scott reaction-diffusion equations is presented in Section 2.2. A detailed description of the DG spatial discretization in one-dimensional space is illustrated in Section 2.3. We apply the proposed numerical method to some test problems of the Gray–Scott model, and their wave patterns evolution is discussed in Section 2.4. Finally, the concluding remarks are discussed in Section 2.5.

2.2 GRAY–SCOTT REACTION-DIFFUSION MODEL

The Gray–Scott reaction-diffusion system models the various spatiotemporal patterns, including, self-replicating and pulse-splitting waves, spots, etc., appearing in nature. The irreversible Gray–Scott model governs the chemical reactions (Gray and Scott, 1983)

$$W_1 + 2W_2 \rightarrow P,$$
$$W_2 \rightarrow P$$

(2.1)

in a gel reactor, where W_2 catalyzes its own reaction with W_1 and P is an inert product. To study the wave pattern evolution mathematically, we analyze the irreversible Gray–Scott reaction-diffusion model in one-dimensional space on the infinite line written as

$$\frac{\partial w_1}{\partial t} = \mu_1 \frac{\partial^2 w_1}{\partial z^2} - w_1 w_2^2 + F(1 - w_1) \equiv \mu_1 \frac{\partial^2 w_1}{\partial z^2} + f(w_1, w_2),$$

$$\frac{\partial w_2}{\partial t} = \mu_2 \frac{\partial^2 w_2}{\partial z^2} + w_1 w_2^2 - (F + k) w_2 \equiv \mu_2 \frac{\partial^2 w_2}{\partial z^2} + g(w_1, w_2),$$

$$(2.2)$$

where $w_1(z, t)$ and $w_2(z, t)$ denote the concentrations of two chemical species W_1 and W_2, respectively. The constant F indicates the rate at which W_1 is fed from the reservoir into the reactor, and W_2 is the overall rate of decay of W_2. To guarantee that the localized pulses can propagate, the choice of F and k is small and $0 \leq \delta^2 \ll 1$. The Gray–Scott model (2.2) can be written in a generalized form as

$$\frac{\partial \mathbf{W}}{\partial t} = \mathbf{D} \nabla^2 \mathbf{W} + \mathbf{R}(\mathbf{W}), \qquad (2.3)$$

with

$$\mathbf{W} = \begin{bmatrix} w_1 \\ w_2 \end{bmatrix}, \ \mathbf{D} = \begin{bmatrix} \mu_1 & 0 \\ 0 & \mu_2 \end{bmatrix}, \ \mathbf{R}(\mathbf{W}) = \begin{bmatrix} f(w_1, w_2) \\ g(w_1, w_2) \end{bmatrix}. \qquad (2.4)$$

Here, $\mathbf{D}$ is the diffusion constant matrix and $\nabla^2 \equiv \frac{\partial^2}{\partial z^2}$ is the Laplacian operator. $\mathbf{R}(\mathbf{W})$ denotes the reaction matrix of nonlinear reaction kinetics f and g.

2.3 MIXED MODAL DG-FEM SCHEME

To simulate the Gray–Scott reaction-diffusion model in one-dimensional space, a mixed modal DG-FEM scheme is employed in this section. In this method, an auxiliary variable $\mathbf{S}$ is added in this formulation to address the higher-order derivatives which are considered the derivative of the solution variable (Singh, 2018). Thus, the Gray–Scott model (2.3) is rewritten as a coupled system of $\mathbf{W}$ and $\mathbf{S}$ for the mixed DG-FEM construction,

$$\mathbf{S} - \nabla \mathbf{W} = 0,$$

$$\frac{\partial \mathbf{W}}{\partial t} + \nabla \mathbf{F}(\mathbf{S}) = \mathbf{R}(\mathbf{W}),$$

$$(2.5)$$

where $\nabla F(S) = -D\nabla W$. To formulate the mixed modal DG-FEM scheme, the following Sobolev space

$$V_h^l = \left\{ v_h \in L_2(\Omega) : \ v_h \mid_{I_n} \in P^k(I_n) \right\}, \tag{2.6}$$

where $I_n = [z_{n-1/2}, z_{n+1/2}]$, $n \in N$ is the local element, $\Omega = [z_L, z_R]$ is the computational domain, v_h is the one component for the vector-valued v_h, $P^k(I_n)$ is the space consisting of discontinuous Galerkin polynomial functions of degree at most k. Now the exact solution of W and S are approximated by the DG-FEM polynomial approximation of $W_h \in V_h^l$ and $S_h \in V_h^l$, respectively as

$$S_h(z,t) = \sum_{i=0}^{N_k} S_h^i(t) b_i(z),$$

$$\tag{2.7}$$

$$W_h(z,t) = \sum_{i=0}^{N_k} Z_h^i(t) b_i(z),$$

where Z_h^i and S_h^i denote the modal coefficients of W and S, and $b_i(z)$ represents the basis function of degrees k. Here, the orthogonal scaled Legendre polynomials for $b_i(z)$ are adopted in the computational element $I_n = [-1, 1]$.

$$b_n(\xi) = \frac{2^i (n!)^2}{(2n)!} P^{0,0}(\xi), \quad n \geq 0,$$

$$\tag{2.8}$$

$$\xi = \frac{2(z - z_i)}{\Delta z}, \quad -1 \leq \xi \leq 1,$$

where $P^{0,0}(\xi)$ is the Legendre polynomial function and z_i is the elemental center, respectively. For the numerical simulations, the first four polynomial functions, which correspond to the third-order DG-FEM approximation, are used in this work. These basis functions are defined globally, implying that they are used by all local elements. The DG discretization of the coupled system of Gray–Scot model (2.5) can be obtained by replacing the exact solutions with the corresponding approximations defined in Equation (2.7) and multiplying by a polynomial function $b_h(z)$ and then integrated by parts over the local element I_n. Taking $b_h \in V_h^l$, $W_h \in V_h^l$, and $S_h \in V_h^l$, we obtain

$$\int_{\Omega_e} S_h b_h \, dV + \int_{\Omega_e} \nabla b_h \cdot S_h \, dV - \int_{\partial\Omega_e} b_h S_h \cdot n \, d\Gamma = 0,$$

$$\frac{\partial}{\partial t} \int_{\Omega_e} W_h b_h \, dV - \int_{\Omega_e} \nabla b_h \cdot F(S_h) \, dV + \int_{\partial\Omega_e} b_h F(S_h) \cdot n \, d\Gamma = \int_{\Omega_e} b_h R(W_h) \, dV.$$

$$\tag{2.9}$$

Here n denotes the outward unit normal vector; Γ and V are the boundary and volume of the element Ω_e, respectively. The interface fluxes are not uniquely defined because of the discontinuity in the solution $\mathbf{W}_h$ and $\mathbf{S}_h$ at the elemental interfaces. The functions $\mathbf{F}(\mathbf{S}_h) \cdot n$ and $\mathbf{W}_h \cdot n$ emerging in Equation (2.9) can be substituted by the numerical fluxes at the elemental interfaces denoted by $\mathbf{H}^{BR1}$ and $\mathbf{H}^{aux}$, respectively. Here, the central flux or Bassi-Rebay (BR1) scheme is employed for the viscous and auxiliary numerical fluxes to calculate the flux at elemental interfaces.

$$\mathbf{H}^{aux}\left(\mathbf{W}_h^{int}, \mathbf{W}_h^{ext}\right) = \frac{1}{2}\left[\mathbf{W}_h^{int} + \mathbf{W}_h^{ext}\right],$$

$$\mathbf{H}^{BR1}\left(\mathbf{S}_h^{int}, \mathbf{S}_h^{ext}\right) = \frac{1}{2}\left[\mathbf{F}\left(\mathbf{S}_h^{int}\right) + \mathbf{F}\left(\mathbf{S}_h^{ext}\right)\right]. \tag{2.10}$$

Here, the superscripts (int) and (ext) deliver the interior and exterior states of the elemental interface. As a result, the DG-FEM weak formulation is obtained as

$$\int_{\Omega_e} \mathbf{S}_h\, b_h\, dV + \int_{\Omega_e} \nabla b_h \cdot \mathbf{S}_h\, dV - \int_{\partial\Omega_e} b_h\, \mathbf{H}^{aux}\, d\Gamma = 0,$$

$$\frac{\partial}{\partial t}\int_{\Omega_e} \mathbf{W}_h\, b_h\, dV - \int_{\Omega_e} \nabla b_h \cdot \mathbf{F}(\mathbf{S}_h)\, dV + \int_{\partial\Omega_e} b_h\, \mathbf{H}^{BR1}\, d\Gamma = \int_{\Omega_e} b_h\, \mathbf{R}(\mathbf{W}_h)\, dV. \tag{2.11}$$

In the abovementioned expression, the emerging volume and surface integrals are approximated by the Gaussian–Legendre quadrature rule within the elements to ensure high order accuracy. Finally, the DG-FEM spatial discretization (2.11) may be expressed in semi-discrete ODEs form as

$$\mathbf{M}\frac{d\mathbf{W}_h}{dt} = \mathbf{L}(\mathbf{W}_h), \tag{2.12}$$

where $\mathbf{M}$ and $\mathbf{L}(\mathbf{W}_h)$ are the orthogonal mass matrix, and the residual function, respectively. Here, an explicit form of Strongly Stability Preserving Runge-Kutta method with third-order accuracy (Shu and Osher, 1988) is adopted as the time marching scheme.

$$\mathbf{W}_h^{(1)} = \mathbf{W}_h^n + \Delta t\, \mathbf{M}^{-1}\mathbf{L}(\mathbf{W}_h),$$

$$\mathbf{W}_h^{(2)} = \frac{3}{4}\mathbf{W}_h^n + \frac{1}{4}\mathbf{W}_h^{(1)} + \frac{1}{4}\Delta t\, \mathbf{M}^{-1}\mathbf{L}\left(\mathbf{W}_h^{(1)}\right), \tag{2.13}$$

$$\mathbf{W}_h^{n+1} = \frac{1}{3}\mathbf{W}_h^n + \frac{2}{3}\mathbf{W}_h^{(2)} + \frac{2}{3}\Delta t\, \mathbf{M}^{-1}\mathbf{L}\left(\mathbf{W}_h^{(2)}\right),$$

where $\mathbf{L}\left(\mathbf{W}_h^n\right)$ is the residual approximation at time t_n, and Δt is the suitable time-step value.

2.4 RESULTS AND DISCUSSION

In this section, we present the wave pattern evolution of the Gray–Scott model, which can be easily seen in real life, such as butterfly wings, damping, gastrulation, embryos, multiple spots, Turing patterns, and many more. For this purpose, three test problems of Gray–Scott model are examined by simulating the numerical experiments with the third-order mixed modal DG-FEM scheme. All the following simulations are done with grid points of 301, and time step of $\Delta z = 0.01$.

Problem 2.1

As the first test problem of the Gray–Scott model, the following initial conditions are considered:

$$w_1(z,0) = 1 - 0.5\sin^{100}(\pi z/L),$$
$$w_2(z,0) = 0.25\sin^{100}(\pi z/L). \tag{2.14}$$

in the limited interval $[0, L]$ which is considered long enough to ensure that the wave dynamics of the model are not influenced by boundaries. The homogenous Neumann conditions are chosen for the boundary conditions. The simulation parameters are taken as

$$\mu_1 = 1, \ \mu_2 = 0.01, \ F = 0.01, \text{ and } k = 0.047, 0.053, 0.060. \tag{2.15}$$

Figure 2.1 illustrates the wave pattern evolution of the Gray–Scott model for the parameters $\mu_1 = 1$, $\mu_2 = 0.01$, $F = 0.01$, $k = 0.053$. From the wave profiles, it can be observed that the two solitary pulses generated from the beginning circumstance was broke into further four pulses and moved apart as time progressed until the equilibrium state was obtained. As the k value decreases, there are fewer peaks over the domain. The parameters F and k are frequently used to assess how many peaks a domain can support. We can observe more peaks if we expand the domain to $[0,400]$. The initial conditions of the simulations differ from those in Figure 2.1. We run the simulation of the Gray–Scott model till time $t = 15,000$ to see the traveling pulses' self-replications on a global scale. Figure 2.2 shows the space-time plots of the concentrations w_1, and w_2 in the Gray–Scott model for different k values. For $k = 0.047$, 0.053, and 0.060, the pulses can grow into 14, 9, and 2 peaks, respectively.

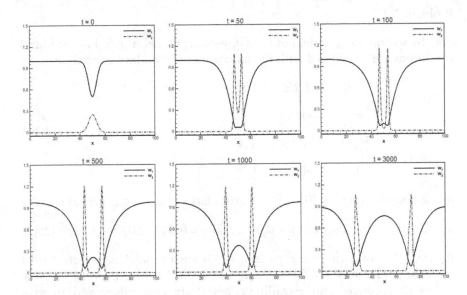

Figure 2.1 Wave pattern evolution of the Gray–Scott model for Problem 2.1 with the parameters $\mu_1 = 1$, $\mu_2 = 0.01$, $F = 0.01$, $k = 0.053$ at different time intervals.

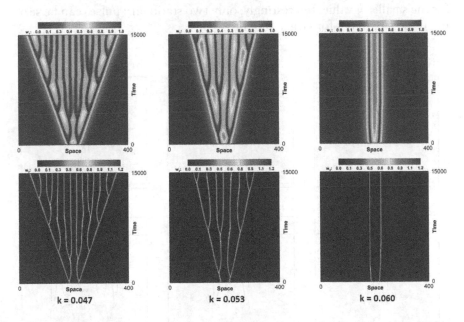

Figure 2.2 Space-time plots of concentrations w_1 and w_2 in the Gray–Scott model for Problem 2.1 with the parameters $\mu_1 = 1$, $\mu_2 = 0.01$, $F = 0.01$ with different k values $(k = 0.047, 0.053, 0.060)$.

Problem 2.2

For the second test problem of the Gray–Scott model, the following initial conditions are considered with the homogenous boundary conditions:

$$w_1(z,0) = \begin{cases} 0.5, & 1.2 \le z \le 1.3 \\ 1, & \text{else.} \end{cases}$$

$$w_2(z,0) = \begin{cases} 0.25, & 1.2 \le z \le 1.3 \\ 0, & \text{else.} \end{cases}$$

(2.16)

in the bounded interval $[0, 2.5]$. The simulation parameters are taken as

$$\mu_1 = 2 \times 10^{-5}, \ \mu_2 = 1 \times 10^{-5}, \ F = 0.023, \text{ and } k = 0.047, 0.050, 0.053. \quad (2.17)$$

Figure 2.3 shows the wave pattern evolution of the Gray–Scott model with the parameters $\mu_1 = 2 \times 10^{-5}$, $\mu_2 = 1 \times 10^{-5}$, $F = 0.023$, $k = 0.050$. New pulses are produced on the trailing edges of previous pulses, and growth-wave patterns and self-splitting pulses have been observed. Figure 2.4 displays the space-time plots of the considered concentrations in the Gray–Scott model for the different k values $(k = 0.047, 0.050, 0.053)$. When the simulations are repeated for a longer time, it is clear that no pattern emerges for the smaller k value. Interestingly, only two stationary pulses can be seen in periodic patterns for larger k value.

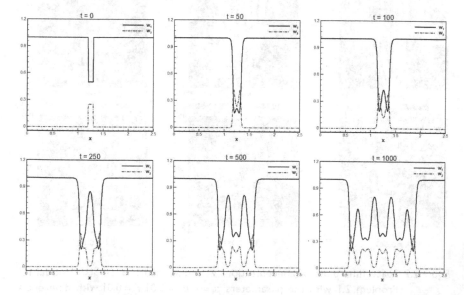

Figure 2.3 Wave pattern evolution of the Gray–Scott model for Problem 2.2 with parameters $\mu_1 = 2 \times 10^{-5}$, $\mu_2 = 1 \times 10^{-5}$, $F = 0.023$, $k = 0.050$ at different time intervals.

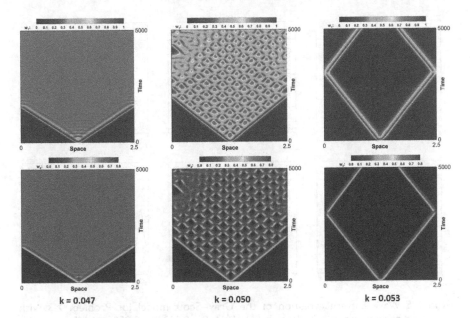

Figure 2.4 Space-time plots of concentrations w_1 and w_2 in the Gray–Scott model for Problem 2.2 with parameters $\mu_1 = 2 \times 10^{-5}$, $\mu_2 = 1 \times 10^{-5}$, $F = 0.023$ with different k values $(k = 0.047, 0.050, 0.053)$.

Problem 2.3

Finally, the last test problem of the Gray–Scott model is considered with the following initial conditions in the bounded interval $[0, 2.5]$:

$$w_1(z,0) = \begin{cases} 0.5, & 0.6 \leq z \leq 0.7 \\ 0.25, & 1.8 \leq z \leq 1.9 \\ 1, & \text{else} \end{cases}$$

$$w_2(z,0) = \begin{cases} 0.25, & 0.6 \leq z \leq 0.7 \\ 0.75, & 1.8 \leq z \leq 1.9 \\ 0, & \text{else.} \end{cases}$$

(2.18)

The simulations are run using Neumann boundary conditions that are homogeneous, and the parameters are set similarly to Problem 2.2. The wave pattern evolutions of the Gray–Scott model for the parameters $\mu_1 = 2 \times 10^{-5}$, $\mu_2 = 1 \times 10^{-5}$, $F = 0.023$, $k = 0.050$ are displayed in Figure 2.5. At longer time, both initial waves are merged and make a more complex pattern. Simulations are run till $t = 5,000$, and the corresponding space-time plots for different k values of the model are shown in Figure 2.6.

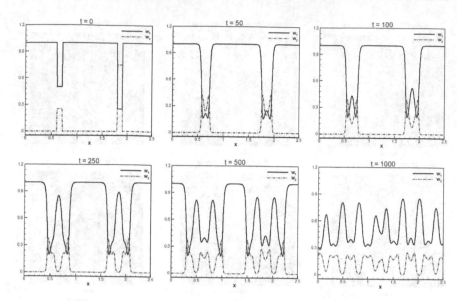

Figure 2.5 Wave pattern evolution of the Gray–Scott model for Problem 2.3 with parameters $\mu_1 = 2 \times 10^{-5}$, $\mu_2 = 1 \times 10^{-5}$, $F = 0.023$, $k = 0.050$ at different time intervals.

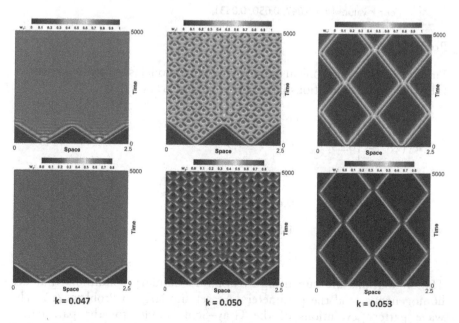

k = 0.047   k = 0.050   k = 0.053

Figure 2.6 Space-time plots of concentrations w_1 and w_2 in the Gray–Scott model for Problem 2.3 with parameters $\mu_1 = 2 \times 10^{-5}$, $\mu_2 = 1 \times 10^{-5}$, $F = 0.023$ with different k values $\left(k = 0.047, 0.050, 0.053 \right)$.

2.5 CONCLUDING REMARKS

This study focuses on the numerical investigation of wave pattern evolution in the Gray–Scott reaction-diffusion model. For this purpose, a mixed modal discontinuous Galerkin finite element method in one-dimensional space is employed to simulate the Gray–Scott model. In this numerical method, a mixed formulation is introduced by adding an extra variable to handle the high-order derivative occurred in diffusion term. For spatial discretization, a hierarchal modal basis function based on scaled Legendre polynomials is utilized, whereas for temporal discretization, an explicit third-order Strongly Stability Preserving Runge-Kutta approach is used. Three different test problems with different initial conditions are adopted to illustrate the wave pattern evolution in the Gray–Scott model. Numerical results reveal that different patterns appear when a modest value in the parameters is changed. The proposed numerical technique, on the other hand, reveals that it is an effective strategy for finding numerical solutions in a wide range of linear and nonlinear physical models. In the future, this method could be used to efficiently solve higher-dimensional reaction-diffusion models.

ACKNOWLEDGMENTS

The author gratefully acknowledges the financial support provided by the Nanyang Technological University, Singapore through the NAP-SUG grant program.

REFERENCES

Doelman, A., Kaper, T.J., & Zegeling, P.A. (1997). Pattern formation in the one-dimensional Gray-Scott model. *Nonlinearity, 10(2)*, 523.

Gray, P., & Scott, S.K. (1983). Autocatalytic reactions in the isothermal, continuous stirred tank reactor: isolas and other forms of multistability. *Chemical Engineering Science, 38(1)*, 29–43.

Harrison, L.G. (1993). *Kinetic Theory of Living Pattern*. Cambridge University Press, Cambridge.

Mazin, W., Rasmussen, K.E., Mosekilde, E., Borckmans, P., & Dewel, G. (1996). Pattern formation in the bistable Gray-Scott model. *Mathematics and Computers in Simulation, 40(3–4)*, 371–396.

Meinhardt, H. (1982). *Models of Biological Pattern Formation*. Academic Press, London.

Murray, J.D. (1989). *Mathematical Biology*. Springer-Verlag, Berlin Hiedelberg.

Pearson, J.E. (1993). Complex patterns in a simple system. *Science, 261(5118)*, 189–192.

Reynolds, W.N., Pearson, J.E., & Ponce, D.S. (1994). Dynamics of self-replicating patterns in reaction-diffusion systems. *Physical Review Letters, 72(17),* 2797.

Sel'Kov, E.E. (1984). Self-oscillations in glycolysis 1. A simple kinetic model. *European Journal of Biochemistry, 4(1),* 79–86.

Shu, W. & Osher, S. (1988). Efficient implementation of essentially non-oscillatory shock-capturing schemes. *Journal of Computational Physics, 77(2),* 439–471.

Singh, S. (2018). *Development of a 3D discontinuous Galerkin method for the second-order Boltzmann-Curtiss-based hydrodynamic models of diatomic and polyatomic gases.* (Doctoral dissertation, Gyeongsang National University South Korea. Department of Mechanical and Aerospace Engineering).

Singh, S. (2021a). Numerical investigation of thermal non-equilibrium effects of diatomic and polyatomic gases on the shock-accelerated square light bubble using a mixed-type modal discontinuous Galerkin method. *International Journal of Heat and Mass Transfer, 169,* 121708.

Singh, S. (2021b). Mixed-type discontinuous Galerkin approach for solving the generalized FitzHugh-Nagumo reaction-diffusion model. *International Journal of Applied and Computational Mathematics, 7,* 207.

Singh, S. (2021c). A mixed-type modal discontinuous Galerkin approach for solving nonlinear reaction-diffusion equations. In *AIP Conference Proceedings* (accepted).

Singh, S. (2022). Computational modeling of nonlinear reaction-diffusion Fisher-KPP equation with mixed modal discontinuous Galerkin scheme. In *Mathematical Modeling for Intelligent Systems,* CRC Press, New York.

Singh, S., & Battiato, M. (2021). An explicit modal discontinuous Galerkin method for Boltzmann transport equation under electronic nonequilibrium conditions. *Computers & Fluids, 224,* 104972.

Singh, S., Karchani, A., Chourushi, T., & Myong, R.S. (2022). A three-dimensional modal discontinuous Galerkin method for second-order Boltzmann-Curtiss constitutive models of rarefied and microscale gas flows. *Journal of Computational Physics, 457,* 111052.

Tok-Onarcan, A., Adar, N., & Dag, I. (2019). Wave simulations of Gray-Scott reaction-diffusion system. *Mathematical Methods in the Applied Sciences, 42(16),* 5566–5581.

Turing, A.M. (1952). The chemical basis of morphogenesis. *Philosophical Transactions of the Royal Society B, 237,* 37–72.

Chapter 3

Modeling the impact of pre-exposure prophylaxis and male circumcision on HIV/AIDS

Prakash Narayan
Pratap University

Kunwer Singh Mathur
Dr. Harisingh Gour Vishwavidyalaya

CONTENTS

NOMENCLATURE

$S(t)$	Susceptible individuals at time 't'
$I(t)$	HIV infected individuals at time 't'
$A(t)$	Individuals with full-blown AIDS at time 't'
$N(t)$	Total population size at time 't'
Λ	Recruitment rate of the individuals
m	Male circumcision rate
ν	Pre-exposure prophylaxis rate
μ	Natural death rate
α_1	The rate at which susceptible individuals become AIDS patients
α_2	Disease induced death rate
β	Transmission rate of HIV virus

DOI: 10.1201/9781003367420-3

δ The rate at which HIV patients become AIDS patients
$S(0)$ Susceptible individuals at time $t = 0$
$I(0)$ Infected individuals at time $t = 0$
$A(0)$ Individuals with full-blown AIDS at time $t = 0$
Ω The region of consideration
ε_0 The infection-free equilibrium point for the model (3.1)
I_0 Infected individuals are not available in the population
S_0 Susceptible individuals when $I_0 = 0$
A_0 Individuals with full-blown AIDS when $I_0 = 0$
ε^* Endemic equilibrium point for the model (3.1)
I^* Infected individuals that are available in the population at $I_0 \neq 0$
S^* Co-existence of susceptible individuals with infected patients
A^* Co-existence of individuals with full-blown AIDS with infected patients
$J[\varepsilon]$ Jacobian matrix of the model
$J[\varepsilon_0]$ Jacobian matrix evaluated at ε_0
$\lambda_{1,2,3}$ Eigenvalues of the jacobian matrix $J[\varepsilon_0]$
R_0 Basic reproduction number for the model (3.1)
Z, W Lyapunov functions
s Small perturbation in S
i Small perturbation in I
a Small perturbation in A
p_1, p_2 Positive constants
$\mathbb{R}^1, \mathbb{R}^2$ One and two-dimensional spaces, respectively
q_1, q_2 Positive constants
E_0 Disease free equilibrium point for the model (3.17)
E^* Endemic equilibrium point for the model (3.17)
R_1 Basic reproduction number for the model (3.17)
x_i State variables
y_j Parameters
$\Gamma^{x_i}_{y_j}$ The sensitivity indices of state variables x_i with respect to the parameters y_j

3.1 INTRODUCTION

A virus known as the human immunodeficiency virus (HIV) causes acquired immunodeficiency syndrome (AIDS). From the beginning, HIV/AIDS has been one of the most deadly diseases in the world, and it continues to be a major global public health issue, having claimed more than 35 million lives so far. In 2017, nearly 9,40,000 people died due to HIV-related causes globally (WHO, 2019). However, the origin of HIV has been a subject of scientific research and debate since the virus was identified in the 1980s. It is also one of the most dangerous sexually transmitted diseases (STDs), with a high impact on the population of various countries in the world. There has been no permanent cure for this disease till now, but some combinations of medicines are used to treat the disease, which is known as highly active antiretroviral therapy (HAART) (Granich et al., 2010). In the history of this infection, there are only two people who are completely cured of this disease. Timothy Ray Brown, an HIV/AIDS patient, had been treated successfully for HIV in 2007, and more than a decade later, the second case was reported on March 6, 2019 in London. The patient was treated with stem cell transplant from donors carrying a genetic mutation that prevent the expression of an HIV receptor CCR5, but this approach is not standard for eradication of the infection. AIDS is the leading cause of death in Sub-Saharan Africa, and several countries of this region have failed to bring the level of the epidemic under control. Therefore, Sub-Saharan Africa has the maximum number of HIV-infected individuals in comparison to other parts of the world. ART is used by the patient as a medication to treat the disease significantly (Arts & Hazuda, 2012). Oral pre-exposure prophylaxis is used daily by HIV-negative people to block the acquisition of the virus. According to the World Health Organization's key facts (WHO, 2019), more than ten randomized controlled studies have demonstrated the effectiveness of pre-exposure prophylaxis in reducing HIV transmission among a range of populations, including men who have sex with men, transgender women, high-risk heterosexual couples, and people who inject drugs. Medical male circumcision reduces the risk of heterosexually acquired HIV infection in men by approximately 60% (WHO, 2019).

Mathematical models and computer simulations have now become useful experimental tools for building and testing theories, assessing quantitative conjectures, answering specific questions, determining sensitivities to changes in parameter values, and estimating critical parameters from data (Hethcote, 2000). A review of the literature (Hethcote et al., 1991; Perelson et al., 1993; Garnett & Anderson, 1996; Kirschner & Webb, 1996; Nowak et al., 1997; Jacquez et al., 1988; Nagelkerke et al., 2002; Brauer, 2017) has shown the speedy growth in the mathematical modeling of sexually transmitted diseases, in particular, HIV/AIDS. The heart of the subject lies in the books (Brauer et al., 2012; Martcheva, 2015) on mathematical modeling in epidemiology. Kermack and McKendrick proposed a basic mathematical model in the epidemiology of infectious diseases (Kermack & McKendrick, 1927). Roy M. Anderson discussed the role of mathematical models in the study of HIV transmission and the epidemiology of AIDS (Anderson, 1988). Since HIV emerged in 1981, several studies, including mathematical modeling, have been devoted to understand the transmission of infection and its outbreak and have obtained some fruitful results, which help us to minimize the burden of infection (Huo et al., 2016; Greenhalgh & Hay, 1997; Naresh et al., 2006; Yusuf & Benyah, 2012; Mathur & Narayan, 2018). There are two major types of HIV/AIDS modeling; first one is in human (Cai et al., 2009; Zhang et al., 2011; Cai et al., 2014; Silva & Torres, 2017) and the second one is in vivo (Perelson et al., 1993; Culshaw et al., 2003; Smith & Wahl, 2004; Song & Cheng, 2005; Rong et al., 2007; Srivastava & Chandra, 2010; Perelson & Ribeiro, 2013; Nowak & Bangham, 1996). In addition, male circumcision (Nagelkerke et al., 2007; Reynolds et al., 2004), use of condoms (Greenhalgh et al., 2001), injecting drug users (Greenhalgh & Hay, 1997; Iannelli et al., 1997) and HAART (Carvalho & Pinto, 2016) are separately used by several researchers as control strategies for the disease in their proposed mathematical models. Moreover, some different combinations of these strategies are also used by researchers (Malunguza et al., 2010; Mukandavire et al., 2007; Mhawej et al., 2010).

Several authors also studied the effects of male circumcision and pre-exposure prophylaxis on HIV/AIDS theoretically and mathematically (Nagelkerke et al., 2007, Doyle et al., 2010, Saha & Samanta, 2019). Nagelkerke and his colleagues modeled the public health impact of male circumcision for HIV prevention in high prevalence areas of Africa. They also explored the effect of male circumcision using a compartmental model (Nagelkerke et al., 2007). They found that the large-scale uptake of male circumcision services in African countries with high HIV prevalence led to substantial reductions in HIV transmission over time among both men and women, in which men benefited somewhat more than women, but prevalence among women was also reduced substantially. Reynolds et al. investigated the relationship between male circumcision and the risk of HIV infection and other sexually transmitted infections in India and discovered that circumcised men have a lower risk of HIV infection than uncircumcised men, and laboratory findings suggest that the foreskin is enriched with HIV (Reynolds et al., 2004). Doyle et al. reviewed the literature on adult male circumcision in the prevention of HIV and studied the impact of male circumcision on HIV theoretically and obtained that male circumcision is a highly effective HIV prevention strategy (Doyle et al., 2010). They also concluded that, as a part of a comprehensive HIV prevention package, male circumcision services should be made available to regions with a high HIV incidence. Vissers et al. studied the impact of pre-exposure prophylaxis on HIV in Africa and India and obtained that it reduces the burden of HIV infection (Vissers et al., 2008). Saha and Samanta developed a mathematical model with optimal control of HIV/AIDS prevention through pre-exposure prophylaxis and limited treatment (Saha & Samanta, 2019). They have obtained that pre-exposure prophylaxis potentially reduces the number of new HIV infections. They also observed that the effect of information regarding pre-exposure prophylaxis plays an important role in reducing the disease burden.

Various researchers studied the impact of male circumcision and pre-exposure prophylaxis separately, which is discussed above. According to the available literature, no researcher has used both strategies together in their mathematical models. The current study is different from the previous studies in the sense that we are using pre-exposure prophylaxis and male circumcision to develop a more realistic mathematical model. Therefore, in this chapter, we have developed a simple HIV/AIDS human epidemic model by incorporating these two concepts. The exact modeling of AIDS, however, is not possible due to the variability of the HIV virus. Therefore, we have attempted to propose an epidemic model having more biological meaning and accuracy, which may help in the control or eradication of the disease from the population. In the mathematical model, pre-exposure prophylaxis is given to all those people who have not yet been exposed to disease caused by HIV. Moreover, male circumcision is used for all those individuals who are living in the region of consideration. Hence, the pre-exposure prophylaxis is given to the susceptible individuals with the rate v and male circumcision strategy is used with the rate m. This is the main contribution in the current study. The literature suggests that these strategies are very useful to overcome HIV/AIDS. We have also obtained some beneficial results through the mathematical model, which makes this study more notable.

This chapter is arranged as follows: The mathematical model is proposed in Section 3.2. In Section 3.3, the positivity and boundedness are obtained, and in Section 3.4, we describe the existence of equilibrium points of the model. In Section 3.5, stability analysis of the equilibrium points is determined. Numerical simulation and discussion are given in Section 3.6, and sensitivity analysis is obtained in Section 3.7. Finally, we conclude this chapter in Section 3.8.

3.2 THE MATHEMATICAL MODEL

It is challenging to evaluate the complete dynamics of HIV/AIDS because it is too complex, and we couldn't able to analyze it all at once (Huo & Feng, 2013). Various researchers, mathematicians, and scientists tried to develop some mathematical models in their own ways in the sense that these models are to be more realistic and include most of the aspects related to sexually transmitted diseases. Also, the models can be analyzed easily in terms of doing mathematical calculations to give crucial findings to reduce or eradicate the disease. Hence, a simple SIA mathematical model for the transmission of HIV/AIDS infection is proposed in this section. In the modeling process, we divide the total population size (i.e., $N(t)$) into three mutually exclusive compartments, viz., susceptible $(S(t))$, infectious $(I(t))$ and full-blown AIDS $(A(t))$ at time t, such that $N(t) = S(t) + I(t) + A(t)$. The model is proposed under the following assumptions:

(A1) The model assumes a simple demographic progression in which recently recruited individuals (such as newborns) come into the population at a rate Λ and go out (due to natural death) at the rate μ.

(A2) Infants are not HIV positive, and there is no vertical transmission of the disease.

(A3) We suppose the population is mixing and interacting equivalently. Also, the movement of the population, i.e., the emigration and immigration in the region of consideration is not considered.

(A4) Male circumcision and pre-exposure prophylaxis are used at that time when it is recommended by the counselor or expert of the STD, in particular, HIV/AIDS. Also, their rates must satisfy the condition $0 < (m+v) < 1$.

(A5) The transmission rate of HIV infection (β) is larger than the rate at which HIV patients enter the AIDS class (i.e., $\beta > \delta$). Moreover, the disease-induced death rate (α_2) is smaller than δ (i.e., $\delta > \alpha_2$).

(A6) Some part of susceptible individuals directly becomes AIDS patients instead of primarily joining the infectious class due to discontinuation of pre-exposure prophylaxis or lack of awareness.

(A7) Infectious individuals include all possible ways by which susceptible become infected.

Hence, the graphical representation of the model is shown below in Figure 3.1.

Keeping in mind the above assumptions and the transition diagram (Figure 3.1), the mathematical model reveals the following system of ordinary differential equations:

$$\frac{dS}{dt} = \Lambda - \Lambda(m+v)S - \beta SI - (\mu + \alpha_1)S$$

$$\frac{dI}{dt} = \beta SI - (\delta + \mu)I \qquad , \qquad (3.1)$$

$$\frac{dA}{dt} = \delta I + \alpha_1 S - (\alpha_2 + \mu)A,$$

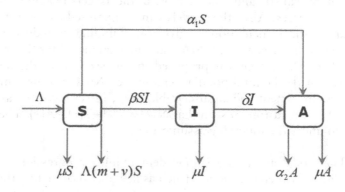

Figure 3.1 Transition diagram of the SIA model.

Table 3.1 Description of Variables and Parameters

Variables/Parameters	Description	Unit
S	Susceptible individuals	Individual
I	HIV infected individuals	Individual
A	Individuals with full-blown AIDS	Individual
N	Total population size	Individual
Λ	Recruitment rate of the individuals	Individual × Time^{-1}
m	Male circumcision rate	Individual^{-1}
v	Pre-exposure prophylaxis rate	Individual^{-1}
μ	Natural death rate	Time^{-1}
α_1	Rate at which susceptible individuals become AIDS patients	Time^{-1}
α_2	Disease-induced death rate	Time^{-1}
β	Transmission rate of HIV virus	Individual^{-1} × Time^{-1}
δ	Rate at which HIV patient becomes AIDS patient	Time^{-1}

with the initial condition: $S(0) > 0$, $I(0) > 0$ and $A(0) \geq 0$, where all the parameters used in the model (3.1) are non-negative and their description is given in Table 3.1.

3.3 POSITIVITY AND BOUNDEDNESS

Non-negativity and boundedness of the solutions will be shown in this section, which are vital for the plausibility of the model. From the second equation of the system (3.1), we get

$$\frac{dI}{dt} \geq -(\delta + \mu) I,$$

which implies that $I(t) \geq I(0) e^{-(\delta + \mu)t} > 0$. Likewise, the third equation can be written as

$$\frac{dA}{dt} \geq -(\alpha_2 + \mu) A,$$

which gives

$$A(t) \geq A(0) e^{-(\alpha_2 + \mu)t} \geq 0.$$

Finally, the first equation gives

$$\frac{dS}{dt} \geq -\left(\Lambda(m + v) + \mu + \alpha_1 + \beta I\right) S.$$

Hence, $S(t) \geq S(0)e^{\left(\int_0^t -(\Lambda(m+v)+\mu+\alpha_1+\beta I(\xi))d\xi\right)} > 0.$

Moreover, we have

$$\frac{dN}{dt} = \Lambda - \Lambda(m+v)S - \alpha_2 A - \mu N \leq \Lambda - \mu N.$$

Clearly, $N(t) \leq \dfrac{\Lambda}{\mu}$ as $t \to \infty$. Therefore, the positivity and boundedness of the model (3.1) are given in the following theorem:

Theorem 3.1

Let $S(0) > 0$, $I(0) > 0$ and $A(0) \geq 0$. Then all the solutions of the system (3.1) positive for all time $t \geq 0$ and bounded in the following feasible region defined as

$$\Omega = \left\{ (S(t),\, I(t),\, A(t)): \; 0 < N(t) = S(t) + I(t) + A(t) \leq \frac{\Lambda}{\mu} \right\}.$$

3.4 THE EXISTENCE OF EQUILIBRIUM POINTS

It is very challenging to eradicate HIV/AIDS infection completely. However, once infected individuals are identified, they can be stopped before they get into the AIDS compartment. Therefore, in this chapter, infection-free solutions mean that only the infected population will be eliminated, and the full-blown AIDS population may still exist. Here, it is obtained that the model (3.1) has the following disease-free equilibrium (DFE), say $\varepsilon_0 = (S_0, I_0, A_0)$, where

$$S_0 = \frac{\Lambda}{\Lambda(m+v)+(\mu+\alpha_1)}, \quad I_0 = 0, \quad A_0 = \frac{\alpha_1 \Lambda}{(\alpha_2+\mu)(\Lambda(m+v)+\mu+\alpha_1)}.$$

Moreover, the unique endemic equilibrium point (EEP), say $\varepsilon^* = (S^*, I^*, A^*)$ exists with

$$S^* = \frac{(\delta+\mu)}{\beta},$$

$$I^* = \frac{(\Lambda(m+v)+\mu+\alpha_1)}{\beta}(R_0 - 1),$$

$$A^* = \frac{\delta\left(\Lambda(m+v)+\mu+\alpha_1\right)(R_0-1)+\alpha_1(\delta+\mu)}{\beta(\alpha_2+\mu)},$$

where $R_0 = \dfrac{\beta\Lambda}{\left(\Lambda(m+v)+\mu+\alpha_1\right)(\delta+\mu)}$ is the basic reproduction number, which defines the total number of secondary infections produced by a single infectious individual in an entirely susceptible population in their entire life span.

3.5 STABILITY ANALYSIS

In this section, the local and global stability of all equilibria of the model system (3.1) is carried out.

3.5.1 Local stability

The Jacobian matrix of the model system (3.1) is given by the following matrix

$$J[\varepsilon] = \begin{pmatrix} -\Lambda(m+v)-\beta I-(\mu+\alpha_1) & -\beta S & 0 \\ \beta I & \beta S-(\delta+\mu) & 0 \\ \alpha_1 & \delta & -(\alpha_2+\mu) \end{pmatrix}.$$

Therefore, the Jacobian evaluated at disease-free equilibrium is

$$J[\varepsilon_0] = \begin{pmatrix} -\Lambda(m+v)-(\mu+\alpha_1) & -\beta S_0 & 0 \\ 0 & \beta S_0-(\delta+\mu) & 0 \\ \alpha_1 & \delta & -(\alpha_2+\mu) \end{pmatrix}.$$

From the above matrix, two eigenvalues $\lambda_1 = -\Lambda(m+v)-(\mu+\alpha_1)$ and $\lambda_2 = -(\alpha_2+\mu)$ are negative. For the stability of DFE, the third eigenvalue must be negative, i.e.,

$$\lambda_3 = \beta S_0-(\delta+\mu) < 0, \text{ which gives } \frac{\beta\Lambda}{\left(\Lambda(m+v)+\mu+\alpha_1\right)(\delta+\mu)} < 1.$$

Therefore, if $R_0 < 1$, then $\lambda_3 < 0$ and hence the DFE is locally asymptotically stable otherwise it is unstable.

Further, the local stability of endemic equilibrium (EE) using the same approach as in (Misra et al., 2015) is given in the following theorem:

Theorem 3.2

The EE exists for $R_0 > 1$ and it is locally asymptotically stable provided that the following condition is satisfied:

$$\frac{1}{2(\alpha_2 + \mu)} < \frac{\Lambda(m+v) + \beta I^* + (\alpha_1 + \mu)}{\alpha_1^2}. \tag{3.2}$$

Proof. Consider the following Lyapunov function

$$Z = \frac{1}{2}s^2 + \frac{p_1}{2}i^2 + \frac{p_2}{2}a^2, \tag{3.3}$$

where p_1 and p_2 are positive constants to be chosen appropriately, while s, i and a are small perturbations in S, I and A, respectively.

Hence, we write $S = S^* + s$, $I = I^* + i$, and $A = A^* + a$.

The derivative of 'Z' with respect to time 't' is

$$\frac{dZ}{dt} = s\frac{ds}{dt} + p_1 i\frac{di}{dt} + p_2 a\frac{da}{dt}. \tag{3.4}$$

After using linearized system of model (3.1) corresponding to endemic equilibrium point ε^*, we get

$$\frac{dZ}{dt} = -\left(\Lambda(m+v) + \beta I^* + \alpha_1 + \mu\right)s^2 - p_1\left((\delta + \mu) - \beta S^*\right)i^2 - p_2(\alpha_2 + \mu)a^2$$

$$+ p_1\left(\beta I^*\right)si + p_2(\delta)ai - \beta S^* si + p_2(\alpha_1)sa. \tag{3.5}$$

Suppose $p_1 = \dfrac{S^*}{I^*}$. then Equation (3.5) reduces to the following equation

$$\frac{dZ}{dt} = -\left(\Lambda(m+v) + \beta I^* + \alpha_1 + \mu\right)s^2 - \frac{S^*}{I^*}\left((\delta + \mu) - \beta S^*\right)i^2$$

$$- p_2(\alpha_2 + \mu)a^2 + p_2(\delta)ai + p_2(\alpha_1)sa. \tag{3.6}$$

The above function dZ/dt will be negative definite provided the following inequalities are satisfied,

$$p_2\alpha_1^2 < 2(\alpha_2 + \mu)\left(\Lambda(m+v) + \beta I^* + \alpha_1 + \mu\right), \tag{3.7}$$

$$p_2\delta^2 < \frac{2S^*}{I^*}(\alpha_2 + \mu)\left((\delta + \mu) - \beta S^*\right). \tag{3.8}$$

Now, since we consider here the co-existence of equilibriums therefore, $I^* \neq 0$ and hence from the second equation of model system (3.1) we get

$(\delta + \mu) = \beta S^*$. Therefore, we obtain the only inequality (3.7) because p_2 is assumed to be positive.

Hence, we can select p_2 in such a way that inequality (3.2) holds.

3.5.2 Global stability

Using a similar approach as in (Castillo-Chavez et al., 2002), the global stability of the DFE is determined. First we divide the model system (3.1) respectively into two general compartments, say $X \in \mathbb{R}^1$ and $Y \in \mathbb{R}^2$, viz., persons without infection and persons with infection (infectious and AIDS patients).

$$\frac{dX}{dt} = \mathbb{F}(X, I),$$

$$\frac{dY}{dt} = \mathbb{G}(X, Y). \tag{3.9}$$

Let $X = (S)$ and $Y = (I, A)$. Then $\mathbb{U}_0 = (X_0, Y_0)$ is the DFE, where $X_0 = S_0$ and $Y_0 = (0, A_0)$.

At $Y = Y_0$, $\mathbb{G}(X, Y_0) = (S_0, 0, A_0)$.

We have

$$\frac{dX}{dt} = \mathbb{F}(X, I) = \Lambda - \Lambda(m+v)S - \beta SI - (\mu + \alpha_1)S.$$

Now, $\dfrac{dX}{dt} = \mathbb{F}(X, 0) = \Lambda - \left(\Lambda(m+v) + \mu + \alpha_1\right)S.$

It is easy to show that $S(t) \to S_0$ as $t \to \infty$.

Hence, $X = X_0 \left(= S_0 = \dfrac{\Lambda}{\Lambda(m+v) + \alpha_1 + \mu} \right)$ is globally asymptotically stable (GAS). Thus, condition (H1) of (Castillo-Chavez et al., 2002) is held.

From equations second and third of the model (3.1), we can write

$$\frac{dY}{dt} = \mathbb{G}(X, Y) = PY - \tilde{Q}(X, Y),$$

where

$$P = \begin{pmatrix} \beta S & 0 \\ \delta + \dfrac{\alpha_1 S}{I} & -(\alpha_2 + \mu) \end{pmatrix} \text{ and } \tilde{Q}(X, Y) = \begin{pmatrix} (\delta + \mu)I \\ 0 \end{pmatrix}.$$

Clearly, P is an M-matrix as off-diagonal elements of it are non-negative and $\hat{Q}(X, Y) > 0$ for $(X, Y) \in \Omega$, where the system has biological meaning.

Thus, both conditions (H1) and (H2) of (Castillo-Chavez et al., 2002) are satisfied. Therefore, the DFE is GAS if $R_0 < 1$.

For the global stability of EE we have the following theorem:

Theorem 3.3

The EE, ε^*, is non-linearly stable if the following inequality hold,

$$\frac{1}{2(\alpha_2 + \mu)} < \frac{\Lambda(m+v) + \mu + \alpha_1 + \beta I}{\alpha_1^2}. \tag{3.10}$$

Proof. Define the following Lyapunov function

$$W = \frac{(S - S^*)^2}{2} + \frac{q_1}{2}(I - I^*)^2 + \frac{q_2}{2}(A - A^*)^2, \tag{3.11}$$

where q_1 and q_2 are positive constants to be chosen appropriately. Taking the derivative of the above equation with respect to 't' along with the solution of model (3.1) and after some algebraic manipulations, we obtain

$$\frac{dW}{dt} = -\left(\Lambda(m+v) + \mu + \alpha_1 + \beta I\right)(S - S^*)^2 - q_1\left((\delta + \mu) - \beta S^*\right)(I - I^*)^2$$

$$- q_2(\alpha_2 + \mu)(A - A^*)^2 - \beta S^*(S - S^*)(I - I^*) + q_1\beta I(S - S^*)(I - I^*)$$

$$+ \alpha_1 q_2(S - S^*)(A - A^*) + q_2\delta(A - A^*)(I - I^*). \tag{3.12}$$

If we take $q_1 = \dfrac{S^*}{I}$ $(\because I \neq 0)$ then the above equation takes the form,

$$\frac{dW}{dt} = -\left(\Lambda(m+v) + \mu + \alpha_1 + \beta I\right)(S - S^*)^2 - \frac{S^*}{I}\left((\delta + \mu) - \beta S^*\right)(I - I^*)^2$$

$$- q_2(\alpha_2 + \mu)(A - A^*)^2 + \alpha_1 q_2(S - S^*)(A - A^*)$$

$$+ q_2\delta(A - A^*)(I - I^*). \tag{3.13}$$

The derivative function dW/dt defined above will be negative definite inside the region Ω, if the following inequalities are satisfied:

$$q_2\alpha_1^2 < 2(\alpha_2 + \mu)\left(\Lambda(m+v) + (\mu + \alpha_1) + \beta I\right), \tag{3.15}$$

$$q_2\delta^2 < \frac{2(\alpha_2+\mu)\big((\delta+\mu)-\beta S^*\big)S^*}{I}.$$ (3.16)

As in the previous case, due to the co-existence of equilibria, inequality (3.16) doesn't hold and hence we obtain (3.15).

Thus, we can easily find positive q_2 from the inequality (3.15) if (3.10) is satisfied.

Remark 3.1

Theorems 3.2 and 3.3 suggest that if male circumcision and pre-exposure prophylaxis are used at the same time, these control strategies produce a more favorable outcome for preventing the individual from contracting HIV/AIDS. The burden of disease can be minimized only by reducing the value of the basic reproduction number R_0 from one.

3.6 NUMERICAL SIMULATION AND DISCUSSION

A deterministic HIV/AIDS epidemic model using control strategies in the form of male circumcision and pre-exposure prophylaxis is analyzed. The parameter $\alpha_1 = 0$ can be taken only when all the people go through a medical checkup, including any kind of tests, so that there is no confusion or uncertainty about the unknown person with HIV/AIDS. Therefore, the model system (3.1) takes the following form:

$$\frac{dS}{dt} = \Lambda - \Lambda(m+v)S - \beta SI - \mu S,$$

$$\frac{dI}{dt} = \beta SI - (\delta + \mu)I,$$ (3.17)

$$\frac{dA}{dt} = \delta I - (\alpha_2 + \mu)A.$$

In this case, system (3.17) has disease-free and endemic equilibria as:

$$E_0 = \left(\frac{1}{\mu + \Lambda(m+v)}, 0, 0\right),$$

$$E^* = \left(\frac{(\delta+\mu)}{\beta}, \frac{\big(\mu+\Lambda(m+v)\big)}{\beta}(R_1-1), \frac{\delta\big(\mu+\Lambda(m+v)\big)}{\beta(\alpha_2+\mu)}(R_1-1)\right),$$

where $R_1 = \dfrac{\beta \Lambda}{(\delta + \mu)(\mu + \Lambda(m + v))}$ is the basic reproduction number for the model (3.17). Clearly, all the results obtained in Section 3.5 for model (3.1) are also satisfied here for model (3.17). The role of R_0 for model (3.1) is the same as the role of R_1 for model system (3.17).

When $\alpha_1 \neq 0$, the analytical findings can be validated by performing some numerical simulation for a set of parameters used in the model (3.1). The values of the parameters like β, α_2, and δ are calculated by using the data provided by the National AIDS Control Organization (NACO, 2016). In contrast, the values of other parameters are chosen intuitively. Here, we perform numerical simulation only for the model (3.1) while a similar approach can be made for the system (3.17). The estimated and assumed values of the parameters are given as follows:

$$\Lambda = 0.956, \quad m = 0.53, \quad v = 0.42, \quad \mu = 0.0067, \quad \beta = 0.9342,$$
$$\delta = 0.8577, \quad \alpha_1 = 0.125, \quad \alpha_2 = 0.6073 \tag{3.18}$$

It is clear that the basic reproduction number $R_0 = 0.993554$ which is less than one and hence only the DFE exists, as shown in Table 3.2. Here, $I_0 = 0$ and in this case, we got $I^* = -0.00717548$, which simply shows that the disease is absent from the population under the condition $R_0 < 1$. Moreover, the DFE is locally asymptotically stable, which is shown in Figure 3.2.

The global stability of DFE is shown in Figure 3.3, which is obtained by using the following set of initial conditions:

$$S(0) = 0.8, \quad I(0) = 0.53, \quad A(0) = 0.3$$
$$S(0) = 0.7, \quad I(0) = 0.5, \quad A(0) = 0.2$$
$$S(0) = 0.809, \quad I(0) = 0.8532, \quad A(0) = 0.7788$$
$$S(0) = 1.489, \quad I(0) = 0.3896, \quad A(0) = 1.8$$
$$S(0) = 3.2, \quad I(0) = 0, \quad A(0) = 0.4$$
$$S(0) = 0.92, \quad I(0) = 0.68, \quad A(0) = 0.3$$
$$S(0) = 1.9, \quad I(0) = 0.1, \quad A(0) = 1.8$$
$$S(0) = 3.195, \quad I(0) = 0, \quad A(0) = 1.604$$

Table 3.2 Numerical Values of the State Variables

Variables	Values	Variables	Values
S_0	0.919319	A^*	0.178349
A_0	0.187158	I^*	-0.00717548
S^*	0.925284	R_0	0.993554

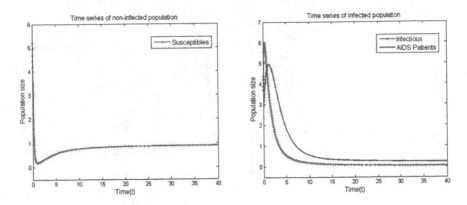

Figure 3.2 Local stability of disease-free equilibrium when $R_0 < 1$.

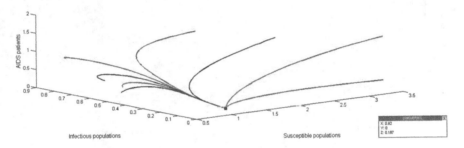

Figure 3.3 Global stability of disease-free equilibrium when $R_0 < 1$.

Table 3.3 Numerical Values of the State Variables

Variables	Values	Variables	Values
S_0	0.963613	A^*	0.249824
A_0	0.196175	I^*	0.0439913
S^*	0.925284	R_0	1.04142

If all the parameters defined in (3.18) remain the same except that $m = 0.5$ and $v = 0.4$, then we obtain the state variables given in Table 3.3. It is obvious from this table that when $R_0 > 1$, then there exists an endemic equilibrium $\varepsilon^* = (S^*, A^*, I^*)$. The EE is locally asymptotically stable if $R_0 > 1$, which is shown in Figure 3.4.

The global stability of the endemic equilibrium is shown in Figure 3.5. The graph is obtained by using the following set of initial conditions:

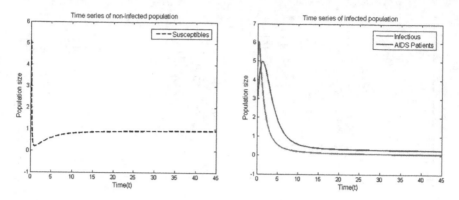

Figure 3.4 Local stability of endemic equilibrium when $R_0 > 1$.

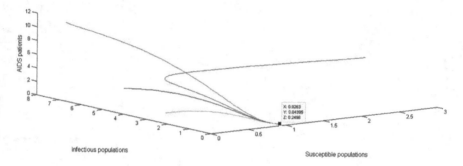

Figure 3.5 Global stability of endemic equilibrium when $R_0 > 1$.

$$S(0) = 2.92, \quad I(0) = 3.04, \quad A(0) = 5.24$$
$$S(0) = 0.18, \quad I(0) = 4.62, \quad A(0) = 3.15$$
$$S(0) = 0.20, \quad I(0) = 7.3896, \quad A(0) = 10.8$$
$$S(0) = 0.22, \quad I(0) = 2.8532, \quad A(0) = 0.7788$$

It has been observed that the basic reproduction number has a great impact on the behavior of the disease. The burden of disease can be minimized only by reducing the value of the basic reproduction number R_0 from one.

The dynamics of R_0 corresponding to m is given in Figure 3.6, which shows that when the value of m is taken to be higher and higher, then the value of R_0 becomes closure and closure to zero. Therefore, R_0 is a decreasing function of m. To control the disease, male circumcision is required higher as much as possible. The graph of v and R_0 is given in Figure 3.7. This graph shows that if we increase the value of v, then the value of R_0 gets nearer to zero and therefore, R_0 is a decreasing function of v. Hence, the

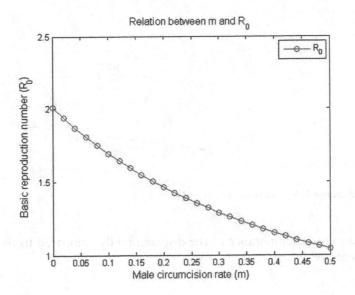

Figure 3.6 The behavior of m versus R_0.

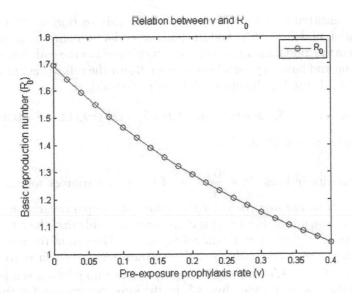

Figure 3.7 The behavior of v versus R_0.

impact of the disease can be controlled by the parameter v, i.e., by increasing the pre-exposure prophylaxis rate. Figure 3.8 depicts the impact of m and v on R_0 at the same time.

Thus, these numerical simulations show that the simultaneous use of male circumcision and pre-exposure prophylaxis is a more efficient way to

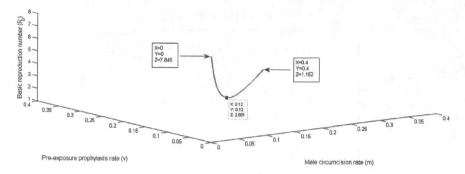

Figure 3.8 Change in R_0 with respect to m and v.

slow down the encumbrance of the disease rapidly compared to the effort made separately.

3.7 SENSITIVITY ANALYSIS

A basic sensitivity analysis is carried out in this section to observe the behavior of model variables and the basic reproduction number concerning changes in the model's parameters. The sensitiveness at the endemic equilibrium point and basic reproduction number R_0 for the following parameters (see Table 3.1) used in the model (3.1) are performed.

$$\Lambda = 0.956, \quad m = 0.5, \quad v = 0.4, \quad \mu = 0.0067, \quad \beta = 0.9342, \quad \delta = 0.8577,$$

$$\alpha_1 = 0.125, \quad \alpha_2 = 0.6073. \tag{3.19}$$

The sensitivity indices $\Gamma_{y_j}^{x_i} = \dfrac{\partial x_i}{\partial y_j} \times \dfrac{y_j}{x_i}$ of the state variables for the model system (3.1) to parameters y_j for the values of the parameters defined in (3.19) are shown in Table 3.4. It is clear from this table that the parameters μ and δ positively effect to S^* while β negatively. The rest of the parameters has no impact on S^*. So, only these three parameters are sensitive to S^*. The meaning of $\Gamma_{\beta}^{S^*} = -1$, is that decreasing (or increasing) β by a given percentage decreases (or increases) always S^* by the same percentage. On the other hand, I^* is effected by every parameter except α_2. The parameters Λ and β make positive impact on I^* while m, v, μ, δ and α_1 have negative impact. Hence, if we increase β and Λ by 10% then I^* increases by 241.406% and 42.046% respectively. Similarly, if we increase m, v and δ by 10% then I^* decreases by 116.311%, 93.049% and 249.458% respectively. Therefore, we can say that the parameters Λ, m, v, β and δ are more sensitive to I^*

Table 3.4 Sensitiveness of the Parameters of Model System (3.1)

Parameters (y_j)	$\Gamma_{y_j}^{S^*}$	$\Gamma_{y_j}^{I^*}$	$\Gamma_{y_j}^{A^*}$	$\Gamma_{y_j}^{R_0}$
Λ	0	+4.2046	+1.0343	+0.1327
m	0	−11.6311	−2.8610	−0.4818
v	0	−9.3049	−2.2888	−0.3854
μ	+0.0078	−0.3579	−0.0931	−0.0145
β	−1	+24.1406	+5.1841	+1
δ	+0.9922	−24.9458	−5.1420	−0.9922
α_1	0	−3.0416	+0.0058	−0.1260
α_2	0	0	−0.9891	0

than others. A^* is affected by all the parameters defined in (3.19). In particular, the parameters Λ, β, and α_1 have a positive impact on it while the rest of the parameters affect negatively to A^*. More clearly, if we increase m, v and δ by 10% then A^* decreases by 28.610%, 22.888% and 51.420% respectively. On the other hand, if we increase Λ and β by 10%, then A^* increases by 10.343% and 51.841%, respectively. The similar conclusion can be obtained for the remaining parameters. Hence, we can say that the parameters m, v, β, δ, and Λ are highly sensitive to A^* than the rest of the parameters. Finally, the basic reproduction number R_0 affected positively only by Λ and β while negatively by other parameters except for α_2 as it has no impact on R_0. The parameters m, v, δ, Λ and β are more sensible to R_0 because when we increase m, v and δ by 10% then R_0 decreases by 4.818%, 3.458% and 9.922% respectively. Similarly, if we increase Λ and β by 10%, then R_0 increases to 1.327% and 10%, respectively.

Graphically, the sensitiveness of the parameters is shown in Figure 3.9.

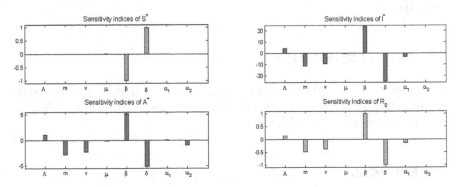

Figure 3.9 Sensitivity analysis of the parameters for EE and R_0 when $R_0 > 1$.

3.8 CONCLUSION

As per the recommendations made by WHO, we have developed and studied an epidemic mathematical model using male circumcision and pre-exposure prophylaxis as two main control strategies. In this chapter, we have obtained two equilibrium points subjected to the value of the basic reproduction number, and we also have determined some sufficient conditions for the local and global stability of both equilibrium points. These conditions show that basic reproduction number R_0 plays a crucial role in the disease outbreak and describes the situation of disease significantly in the population. We have obtained that if $R_0 < 1$, the DFE is locally as well as globally asymptotically stable, and if $R_0 > 1$, the infection may persist in the population before full-blown AIDS. The sensitivity analysis shows that the parameters m and v both are very sensitive to the basic reproduction number R_0 and EE. Moreover, it is shown that both parameters are inversely proportional to R_0. This means that if we increase (or decrease) the value of any one parameter, then the value of R_0 decreases (or increases) accordingly. Therefore, the basic reproduction number R_0 is a decreasing function of both parameters m and v. The parameters m and v make a negative impact on both EE and R_0. Moreover, these analytic findings also reveal that pre-exposure prophylaxis along with male circumcision has a high impact on the spread of HIV/AIDS. Hence, it is concluded that the rates of male circumcision (m) and pre-exposure prophylaxis (v) control the basic reproduction number R_0 and hence they control the spread of infection in the population. Therefore, we can say that m and v are the control parameters for the proposed model (3.1). We may suggest that if male circumcision and pre-exposure prophylaxis are taken at the same time, then these control strategies give more favorable outcomes to save the person from acquiring HIV/AIDS. The simultaneous effect of m and v is demonstrated in Figure 3.8. All the results of this study are in line with the recommendations given by WHO and the results obtained by the researchers in their studies.

ACKNOWLEDGEMENTS

The authors are very glad to thank the anonymous referees for their careful reading, helpful comments, and valuable suggestions, which have helped us to improve the quality of this work.

REFERENCES

Anderson, R. M. (1988). The role of mathematical models in the study of HIV transmission and the epidemiology of AIDS. *Journal of Acquired Immune Deficiency Syndromes, 1*(3), 241–256.

Arts, E. J., & Hazuda, D. J. (2012). HIV-1 antiretroviral drug therapy. *Cold Spring Harbor Perspectives in Medicine*, 2(4), a007161.

Brauer, F. (2017). Mathematical epidemiology: past, present, and future. *Infectious Disease Modelling*, 2(2), 113–127.

Brauer, F., Castillo-Chavez, C., & Castillo-Chavez, C. (2012). *Mathematical Models in Population Biology and Epidemiology* (Vol. 2, p. 508). New York: Springer.

Cai, L., Guo, S., & Wang, S. (2014). Analysis of an extended HIV/AIDS epidemic model with treatment. *Applied Mathematics and Computation*, 236, 621–627.

Cai, L., Li, X., Ghosh, M., & Guo, B. (2009). Stability analysis of an HIV/AIDS epidemic model with treatment. *Journal of Computational and Applied Mathematics*, 229(1), 313–323.

Carvalho, A. R., & Pinto, C. M. (2016). Emergence of drug-resistance in HIV dynamics under distinct HAART regimes. *Communications in Nonlinear Science and Numerical Simulation*, 30(1–3), 207–226.

Castillo-Chavez, C., Feng, Z., & Huang, W. (2002). On the computation of ro and its role on. In *Mathematical Approaches for Emerging and Reemerging Infectious Diseases: An Introduction*, edited by Carlos Castillo-Chavez, Sally Blower, Pauline van den Driessche, Denise Kirschner, & Abdul-Aziz Yakubu (Vol. 1, p. 229). Springer, New York.

Culshaw, R. V., Ruan, S., & Webb, G. (2003). A mathematical model of cell-to-cell spread of HIV-1 that includes a time delay. *Journal of Mathematical Biology*, 46(5), 425–444.

Doyle, S. M., Kahn, J. G., Hosang, N., & Carroll, P. R. (2010). The impact of male circumcision on HIV transmission. *The Journal of Urology*, 183(1), 21–26.

Garnett, G. P., & Anderson, R. M. (1996). Sexually transmitted diseases and sexual behavior: insights from mathematical models. *Journal of Infectious Diseases*, 174(Supplement_2), S150–S161.

Granich, R., Crowley, S., Vitoria, M., Lo, Y. R., Souteyrand, Y., Dye, C., ... & Williams, B. (2010). Highly active antiretroviral treatment for the prevention of HIV transmission. *Journal of the International AIDS Society*, 13(1), 1–8.

Greenhalgh, D., Doyle, M., & Lewis, F. (2001). A mathematical treatment of AIDS and condom use. *Mathematical Medicine and Biology: A Journal of the IMA*, 18(3), 225–262

Greenhalgh, D., & Hay, G. (1997). Mathematical modelling of the spread of HIV/AIDS amongst injecting drug users. *Mathematical Medicine and Biology: A Journal of the IMA*, 14(1), 11–38.

Hethcote, H. W. (2000). The mathematics of infectious diseases. *SIAM Review*, 42(4), 599–653.

Hethcote, H. W., Van Ark, J. W., & Longini Jr, I. M. (1991). A simulation model of AIDS in San Francisco: I. Model formulation and parameter estimation. *Mathematical Biosciences*, 106(2), 203–222.

Huo, H. F., Chen, R., & Wang, X. Y. (2016). Modelling and stability of HIV/AIDS epidemic model with treatment. *Applied Mathematical Modelling*, 40(13–14), 6550–6559.

Huo, H. F., & Feng, L. X. (2013). Global stability for an HIV/AIDS epidemic model with different latent stages and treatment. *Applied Mathematical Modelling*, 37(3), 1480–1489.

Iannelli, M., Milner, F. A., Pugliese, A., & Gonzo, M. (1997). The HIV/AIDS epidemics among drug injectors: a study of contact structure through a mathematical model. *Mathematical Biosciences*, 139(1), 25–58.

Jacquez, J. A., Simon, C. P., Koopman, J., Sattenspiel, L., & Perry, T. (1988). Modeling and analyzing HIV transmission: the effect of contact patterns. *Mathematical Biosciences*, 92(2), 119–199.

Kermack, W. O., & McKendrick, A. G. (1927). A contribution to the mathematical theory of epidemics. *Proceedings of the Royal Society of London. Series A, Containing Papers of a Mathematical and Physical Character*, 115(772), 700–721.

Kirschner, D., & Webb, G. F. (1996). A model for treatment strategy in the chemotherapy of AIDS. *Bulletin of Mathematical Biology*, 58(2), 367–390.

Malunguza, N., Mushayabasa, S., Chiyaka, C., & Mukandavire, Z. (2010). Modelling the effects of condom use and antiretroviral therapy in controlling HIV/AIDS among heterosexuals, homosexuals and bisexuals. *Computational and Mathematical Methods in Medicine*, 11(3), 201–222.

Martcheva, M. (2015). *An introduction to mathematical epidemiology* (Vol. 61, pp. 9–31). New York: Springer.

Mathur, K. S., & Narayan, P. (2018). Dynamics of an SVEIRS epidemic model with vaccination and saturated incidence rate. *International Journal of Applied and Computational Mathematics*, 4(5), 1–22.

Mhawej, M. J., Moog, C. H., Biafore, F., & Brunet-François, C. (2010). Control of the HIV infection and drug dosage. *Biomedical Signal Processing and Control*, 5(1), 45–52.

Misra, A. K., Sharma, A., & Shukla, J. B. (2015). Stability analysis and optimal control of an epidemic model with awareness programs by media. *Biosystems*, 138, 53–62.

Mukandavire, Z., Bowa, K., & Garira, W. (2007). Modelling circumcision and condom use as HIV/AIDS preventive control strategies. *Mathematical and Computer Modelling*, 46(11–12), 1353–1372.

NACO (2016). India, H. I. V. Estimations 2015. Technical Report. Available at http://www.naco.gov.in/upload/2015%20MSLNS/HSS/India%20HIV%20Estimations%202015.pdf. Accessed on 20 April, 2016.

Nagelkerke, N. J., Jha, P., Vlas, S. J. D., Korenromp, E. L., Moses, S., Blanchard, J. F., & Plummer, F. A. (2002). Modelling HIV/AIDS epidemics in Botswana and India: impact of interventions to prevent transmission. *Bulletin of the World Health Organization*, 80, 89–96.

Nagelkerke, N. J., Moses, S., de Vlas, S. J., & Bailey, R. C. (2007). Modelling the public health impact of male circumcision for HIV prevention in high prevalence areas in Africa. *BMC Infectious Diseases*, 7(1), 1–15.

Naresh, R., Tripathi, A., & Omar, S. (2006). Modelling the spread of AIDS epidemic with vertical transmission. *Applied Mathematics and Computation*, 178(2), 262–272.

Nowak, M. A., & Bangham, C. R. (1996). Population dynamics of immune responses to persistent viruses. *Science*, 272(5258), 74–79.

Nowak, M. A., Bonhoeffer, S., Shaw, G. M., & May, R. M. (1997). Anti-viral drug treatment: dynamics of resistance in free virus and infected cell populations. *Journal of Theoretical Biology*, 184(2), 203–217.

Perelson, A. S., Kirschner, D. E., & De Boer, R. (1993). Dynamics of HIV infection of CD4+ T cells. *Mathematical Biosciences*, 114(1), 81–125.

Perelson, A. S., & Ribeiro, R. M. (2013). Modeling the within-host dynamics of HIV infection. *BMC Biology*, 11(1), 1–10.

Reynolds, S. J., Shepherd, M. E., Risbud, A. R., Gangakhedkar, R. R., Brookmeyer, R. S., Divekar, A. D., ... & Bollinger, R. C. (2004). Male circumcision and risk of HIV-1 and other sexually transmitted infections in India. *The Lancet, 363*(9414), 1039–1040.

Rong, L., Feng, Z., & Perelson, A. S. (2007). Emergence of HIV-1 drug resistance during antiretroviral treatment. *Bulletin of Mathematical Biology, 69*(6), 2027–2060.

Saha, S., & Samanta, G. P. (2019). Modelling and optimal control of HIV/AIDS prevention through pre-exposure prophylaxis and limited treatment. *Physica A: Statistical Mechanics and its Applications, 516*, 280–307.

Silva, C. J., & Torres, D. F. (2017). A SICA compartmental model in epidemiology with application to HIV/AIDS in Cape Verde. *Ecological Complexity, 30*, 70–75.

Smith, R. J., & Wahl, L. M. (2004). Distinct effects of protease and reverse transcriptase inhibition in an immunological model of HIV-1 infection with impulsive drug effects. *Bulletin of Mathematical Biology, 66*(5), 1259–1283.

Song, X., & Cheng, S. (2005). A delay-differential equation model of HIV infection of CD4+ T-cells. *Journal of the Korean Mathematical Society, 42*(5), 1071–1086.

Srivastava, P. K., & Chandra, P. (2010). Modeling the dynamics of HIV and CD4+ T cells during primary infection. *Nonlinear Analysis: Real World Applications, 11*(2), 612–618.

Vissers, D. C., Voeten, H. A., Nagelkerke, N. J., Habbema, J. D. F., & de Vlas, S. J. (2008). The impact of pre-exposure prophylaxis on HIV epidemics in Africa and India: a simulation study. *PLoS One, 3*(5), e2077.

WHO. (2019). World Health Organization: Key Facts. https://www.who.int/news-room/fact-sheets/detail/hiv-aids.

Yusuf, T. T., & Benyah, F. (2012). Optimal strategy for controlling the spread of HIV/AIDS disease: a case study of South Africa. *Journal of Biological Dynamics, 6*(2), 475–494.

Zhang, T., Jia, M., Luo, H., Zhou, Y., & Wang, N. (2011). Study on a HIV/AIDS model with application to Yunnan province, China. *Applied Mathematical Modelling, 35*(9), 4379–4392.

Chapter 4

Modeling the effect of media awareness campaigns on the spread of HIV/AIDS

Agraj Tripathi

Pranveer Singh Institute of Technology

Ram Naresh

Harcourt Butler Technical University

CONTENTS

4.1 INTRODUCTION

The HIV (Human Immuno-deficiency Virus) infection has been a very serious public health hazard since its inception and has increased the economic burden globally. By the end of 2019, approximately 38 million people worldwide lived with HIV, while 25.4 million people received antiretroviral therapy (ART). In 2019, 690,000 people died from AIDS-related illnesses, while 32.7 million people have died in total since the epidemic began. In 2019, 1.7 million new cases of HIV infection were reported worldwide. There were 1.7 million new HIV infections worldwide in 2019. Since 1998, the number of new infections has declined by 40%. Since 2010, there have been 2.1 million new HIV infections, but in

DOI: 10.1201/9781003367420-4

2019 there were only 1.7 million. Since 2004, AIDS-related deaths have declined by 60% (UNAIDS, 2020).

Though the survival period of infected persons has increased significantly due to advancements on biomedical front, the behavior change is alternatively accepted as an effective precaution strategy to curtail the spread of the disease. Mass media campaigns can be effective tools to create awareness amongst susceptibles which may motivate them to modify their behavior and to keep them away from risky sexual behavior, bringing down the prevalence of the disease.

In order to restrict the spread of infectious illnesses, the media plays a vital role and is a potent instrument. The mass media campaigns assist in inducing behaviour adjustment as a result of the awareness impact, resulting in a reduction in unsafe sexual engagement. In recent years, several mathematical models have been proposed to investigate the impact of knowledge on infectious disease control (Anderson et al., 1986, Anderson, 1988, Bakare et al., 2014, Dubey et al., 2016, Funk et al., 2010, Kiss et al., 2010, Liu et al., 2007, Liu, 2013, Liu & Cui, 2008, May & Anderson, 1987, Mukandavire & Garira, 2006, Naresh et al., 2011, Rahman & Rahman, 2007, Sahu & Dhar, 2015, Samanta et al., 2013, Sharma & Mishra, 2014, Sharma, 2014, Tchuenche et al., 2011, Tchuenche & Bauch, 2012, Tripathi et al., 2007). In particular, Cui et al. (2008) constructed a mathematical model for infectious illnesses such as SARS in an area under the influence of media awareness and demonstrated the role of media programmes. Liu and Cui, (2008), presented a compartmental model to examine the influence of media coverage on infectious disease spread and its control in a specific place. Using a SIRS model, Funk et al., (2010), studied the connection between infectious illness transmission and awareness in a well-mixed society. They divided the population into six groups based on illness awareness (aware, ignorant) and infection stage (susceptible, infected, recovered), reasoning that the uninformed people are more responsible for disease spread. It has been observed that raising public knowledge lowers the burden of sickness.

In most of the above studies, the transmission rates have been considered as a decreasing function of infected individuals due to media alerts. The cumulative density of awareness programs is assumed to be constant, but in a real situation, it is varying and should be considered in the modeling process. Some mathematical models have been developed by incorporating this idea (Cui et al., 2008, Driessche & Watmough, 2002, Joshi et al., 2015, May & Anderson, 1987, Mukandavire & Garira, 2006, Sahu & Dhar, 2015, Sharma & Mishra, 2014). In particular, Misra et al. (2011), proposed a nonlinear SIS model to study the effect of media awareness programs on the spread of infectious diseases. They suggested that awareness programs can be used to effectively control the spread of disease. Tchuenche and Bauch (2012), developed a Susceptible-Infected-Vaccinated-Recovered (SIVR) epidemic model to study the effect of media broadcasting on the spread and control over an Influenza outbreak. Using optimal control theory, they obtained the effect of costs due to media coverage. Samanta et al. (2013), proposed and analyzed a mathematical model to assess the effect of awareness programs by media on the prevalence of infectious diseases. They concluded that although media awareness can have a substantial effect in controlling disease prevalence, the system shows limit cycle oscillations above a threshold value of their execution rate. Greenhalgh et al. (2015), formulated and analyzed a mathematical model to study the impact of awareness programs on an infectious disease outbreak. These programs induce behavioral changes in the population, wherein the susceptible class was divided into two subclasses namely, aware susceptible and unaware susceptible. Dubey et al. (2016), proposed a model to show the impact of awareness programs as well as treatment in an SIR model. They concluded that if the exposure to the awareness programs is high and adequate treatment is available, the infection can be eliminated.

All the above studies, where the effect of media awareness campaigns has been taken into account, are for curable infectious diseases. It is important to note that in the absence of vaccination, the available treatment of HIV infection does not provide complete immunity to the patients. Thus, a drive of awareness campaigns through mass media is required to be implemented so that people become aware of the disease and its preventive measures. A few mathematical models have been proposed to understand the epidemiology of HIV/AIDS and to help improve our understanding of various factors and intervention strategies like screening of infectives, contact tracing, quarantine, treatment, vaccination etc. affecting the disease spread (Anderson & May, 1991, Driessche & Watmough, 2002, Greenhalgh et al., 2015, Hethcote, 2000, Kumar et al., 2017, Mastahun & Abdurahman, 2017, Misra et al., 2011, 2015, Naresh et al., 2008, 2009, Tchuenche & Bauch, 2012). In particular, Tripathi et al. (2007), proposed a model for HIV infection to study the effect of screening of unaware infectives in a homogeneous population and observed that the endemicity of the infection is reduced when people after becoming aware of their infection do not take part in sexual interaction. The impact of vaccination on the spread of HIV/AIDS in a homogeneously mixing population of variable size structure has also been studied by Naresh et al. (2008). They have shown that if the vaccine efficacy is not high enough, no amount of vaccination could lead to HIV eradication.

These studies do not take into account the impact of media awareness as a preventive measure to stop spreading the HIV/AIDS. Few mathematical models for HIV/AIDS were developed by incorporating a separate class of media awareness programmes. Rahman and Rahman (2007), found that education, occupation, socioeconomic status, the status of household food consumption, area of residence and media exposure have significant contribution in determining HIV/AIDS awareness level. They concluded that it is most urgent to give emphasis on education, alleviation of poverty, ensuring electronic media exposure, head-to-head communication program, institutional-based sex education and necessary information to learn about HIV/

AIDS for the young, adult, and adolescents all over the country. Tripathi and Naresh (2019), established a nonlinear mathematical model to evaluate the impact of media campaigns on lowering HIV/AIDS transmission, concluding that the spread of HIV/AIDS can be greatly decreased when media awareness initiatives are successfully targeted at the susceptible population.

Based on the investigation of the aforementioned studies, successful media campaigns can be highly useful in educating people, who may then choose to avoid contact with infectious individuals. In light of the foregoing, we present a nonlinear mathematical model to investigate the impact of media awareness initiatives on HIV transmission in a population with continual recruitment of susceptibles, in this work. The role of media is incorporated explicitly by assuming the cumulative density of awareness programs as a dynamical variable whose growth depends on the size of HIV-infected individuals. The susceptibles are assumed to become aware upon exposure with media awareness programs, modeled using Holling-type interaction. Numerical studies were also carried out to back up the model system's analytical results.

The design of the chapter is as follows, Section 4.1 contains a general introduction and literature survey. The model formulation is given in Section 4.2 with the region of biologically feasible solutions in Sections 4.3 and 4.4. Existence of equilibria is given in Sections 4.5 and 4.6 and the stability (local and nonlinear) analysis of the steady state is provided in Section 4.7. In Sections 4.8 and 4.9, an optimal control system and its characterization is discussed followed by numerical simulation in Section 4.10. Conclusions are given in Section 4.11.

4.2 FORMULATION OF MATHEMATICAL MODEL

For model building, we have subdivided the total population $N(t)$ at time t, with constant immigration rate Q, into four compartments namely, susceptibles $X(t)$, HIV infectives $Y(t)$, aware susceptibles $X_M(t)$ and AIDS population $A(t)$. The variable $M(t)$ represents the media class. The susceptible individuals $X(t)$ become HIV infected on contact with HIV infectives at a rate β. It is assumed that anti-HIV treatment is not available in the community and therefore, all the HIV infectives are bound to develop full-blown AIDS at a rate δ. The susceptibles join the aware susceptible class $X_M(t)$, when they come into contact with media at a rate λ. Some of the previously conscious susceptible individuals may become susceptible again at a rate ϑ due to fading of the effect of media awareness. Thus, the spread of the disease is assumed to be governed by the following nonlinear differential equations,

$$\frac{dX(t)}{dt} = Q - \frac{\beta XY}{N} - dX - f(M)X + \vartheta X_M \qquad (4.1)$$

$$\frac{dY(t)}{dt} = \frac{\beta XY}{N} - \delta Y - dY \tag{4.2}$$

$$\frac{dX_M(t)}{dt} = f(M)X - \vartheta X_M - dX_M \tag{4.3}$$

$$\frac{dA(t)}{dt} = \delta Y - dA - \alpha A \tag{4.4}$$

$$\frac{dM(t)}{dt} = \mu Y - \sigma M \tag{4.5}$$

where d is the natural mortality rate constant and α is disease-induced death rate constant. The constant μ represents the rate by which awareness programs are being implemented and is assumed to be directly proportional to the HIV infective population. The constant σ represents the depletion rate of these programs due to ineffectiveness, social problems in the population, etc.

The term $f(M) = \dfrac{\lambda M}{1 + \gamma M}$ is used for the effect of media coverage on susceptible population where λ represents the dissemination rate of awareness programs among susceptibles where γ limits the effect of awareness programs on susceptibles. Thus $\dfrac{\lambda}{\gamma}$ is the uppermost effect that media can put on susceptibles (Dubey et al., 2016).

It is assumed that all the dependent variables and parameters of the model system (4.1)–(4.5) are non-negative. Since variable A is not appearing in the above three equations of the model system, we can analyze the following sub-system given by (4.6)–(4.9) using,

$$N(t) = X(t) + Y(t) + X_M(t).$$

$$\frac{dN(t)}{dt} = Q - dN - \delta Y \tag{4.6}$$

$$\frac{dY(t)}{dt} = \frac{\beta(N - Y - X_M)Y}{N} - (\delta + d)Y \tag{4.7}$$

$$\frac{dX_M(t)}{dt} = \frac{\lambda(N - Y - X_M)M}{1 + \gamma M} - (\vartheta + d)X_M \tag{4.8}$$

$$\frac{dM(t)}{dt} = \mu Y - \sigma M \tag{4.9}$$

Continuity of the right-hand side of the system of equations and its derivative implies that the model is well posed for $N > 0$ and $\beta > (\delta + d)$. We may define the threshold, $R_0 = \dfrac{\beta}{\delta + d}$ as the basic reproductive number (Driessche & Watmough, 2002). The basic reproductive number defines the average number of new infections produced by one primary infected individual in a wholly susceptible population (Anderson et al., 1986). From the model, it is noted that in the absence of infection the population size approaches the steady state $\dfrac{Q}{d}$.

4.3 POSITIVITY OF SOLUTIONS

In this section, we prove that all solutions of the system (4.6)–(4.9) with positive initial data will remain positive for all time $t > 0$.

Lemma 4.1

Let the initial data be $N(0) = N_0 > 0$, $Y(0) = Y_0 > 0$, $X_M(0) = X_{M0} > 0$ and $M(0) = M_0 > 0$ for all t. Then the solution $(N(t), Y(t), X_M(t), M(t))$ of the model remains positive for time $t > 0$.

   Proof. From Equation (4.7),

$$\frac{dY(t)}{dt} \geq -(\delta + d)Y \qquad (4.10)$$

   Thus,

$$Y(t) \geq c_1 e^{-(\delta + d)t} > 0 \qquad (4.11)$$

Here c_1 is a constant of integration. In a similar pattern, the remaining equations show that they are always positive for $t > 0$.

4.4 INVARIANT REGION

To explore the stability of equilibria, we need the bounds of dependent variables of the model system (4.6)–(4.9). The invariant region is stated in the form of the following lemma.

Lemma 4.2

The closed set

$$\Omega = \left\{ (N, Y, X_M, M) : \frac{Q}{\delta + d} < N < \frac{Q}{d}, 0 < Y + X_M \leq N, 0 < M < \frac{\mu}{\sigma} \frac{Q}{d} \right\}$$

is the invariant region which attracts all solutions initiating in the interior of the positive octant.

4.5 EXISTENCE OF EQUILIBRIA

The model system (4.6)–(4.9) has two equilibria as listed below,

i. $E_0\left(\dfrac{Q}{d},0,0,0\right)$, the disease-free equilibrium,

ii. $E^*\left(N^*,Y^*,X_M^*,M^*\right)$, the endemic equilibrium in presence of media awareness campaigns.

The existence of E_0 is obvious. In the following, we prove the existence of E^*.

4.6 EXISTENCE OF ENDEMIC EQUILIBRIUM E*(N*, Y*, X*_M, M*)

Here N^*, Y^*, X_M^*, and M^* are positive solutions of the following system of algebraic equations obtained by putting right-hand side of model Equations (4.6)–(4.9) to zero,

$$Q - dN - \delta Y = 0$$

$$\frac{\beta(N - Y - X_M)Y}{N} - (\delta + d)Y = 0$$

$$\frac{\lambda(N - Y - X_M)M}{(1 + \gamma M)} - (\vartheta + d)X_M = 0$$

(4.12)

$$\mu Y - \sigma M = 0$$

On solving simultaneous algebraic Equations (4.12) we obtain,

$$Y = \frac{Q - dN}{\delta} = f(N)$$

$$M = \frac{\mu}{\sigma}f(N)$$

(4.13)

$$X_M = \frac{[\beta - (\delta + d)]N - \beta f(N)}{\beta} = g(N)$$

Substituting these values in the third equation of system (4.12) we get,

$$F(N) = \lambda[N - f(N) - g(N)]\frac{\mu}{\sigma}f(N) - (\vartheta + d)\left[1 + \left(\frac{\gamma\mu}{\sigma}\right)f(N)\right]g(N) = 0$$

(4.14)

To show the existence of E^*, it would be sufficient to show that Equation (4.14) has one and only one positive root between $\frac{Q}{\delta + d}$ and $\frac{Q}{d}$. To prove this, from Equation (4.14) we have,

$$F\left(\frac{Q}{\delta + d}\right) = \frac{Q}{\beta}\left[\frac{\lambda}{\sigma}\left(\frac{\mu Q}{(\delta + d)}\right) + (\vartheta + d)\left(1 + \frac{\gamma}{\sigma}\left(\frac{\mu Q}{(\delta + d)}\right)\right)\right] > 0 \qquad (4.15)$$

$$F\left(\frac{Q}{d}\right) = -\frac{Q}{d}(\vartheta + d)\left(1 - \frac{1}{R_0}\right) < 0. \qquad (4.16)$$

since $R_0 = \dfrac{\beta}{(\delta + d)} > 1.$

Also

$$F'(N) = -\lambda[N - f(N) - g(N)]\frac{\mu d}{\sigma\delta} + \lambda\frac{\mu f(N)}{\sigma}\frac{(\delta + d)}{\beta}$$

$$+ (\vartheta + d)\left[\frac{\gamma\mu d}{\sigma\delta}\right]g(N) - (\vartheta + d)\left[\frac{\sigma + \gamma\mu f(N)}{\sigma}\right]\left[\frac{(\beta - \delta)(\delta + d)}{\beta\delta}\right]$$

(4.17)

If

$$\lambda\frac{\mu f(N)}{\sigma}\frac{(\delta + d)}{\beta} + (\vartheta + d)\left[\frac{\gamma\mu d}{\sigma\delta}\right]g(N) < \lambda[N - f(N) - g(N)]\frac{\mu d}{\sigma\delta}$$

$$+ (\vartheta + d)\left[\frac{\sigma + \gamma\mu f(N)}{\sigma}\right]\left[\frac{(\beta - \delta)(\delta + d)}{\beta\delta}\right], \qquad (4.18)$$

then $F'(N) < 0$. It is easy to note by the intermediate property of calculus that if $F'(N) < 0$, then Equation (4.14) has only one root (say N^*) between $\frac{Q}{\alpha + d}$ and $\frac{Q}{d}$. Using N^*, the values of Y^*, X_M^* and M^* can be found easily.

4.7 STABILITY ANALYSIS

The results of stability analysis of the equilibria are given below in the form of the following theorems.

Theorem 4.1

The disease-free equillibrium E_0 is locally asymptotically stable for $R_0 < 1$ and unstable otherwise.

Theorem 4.2

The endemic equilibrium E^* is locally asymptotically stable under the following conditions,

$$\frac{2\lambda M^*}{1+\gamma M^*} < \frac{1}{3}\left(\frac{\lambda M^*}{1+\gamma M^*} + \vartheta + d\right) \tag{4.19}$$

$$\mu^2 M^* < \frac{\beta Y^* \sigma}{N^*} \tag{4.20}$$

Theorem 4.3

The endemic equilibrium E^* is nonlinearly asymptotically stable under the following conditions, in the invariant region Ω.

$$\frac{2\lambda M^*}{1+\gamma M^*} < \frac{1}{3}\left(\frac{\lambda M^*}{1+\gamma M^*} + \vartheta + d\right) \tag{4.21}$$

$$\mu^2 < \frac{\beta\sigma}{N^*} \tag{4.22}$$

(See Appendices A and B for the proof of Theorems 4.1–4.3)

Remark

When the contact rate of susceptibles with media i.e. $\lambda \to 0$ and implementation of media programs i.e. $\mu \to 0$, then the above conditions are satisfied easily showing that these parameters have a destabilizing effect on the model system.

4.8 OPTIMAL CONTROL SYSTEM

To investigate the optimal level of effort that would be needed to control the disease, we give the objective functional J, which is to minimize the number of HIV infectives. In this section, the model system (4.6)–(4.9) is extended for the formulation of the optimal control model,

$$\frac{dN(t)}{dt} = Q - dN - \delta Y$$

$$\frac{dY(t)}{dt} = \frac{\beta(N - Y - X_M)Y}{N} - (\delta + d)Y$$

$$\frac{dX_M(t)}{dt} = \lambda(t)\frac{(N - Y - X_M)M}{1 + \gamma M} - (\vartheta + d)X_M$$

$$\frac{dM(t)}{dt} = \mu(t)Y - \sigma M$$

(4.23)

The control parameters $\lambda(t)$ and $\mu(t)$ are Lebesgue measurable function on a finite interval [0,T]. The objective functional can be defined as follows,

$$J(\lambda,\mu) = \int_0^T \left[AY(t) - \left(B_1 X_M(t) + B_2 \frac{\lambda^2}{2} + B_3 \frac{\mu^2}{2} \right) \right] dt$$

(4.24)

Here the contact rate λ with media and implementation rate of the media awareness programs μ are both functions of time. Our goal is to find the optimal control λ and μ in order to minimize the objective functional given by $J(\lambda,\mu)$.

$$\min J(\lambda,\mu) = J(\lambda^*,\mu^*)$$

(4.25)

where $(\lambda,\mu)\epsilon U$ and A and B_i $(i = 1,\ 2,\ 3)$ are the positive balancing constants which help to balance the units of integrand. The control function λ represents the contact rate of susceptibles with media in the population so that the infected population is reduced, while μ is the corresponding implementation rate of media programs.

The control set U is,

$$U = [(\lambda,\mu) : 0 \leq a_1 \leq \lambda(t) \leq b_1 < 1, 0 \leq a_2 \leq \mu(t) \leq b_2 < 1, 0 \leq t \leq T]\ (4.26)$$

This system satisfies the standard condition for the existence of an optimal control and now we derive the necessary conditions, using Pontryagin's maximum principle.

4.9 CHARACTERIZATION OF OPTIMAL CONTROL FUNCTION

Theorem 4.4

Let λ^* and μ^* be the optimal control parameter and N^*, Y^*, X_M^* and M^* are corresponding optimal state variables of the control system (4.23). The optimal controls (λ^*, μ^*) are given as,

$$\lambda^* = \min\left[b_1, \max\left(a_1, \lambda_3 \frac{(N-Y-X_M)M}{(1+\gamma M)B_2}\right)\right], \mu^* = \min\left[b_2, \max\left(a_2, \lambda_4 \frac{Y}{B_3}\right)\right],$$

Proof. The Hamiltonian of the system (4.23) and (4.24) is

$$H = AY(t) - \left[B_1 X_M(t) + B_2 \frac{\lambda^2(t)}{2} + B_3 \frac{\mu^2(t)}{2}\right]$$

$$+ \lambda_1 [Q - dN - \delta Y] + \lambda_2 \left[\frac{\beta(N-Y-X_M)Y}{N} - (\delta+d)Y\right] \quad (4.27)$$

$$+ \lambda_3 \left[\frac{\lambda(N-Y-X_M)M}{1+\gamma M} - (\vartheta+d)X_M\right] + \lambda_4 [\mu Y - \sigma M]$$

where $\lambda_i(t)(i=1,2,3,4)$ are the adjoint variables. To determine the transversality conditions and the adjoint equations, we use the Hamiltonian. The adjoint system results from Pontryagin's maximum principle.

$$\frac{d\lambda}{dt} = -\frac{\partial H}{\partial N}, \frac{d\lambda_2}{dt} = -\frac{\partial H}{\partial Y}, \frac{d\lambda_3}{dt} = -\frac{\partial H}{\partial X_M}, \frac{d\lambda_4}{dt} = -\frac{\partial H}{\partial M}$$

with $\lambda_i(T) = 0$, for $(i=1,2,3,4)$. In order to obtain the characterization of the control,

$$\lambda_1' = -\frac{\partial H}{\partial N} = -\left[\lambda_2 \frac{\beta(Y+X_M)Y}{N^2} + \lambda_3 \frac{\lambda M}{1+\gamma M} - \lambda_1 d\right] \quad (4.28)$$

$$\lambda_2' = -\frac{\partial H}{\partial Y}$$

$$= -\left[A + \lambda_2 \left[\frac{\beta(N-Y-X_M)}{N} - \frac{\beta Y}{N} - (\delta+d)\right] + \lambda_4 \mu - \lambda_1 \delta - \lambda_3 \frac{\lambda M}{1+\gamma M}\right]$$

$$(4.29)$$

$$\lambda_3' = -\frac{\partial H}{\partial X_M} = -\left[-B_1 - \lambda_2 \frac{\beta Y}{N} - \lambda_3 \frac{\lambda M}{1+\gamma M} - \lambda_3(\vartheta + d)\right] \quad (4.30)$$

$$\lambda_4' = -\frac{\partial H}{\partial M} = -\left[\lambda_3 \frac{\lambda(N-Y-X_M)}{(1+\gamma M)^2} - \lambda_4 \sigma\right] \quad (4.31)$$

The optimality conditions on the interior of the control set of an optimal control pair λ^*, μ^* are $\frac{\partial H}{\partial \lambda} = 0, \; \frac{\partial H}{\partial \mu} = 0.$

$$0 = \frac{\partial H}{\partial \lambda} = -B_2\lambda + \lambda_3\left[\frac{(N-Y-X_M)M}{1+\gamma M}\right] \quad (4.32)$$

$$0 = \frac{\partial H}{\partial \mu} = -B_3\mu + \lambda_4 Y \quad (4.33)$$

Hence using our control bounds,

$$\lambda^* = \min\left[b_1, \max\left(a_1, \lambda_3 \frac{(N-Y-X_M)M}{(1+\gamma M)B_2}\right)\right], \; \mu^* = \min\left[b_2, \max\left(a_2, \lambda_4 \frac{Y}{B_3}\right)\right],$$

4.10 NUMERICAL SIMULATION AND DISCUSSION

To see the dynamical behavior (existence of equilibria and feasibility of stability conditions) of the system, we integrate the system (4.6)–(4.9) by fourth-order Runge-Kutta method using MATLAB. The parameter values used in the model are given in Table 4.1.

The endemic equilibrium values of the model system (4.6)–(4.9) are computed as,

$$N^* = 17313.58547, \; Y^* = 11576.09803, \; X_M^* = 4253.465822,$$

$$M^* = 42831.56273$$

whereas eigenvalues are,

$$-1.9202, -0.0985 + i0.0454, -0.0985 - i0.0454, -0.0999$$

Here all the eigenvalues are negative or have negative real part showing that the endemic equilibrium E^* is locally asymptotically stable.

In Figure 4.1, we have plotted a dimensional plot for population $N(t)$, infective population $Y(t)$ and aware susceptible population $X_M(t)$ for different initial starts and found that all the trajectories approach to its equilibrium point, which shows the nonlinear stability behavior of the system.

Table 4.1 Parameters of the Model

Parameter Symbol	Value
Recruitment rate (Q)	1400/year, Estimated
Transfer rate from Y to $A(\delta)$	0.1/year, Anderson & May (1991)
Mortality rate (d)	0.014/year, Statista (2020)
Contact rate (β)	1.33/year, Tripathi & Naresh (2019)
Contact rate of susceptibles with media (λ)	0.9/year, Estimated
Limiting effect of awareness programs (γ)	1, Estimated
Efficacy of awareness programs (ϑ)	0.3/year, Estimated
Implementation rate of media programs (μ)	0.37/year, Estimated
Depletion of media programs dueto ineffectiveness (σ)	0.01/year, Estimated
Lower and upper bound for control $\lambda(a_1, b_1)$	$(0, 1)$
Lower and upper bound for control $\mu(a_2, b_2)$	$(0, 1)$
Balancing constants (B_1, B_2, B_3)	$\left(10^{-3}, 10^{-2}, 10^{-2}\right)$, Estimated
Balancing constant (A)	1, Estimated

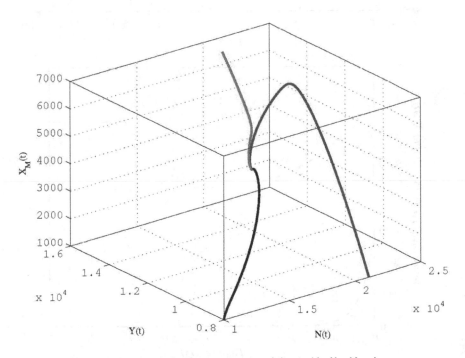

Figure 4.1 Three-dimensional plot for nonlinear stability in $N - Y - X_M$ plane.

In Figures 4.2–4.7, the variation of infective and aware susceptible population is shown with time t for different parameter values without optimal control. From Figure 4.2 it can be seen that as the contact rate of susceptibles

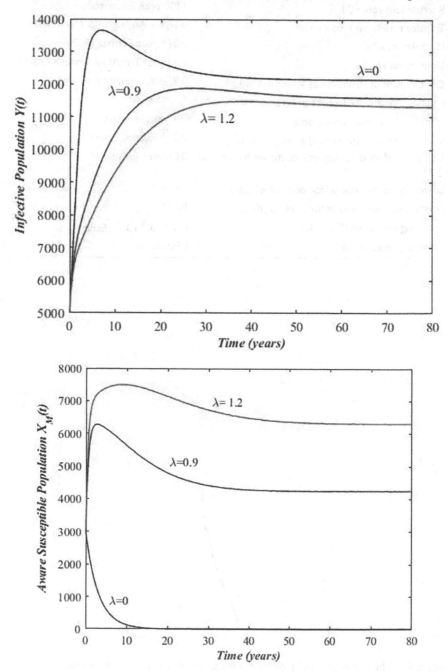

Figures 4.2 and 4.3 Time variations of infective and aware susceptible populations for different values of λ.

with media i.e. λ increases, the HIV infective population declines significantly. This is expected due to an increase in aware susceptible population
as a result of increased media awareness campaigns (Figure 4.3).

Figures 4.4 and 4.5 depict the evolution of the infective and aware susceptible populations over time t for various values of ϑ, the rate at which aware
susceptibles become susceptible again due to the fading of the influence of

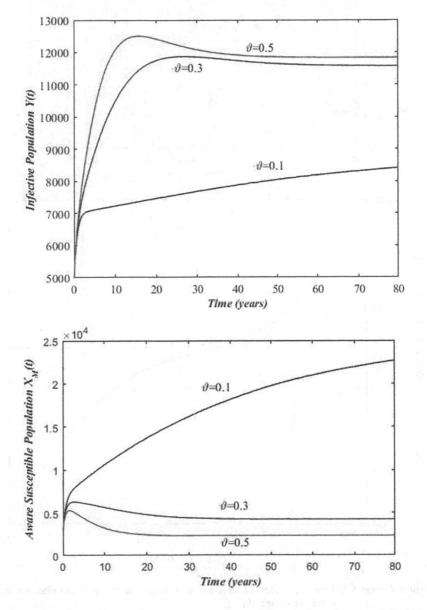

Figures 4.4 and 4.5 Time variations of infective and aware susceptible populations for
different values of ϑ.

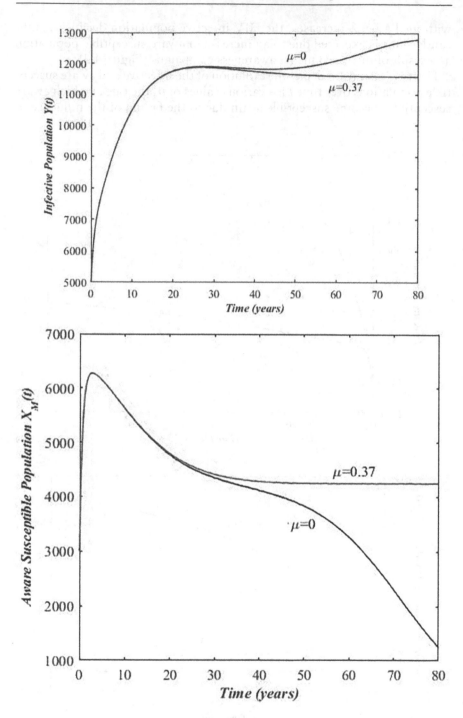

Figures 4.6 and 4.7 Time variations of infective and aware susceptible populations for different values of μ.

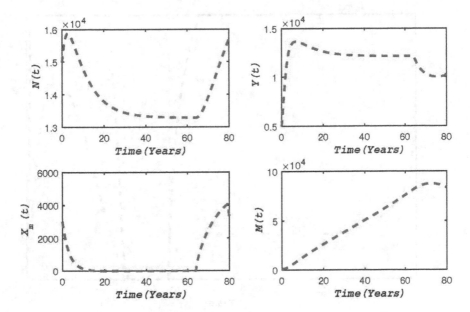

Figure 4.8 Variation of populations with optimal control.

media programs. These results show that when the impact of media awareness efforts diminishes, the HIV infective population grows (Figure 4.4). This rise in infective population is attributed to a decrease in the number of people who are aware that they are infected (Figure 4.5). Thus, if people do not present a positive attitude to modify their behavior as a result of media awareness campaigns, the infection continues to grow.

In Figures 4.6 and 4.7, the variation of infective population and aware susceptible population with time t is shown for different values of μ, the implementation rate of media programs. It is observed that on increasing the implementation rate of media programs, the infective population declines (Figure 4.6) due to an increase in aware susceptible population (Figure 4.7).

The behavior of optimal system has been studied numerically by fourth order Runge-Kutta method using parameter values given in Table 4.1. We investigated the dynamics for a varying combination of control measures λ^* and μ^* and for different balancing weight constants presented in objective function. In Figure 4.8, we again present the plot for different populations with optimal control. From this figure, it is noted that the HIV-infected population increases in beginning of the epidemic while it reduces rapidly after some time due to the corresponding decrease in aware susceptible population in that duration.

In Figures 4.9–4.13, the behavior of infective and aware susceptible populations can be seen for optimal parameter with different balancing constants.

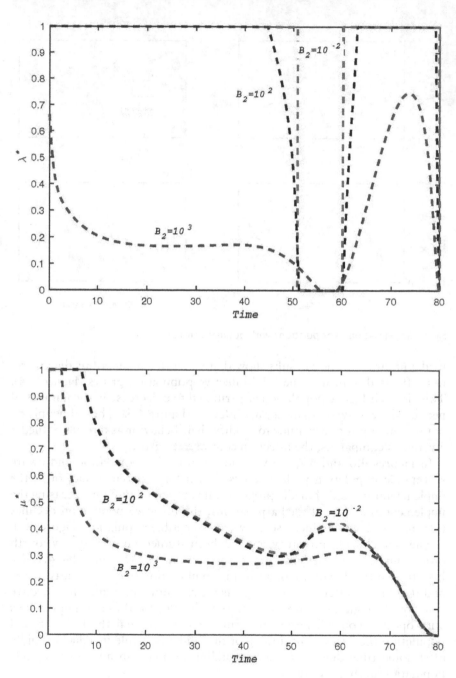

Figures 4.9 and 4.10 Profile of optimal control for different weight constants.

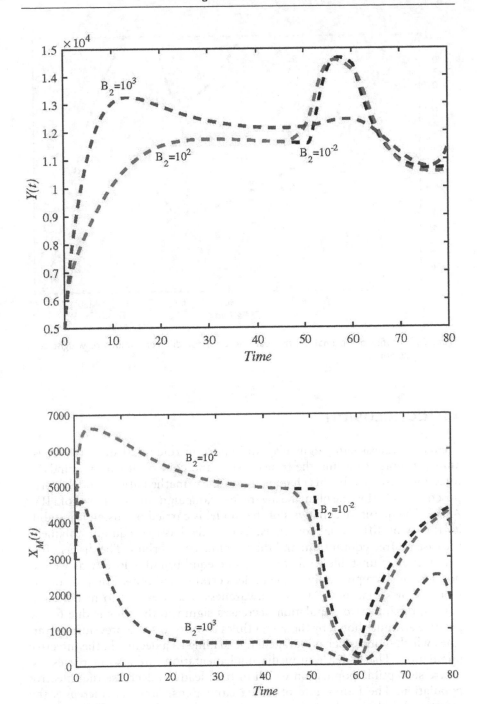

Figures 4.11 and 4.12 Profile of infective and aware susceptible population for different weight constants.

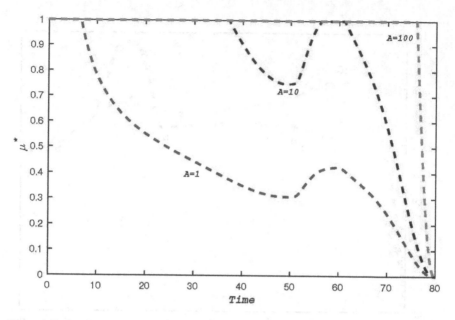

Figure 4.13 Profile of optimal control parameter μ for different values of weight constant A.

4.11 CONCLUSION

Media awareness campaigns play an important role in creating awareness among people regarding the transmission and control of various kinds of infectious diseases. In this chapter, a nonlinear mathematical model is proposed to study the role of media awareness campaigns on the spread of HIV/ AIDS. The qualitative analysis of the model is carried out using the stability theory of differential equations. The model has two equilibria namely, the disease-free equilibrium and the endemic equilibrium. The disease-free equilibrium is unstable while the endemic equilibrium is locally as well as nonlinearly asymptotically stable under certain conditions. Analysis of the model shows that as the level of awareness increases due to media campaigns, the infective population decreases significantly. This is due to the positive attitude shown by the susceptibles towards disease prevention strategies which induces behavioural change leading to a decline in the infective population. The increase in media implementation rate also increases the aware susceptible population which in turn leads to decrease the infective population. The fading rate of media campaigns, however, increases the infective population due to lowering of aware suceptible population. It further investigates the optimal control problem to see its effect on slowing down the disease prevalence. The analysis suggests that the optimal control parameters have a significant impact in reducing the infective population. The extension of the model incorporating diffusion, and pattern formation can be used for future research work.

APPENDIX A

Proof of Theorem 4.1. In order to study the local stability behavior of E_0, the Jacobian matrix is computed at that equilibrium point and the sign of the eigenvalues is determined. We have the following Jacobian matrix evaluated at E_0.

$$J(E_0 - xI) = \begin{bmatrix} -(d+x) & -\delta & 0 & 0 \\ 0 & \left[(\beta - (\delta + d)) - x\right] & 0 & 0 \\ 0 & 0 & -(\vartheta + d + x) & \lambda \dfrac{Q}{d} \\ 0 & 0 & 0 & -(\sigma + x) \end{bmatrix}$$

Here all the eigenvalues of the Jacobian matrix are negative if $\beta < (\delta + d)$ i.e. $R_0 < 1$, thus disease-free equilibrium E_0 is locally asymptotically stable. It is unstable for $R_0 > 1$ and the endemic equilibrium E^* exists.

Proof of Theorem 4.2. To establish the local stability of the endemic equilibrium E^*, we linearize the system using small perturbations n, y, x_m, and m as follows:

$$N = N^* + n, Y = Y^* + y, X_M = X_M^* + x_m, M = M^* + m$$

Let us consider the following positive definite function,

$$U = p_0 \frac{1}{2}n^2 + p_1 \frac{1}{2}y^2 + p_2 \frac{1}{2}x_m^2 + p_3 \frac{1}{2}m^2 \tag{A.1}$$

where p_0, p_1, p_2, and p_3 are positive constants to be chosen appropriately. On differentiating U with respect to t, we get

$$\frac{dU}{dt} = p_0 n \frac{dn}{dt} + p_1 y \frac{dy}{dt} + p_2 x_m \frac{dx_m}{dt} + p_3 m \frac{dm}{dt} \tag{A.2}$$

Now using the linearized system of (4.6)–(4.9) and after some algebraic manipulations, we get

$$\frac{dU}{dt} = -p_0 dn^2 - p_1 B y^2 - p_2 (C + \vartheta + d) x_m^2 - p_3 \sigma M m^2$$

$$+ [p_1 A - p_0 \delta] ny + p_2 Cnx_m - p_1 Byx_m - p_2 Cyx_m \qquad (A.3)$$

$$+ p_2 Dx_m m + p_3 \mu M^* ym$$

Choosing $p_1 = p_3 = 1$, $p_0 = \dfrac{A}{\delta}$, and $0 < p_2 < (C + \vartheta + d) \min \left[\dfrac{3p_0 d}{C^2}, \dfrac{2\sigma M^*}{3D^2} \right]$,

we get $\dfrac{dU}{dt}$ to be negative definite under the conditions stated in the theorem.

APPENDIX B

Proof of Theorem 4.3. To prove this theorem, we consider the following positive definite function,

$$V = \frac{1}{2} k_0 (N - N^*)^2 + k_1 \left(Y - Y^* - Y^* ln \frac{Y}{Y^*} \right) + \frac{1}{2} k_2 (X_M - X_M^*)^2 + \frac{1}{2} k_3 (M - M^*)^2$$

$$(B.1)$$

where k_i ($i = 0, 1, 2, 3$) are positive constants, to be chosen appropriately. Differentiating V with respect to t, we get

$$\frac{dV}{dt} = k_0 \left(N - N^* \right) \frac{dN}{dt} + k_1 \frac{(Y - Y^*)}{Y} \frac{dY}{dt} + k_2 \left(X_M - X_M^* \right) \frac{dX_M}{dt} + k_3 \left(M - M^* \right) \frac{dM}{dt}$$

$$(B.2)$$

After some algebraic manipulations, $\dfrac{dV}{dt}$ can be written as,

$$\frac{dV}{dt} = -k_0 d (N - N^*)^2 - k_1 \frac{\beta}{N^*} (Y - Y^*)^2 - k_2 (C + \vartheta + d)(X_M - X_M^*)^2$$

$$- k_3 \sigma (M - M^*)^2 + \left(k_1 \frac{\beta}{N^*} - k_0 \delta \right) \left(N - N^* \right) \left(Y - Y^* \right)$$

$$+ k_2 C \left(N - N^* \right) \left(X_M - X_M^* \right) - \left(k_1 \frac{\beta}{N^*} + k_2 C \right) \left(Y - Y^* \right) \left(X_M - X_M^* \right)$$

$$- \frac{\beta k_1}{N^*} \left(Y_H - Y_H^* \right) \left(X_M - X_M^* \right) - k_2 \lambda M^* \left(Y_H - Y_H^* \right) \left(X_M - X_M^* \right)$$

$$+ k_2 \frac{\lambda (N - Y - X_M)}{(1 + \gamma M)(1 + \gamma M^*)} \left(X_M - X_M^* \right) \left(M - M^* \right) + k_3 \mu \left(Y - Y^* \right) \left(M - M^* \right)$$

$$(B.3)$$

which is further simplified to,

$$\frac{dV}{dt} \le -k_0 d(N - N^*)^2 - k_1 \frac{\beta}{N^*}(Y - Y^*)^2 - k_2(C + \vartheta + d)(X_M - X_M^*)^2$$

$$- k_3\sigma(M - M^*)^2 + \left(k_1 \frac{\beta}{N^*} - k_0\delta\right)(N - N^*)(Y - Y^*)$$

$$+ k_2 C(N - N^*)(X_M - X_M^*) - \left(k_1 \frac{\beta}{N^*} + k_2 C\right)(Y - Y^*)(X_M - X_M^*) \quad \text{(B.4)}$$

$$- \frac{\beta k_1}{N^*}(Y - Y^*)(X_M - X_M^*) - k_2\lambda M^*(Y - Y^*)(X_M - X_M^*)$$

$$+ k_2 \frac{\lambda N_{max}}{(1 + \gamma M^*)}(X_M - X_M^*)(M - M^*) + k_3\mu(Y - Y^*)(M - M^*)$$

Choosing $\quad k_1 = k_3 = 1, \quad k_0 = \frac{\beta}{\delta N^*}, \quad$ and $\quad 0 < k_2 < (C + \vartheta + d)\min.$

$$\left[\frac{3k_0 d}{C^2}, \frac{2\sigma(1 + \gamma M^*)^2}{3\lambda^2(\frac{Q}{d})^2}\right], \text{ we get } \frac{dV}{dt} \text{ to be negative definite under the condi-}$$

tions stated in theorem.

Here the coefficients A, B, C and D used in the above theorems are defined as follows,

$$A = \beta \frac{(Y^* + X_M^*)Y^*}{N^{*2}} > 0, \quad B = \beta \frac{Y^*}{N^*} > 0, \quad C = \lambda \frac{M^*}{(1 + \gamma M^*)} > 0,$$

$$D = \lambda \frac{(N^* - Y^* - X_M^*)}{(1 + \gamma M^*)^2} > 0.$$

REFERENCES

Anderson R.M. (1988). The role of mathematical models in the study of HIV transmission and the epidemiology of AIDS. *J. AIDS*, 1, 241–256.

Anderson R.M. & May R.M. (1991). *Infectious Diseases of Humans*, Oxford Univesity Press, London/New York.

Anderson R.M., Medley G.F., May R.M., & Johnson, A.M. (1986). A preliminary study of the transmission dynamics of the human immunodeficiency virus (HIV), the causative agent of AIDS. *IMA J. Math. Appl. Med. Biol.*, 3, 229–263.

Bakare E. A., Nwagwo A., & Danso-Addo E. (2014). Optimal control analysis of an SIR epidemic model with constant recruitment. *Int. J. Appl. Math. Res.*, 3(3), 273–285.

Cui J., Sun Y. & Zhu H. (2008). The impact of media on the control of infectious disease. *J. Dynam. Diff. Eqns.*, 20(1), 31–53.

Cui J., Tao X. & Zhu H. (2008). An SIS infection model incorporating media coverage. *Rocky Mt. J. Math.*, 38(5), 1323–1334.

Driessche P. V. & Watmough J. (2002). Reproduction numbers and sub-threshold endemic equilibria for compartmental models of disease transmission. *Math. Biosci.*, 180, 29–48.

Dubey B., Dubey V., & Dubey U.S. (2016). Role of media and treatment on an SIR model. *Nonlinear Anal.: Model. Control*, 21(2), 185–200.

Funk S., Gilad E., & Jansen V.A.A. (2010). Endemic disease, awareness, and local behavioural response. *J. Theor. Biol.*, 264(2), 501–509.

Greenhalgh D., Rana S., Samanta S., Sardar T., Bhattacharya S., & Chattopadhyaya J. (2015). Awareness programs control infectious disease-multiple delay induced model. *Appl. Math. Comp.*, 251, 539–563.

Hethcote H.W. (2000). The mathematics of infectious diseases. *SIAM Rev.*, 42(4), 599–653.

Joshi H., Lenhart S., Hota S., & Augusto F. (2015). Optimal control of an SIR model with changing behavior through and education campaign. *Electron. J. Differ. Equ*, 50, 1–14.

Kiss I. J., Cassell J., Recker M., & Simon P.L. (2010). The impact of information transmission on epidemic outbreaks. *Math. Biosci.*, 255, 1–10.

Kumar A., Srivastava P.K., & Takeuchi Y. (2017). Modelling the role of information and limited optimal treatment on disease prevalence. *J. Theor. Biol.*, 414, 103–119.

Liu R., Wu J., & Zhu H. (2007). Media/psychological impact on multiple outbreaks of emerging infectious diseases. *Comp. Math. Meth. Med.*, 8(3), 153–164.

Liu W. (2013). A SIRS epidemic model incorporating media coverage with random perturbation. *Abstr. Appl. Anal.*, 2013, 792308.

Liu Y. & Cui J. (2008). The impact of media coverage on the dynamics of infectious disease. *Int. J. Biomath.*, 1, 65–74.

Mastahun M. & Abdurahman X. (2017). Optimal control of an HIV/AIDS epidemic model with infective immigration and behavioral change. *Appl. Math.*, 8, 87–105.

May R.M. & Anderson R.M. (1987). Transmission dynamics of HIV infection. *Nature, 326*, 137–142.

Misra A.K., Sharma A., & Shukla J.B. (2011). Modeling and analysis of effects of awareness programs by media on the spread of infectious diseases. *Math. Comp. Model.*, 53, 1221–1228.

Misra A.K., Sharma A., & Shukla J.B. (2015). Stability analysis and optimal control of an epidemic model with awareness programs by media. *Biosys.*, 138, 53–62.

Mukandavire Z. & Garira W. (2006). HIV/AIDS model for assessing the effects of prophylactic sterilizing vaccines, condoms and treatment with amelioration. *J. Biol. Syst.*, 14(3), 323–355.

Naresh R., Tripathi A., Biazar J., & Sharma D. (2008) Analysis of the effect of vaccination on the spread of AIDS epidemic using adomian decomposition method. *J. Nat. Sci. Sus. Tech.*, 2, 183–213.

Naresh R., Tripathi A., & Sharma D. (2009). Modelling and analysis of the spread of AIDS epidemic with immigration of HIV infectives. *Math. Comp. Model.*, 49, 880–892.

Naresh R., Tripathi A., & Sharma D. (2011b). A nonlinear HIV/AIDS model with contact tracing. *Appl. Math. Comp.*, 217, 9575–9591.

Rahman M.S. & Rahman M.L. (2007). Media and education play a tremendous role in mounting AIDS awareness among married couples in Bangladesh. *AIDS Res. Ther.*, 4, 10–16.

Sahu G.P. & Dhar J. (2015). Dynamics of an SEQIHRS epidemic model with media coverage, quarantine and isolation in a community with pre-existing immunity. *J. Math. Anal. Appl.*, 421, 1651–1672.

Sharma R. (2014). Stability analysis of infectious diseases with media coverage and poverty. *Math. Theo. Model.*, 4(4), 107–113.

Sharma A. & Misra A.K. (2014). Modelling the impact of awareness created by media campaigns on vaccination coverage in a variable population. *J. Biol. Syst.*, 22(2), 249–270.

Samanta S., Rana S., Sharma A., Misra A.K, & Chattopadhyaya J. (2013). Effects of awareness programs by media on the epidemic outbreaks: A mathematical model. *Appl. Math. Comp.*, 219, 6965–6977.

Statista (2020). Life expectancy (from birth) in India from 1800 to 2020. www.statista.com/statistics/1041383/life-expectancy-india-all-time/.

Tchuenche J.M. & Bauch C.T. (2012). Dynamics of an infectious disease where media coverage influences transmission. *ISRN Biomath.*, 2012, 581274.

Tchuenche J.M., Dube N., Bhanu C.P., Smith R. J., & Bauch C.T. (2011). The impact of media coverage on the transmission dynamics of human influenza. *BMC Pub. Heal.*, 11(suppl 1), S5.

Tripathi A. & Naresh R. (2019). Modeling the role of media awareness programs on the spread of HIV/AIDS. *World. J. Model. Simul.*, 15(1), 12–24.

Tripathi A., Naresh R., & Sharma D. (2007). Modelling the effect of screening of unaware infectives on the spread of HIV infection. *Appl. Math. Comp.*, 184, 1053–1968.

UNAIDS (2020). UNAIDS fact sheets. www.unaids.org.

Chapter 5

Relaxation oscillation and canard explosion in slow–fast predator–prey systems

Tapan Saha
Presidency University

Pallav Jyoti Pal
Krishna Chandra College

CONTENTS

5.1 INTRODUCTION

Many biological, physical, chemical, pharmacological, neurobiological, and other systems of study involve multiple time scales and are modeled as slow–fast systems which capture amazing predictions of complicated oscillations [1–8]. It has received considerable attention during the last few decades, and important advances related to the mathematical theory were reported [9,10]. However, from a mathematical point of view, slow–fast systems have been studied for a long time. In 1920, Van der Pol [11] converted

DOI: 10.1201/9781003367420-5

a second-order equation into a first-order slow–fast system to examine a vacuum tube triode circuit, and some noteworthy dynamical properties were pointed out. Another very important example of a slow–fast system on neuronal excitability is the FitzHugh–Nagumo (FHN) model [1] developed by Richard FitzHugh which exhibits a wide range of complex dynamics. There are several models of slow–fast systems in the field of biological, physical, chemical, etc. [1,2,5–8] as well as the references therein which capture remarkable predictions of complex oscillations. When analyzing a slow–fast system, the key concept is to isolate the sub-processes working at various time scales, understand them, and then attempt to characterize the entire dynamics of the full slow–fast system based on the subsystems. The above crude notion may be made mathematically rigorous and, in fact, serves as the foundation for the geometric analysis of slow–fast systems. The method of isolation of the sub-processes, their investigation, and the process to characterize the entire dynamics of the full system based on the sub-processes is described in Section 5.2. Neil Fenichel devised a geometric approach [12,13] for studying a slow–fast system using invariant manifold theory in 1970. This approach works well in the case of normally hyperbolic critical manifolds. Later on, Krupa and Szmolyan [5,14], extended the above-mentioned geometrical approach for non-hyperbolic points (fold points and canard points) for planar systems. Their analysis carries over to higher dimensional problems as well [14]. A well-known phenomenon for singular perturbation systems is the occurrence of relaxation oscillations via canard cycles. A relaxation oscillation consisting of a long interval of quasi-steady state followed by a short interval of rapid variation is a special type of periodic solution and a canard is a trajectory of a slow–fast system that remains not only near the attracting but also the repelling slow manifolds for $\mathcal{O}(1)$ time [5]. In the seminal work [15,16], Dumortier and Roussarie first used the blow-up technique for a singular perturbation problem and investigated the presence of canard cycles for the van der Pol equation. The reason why the systems with multiple time scales attract considerable attention amongst researchers for a long time and have been widely applied to various fields of study are the following: (1) the systems with multiple time scales are ubiquitous in nature which provides a more realistic description of the dynamics of systems and (2) several techniques including asymptotic, geometric methods and blow-up technique have been developed to understand even complicated slow–fast dynamics.

For predator–prey models, slow–fast processes are evolved under certain circumstances, e.g., when the growth rate of the prey population is much faster than the predator. Taking the ratio(s) between the birth rate(s) of the predator population(s) to the prey population(s) very small, the separation of timescale(s) is introduced into the predator–prey system as a singular parameter. Many ecological models have made similar assumptions. The first article on a slow–fast predator–prey model was investigated by Rinaldi and Muratori in [17], where they highlighted the cyclical presence of the slow–fast limit cycle. Recently, in [18], the authors analyzed the slow–fast dynamics of a predator–prey system with generalized Holling type III functional response in detail and observed relaxation oscillations, canard cycles, canard explosion, heteroclinic and homoclinic orbits. In [19], the authors gave an elementary proof of the entry–exit theorem based on a direct study of asymptotic formulas of the underlying solutions. Wang and Zhang [18] studied a slow–fast predator–prey system with Holling type III functional response and proved the existence of canard cycles, canard explosion, relaxation oscillations, heteroclinic and homoclinic orbits, cyclicity of slow–fast cycles, etc. In [20], the authors studied bifurcations of canards and homoclinic orbits in a slow–fast modified Leslie–Gower predator–prey model with the Allee effect in prey population growth. They have proved that the cyclicity of canard cycles in this model is at most four. In [21], the authors obtained very rich and complicated dynamical phenomena, including the existence of relaxation oscillation, canard cycles near the Hopf bifurcation point, and the interesting phenomenon of canard explosion for a smooth planar predator–prey system. Recently, in [22], the authors analyzed the slow–fast dynamics of a modified Leslie–Gower model with piecewise-smooth Holling type I functional response. They have studied the existence of saddle-node bifurcation, Bogdanov–Takens bifurcation, bi-stability, singular Hopf bifurcation, canard orbits, multiple relaxation oscillations, saddle-node bifurcation of limit cycle, and boundary equilibrium bifurcations.

In this chapter, we study the dynamics of prey and generalist predator population with Holling type II functional response, where the growth of the generalist predator is logistic-type with prey–dependent environmental carrying capacity and induced by the Allee effect. The primary goal of this chapter is three-folded. The first is to offer an overview of certain relevant slow–fast analytical methodologies in concise and compact form. The second goal is to demonstrate some phenomena that arise from the interplay of two-time scales of an ecological system modeled by slow–fast planar ordinary differential equations. Lastly, the third is to observe how the Allee effect and quantity of additional food sources in the predator population affect the dynamics of a slow–fast system. During such study, we investigate various types of interesting complex oscillatory dynamics exhibited by the model including singular Hopf bifurcation, canard cycle, canard explosion, relaxation oscillation, and singular Homoclinic orbit, and we utilize the methods like geometric singular perturbation theory, blow-up technique, and bifurcation analysis to describe them.

The remaining part of this chapter is organized as follows: In Section 5.2, we present some preliminaries and certain relevant slow–fast analytical methodologies in concise and compact form, including the existence of interesting phenomena of the singular Hopf bifurcation, canard cycle, singular Homoclinic orbit, canard explosion, and relaxation oscillation. Based on the methodology discussed in Section 5.2, as a case study, we investigate the dynamics of a predator–prey model in Section 5.3 where the prey population reproduces much faster than the predator population. Finally, the chapter concludes with some brief conclusions of our findings in Section 5.4.

5.2 PRELIMINARIES

5.2.1 Slow–fast systems

In this subsection, we begin with a quick overview of continuous-time slow–fast systems and their notation. From a mathematical standpoint, one of the easiest approaches to dealing with systems containing multiple time scales is to start with singularly perturbed ordinary differential equations (ODEs) with only two-time scales of the form

$$\dot{x} = f(x, y, \mu, \epsilon), \tag{5.2.1a}$$

$$\dot{y} = \epsilon g(x, y, \mu, \epsilon), \tag{5.2.1b}$$

where f and g are C^p-functions with $p \geq 3$, $x \in \mathbb{R}^m$ and $y \in \mathbb{R}^n$ are called the fast and the slow variables respectively, $\mu \in \mathbb{R}^k$ are parameters, m, n, $k \geq 1$ in general, and $0 < \epsilon \ll 1$ is the singular perturbation parameter that indicates system separation into slow and fast time scales. The over dot $(\cdot)$ denotes differentiation with respect to the fast time $\tau \in \mathbb{R}$.

Switching the fast time to the slow time scale $\tau = \epsilon t$, transforms the system (5.2.1) into the following ODEs

$$\epsilon x' = f(x, y, \mu, \epsilon), \tag{5.2.2a}$$

$$y' = g(x, y, \mu, \epsilon), \tag{5.2.2b}$$

where the prime $(')$ denotes differentiation with respect to $\tau \in \mathbb{R}$. The systems (5.2.1) and (5.2.2) are equivalent for $\epsilon > 0$ and hence have identical phase portraits. However, for the singular case $\epsilon = 0$, they do have a separate set of mathematical equations.

Taking the singular limit, $\epsilon \to 0$, the systems (5.2.1) and (5.2.2) reduce to

$$\dot{x} = f(x, y, \mu, 0), \tag{5.2.3a}$$

$$\dot{y} = 0, \tag{5.2.3b}$$

and

$$0 = f(x, y, \mu, 0), \tag{5.2.4a}$$

$$y' = g(x, y, \mu, 0), \tag{5.2.4b}$$

respectively. The system (5.2.3) is a lower-dimensional differential equation in the fast variable x where the slow variable y is treated as a parameter. The system (5.2.3) is called a fast subsystem or layer system, which explains how x changes whenever y stays constant. The system (5.2.4) is a Differential-Algebraic Equation (DAE) called the slow subsystem or reduced system. The slow subsystem (5.2.4) is defined on the critical set,

$$M_0 = \left\{ x \in \mathbb{R}^m, \; y \in \mathbb{R}^n \middle| f(x, y, \mu, 0) = 0 \right\}, \tag{5.2.5}$$

the points of which represent the equilibrium points for the fast subsystem (5.2.3).

The methods such as asymptotic expansions [23], Geometric Singular Perturbation theory (GSPT) (see [12,13,24] for details) have been developed to analyze the dynamics of a slow–fast system.

5.2.2 Geometric singular perturbation theory

The geometric singular perturbation theory (GSPT), developed by Fenichel in the 1970s and popularized by Jones [25], is a more qualitative and rigorous way of analyzing the dynamics of a slow–fast system. One of the objectives of GSPT is to investigate first the simpler limiting subsystems (5.2.3) and (5.2.4) with separated time scales, and later, we seek to explain the full dynamics of the complex system (5.2.1) or (5.2.2) for $0 < \epsilon \ll 1$. An essential feature of critical manifolds is normal hyperbolicity [10] which is defined below:

Definition 5.1

A point $p = (x, y) \in M_0$ is said to be hyperbolic if the matrix $D_x f(p, \mu, 0) \in \mathbb{R}^{m \times m}$, where D_x is the total derivative with respect to x, has no eigenvalues on the imaginary axis. The critical manifold M_0 is said to be normally hyperbolic if all points $p \in M_0$ are hyperbolic. If the matrix $D_x f(p, \mu, 0) \in \mathbb{R}^{m \times m}$ has a minimum of one eigenvalue on the imaginary axis, then the point p is referred to as non-hyperbolic.

Definition 5.2

A subset $S \subset M_0$ which is normally hyperbolic and is said to be attracting (repelling) if every eigenvalue of $D_x f(p, \mu, 0)$ have negative (positive) real parts for $p \in S$. If S is normally hyperbolic and not attracting or repelling, it is referred to as saddle type.

The techniques required to investigate a slow–fast system depend on whether the critical manifold is normally hyperbolic or not. If it is normal hyperbolic, we use Fenichel's theorem [10] and in the case of loss of normal hyperbolicity of the critical manifold due to the presence of non-hyperbolic points such as fold and canard points, we may use the blow-up technique.

5.2.3 The blow-up method for slow–fast systems

The collapse of normal hyperbolicity of M_0 is a very challenging phenomenon. It is a major obstacle to the use of geometric theory provided by Fenichel theory. We emphasize the importance of the loss of normal hyperbolicity in slow–fast systems because it is linked to dynamic aspects such as relaxation oscillations, canards, mixed-mode oscillations, etc. The blow-up method incorporates a change of coordinates, by which the analysis of a slow–fast system around the non-hyperbolic point is transformed into a series of sub-problems with partly-hyperbolic singularities. Understanding and compiling all the results of the investigation of the sub-problems, one can thereafter draw conclusions about the dynamics of the full slow–fast system. In a seminal work by Dumortier and Roussarie [15], the blow-up technique was first used for the investigation of the dynamics of a slow–fast system. They observed that by inserting an appropriate manifold, such as a sphere at a non-hyperbolic singular point, such singularities may be turned into partly hyperbolic singularities. Different charts are used to study the dynamics of the manifold for the blown-up system. For the details of this technique, readers are referred to [5,10].

5.2.4 Singular Hopf bifurcation, canard cycles, canard explosion and relaxation oscillation

In this section, we briefly review the theory behind the phenomena of singular Hopf bifurcation, canard cycles, canard explosion, and relaxation oscillation. We refer to [5,10] for the details of the theory. We consider a planar slow–fast system of the following form

$$\frac{dx}{dt} = f\left(x,\, y,\, \mu,\, \epsilon\right), \tag{5.2.6a}$$

$$\frac{dy}{dt} = \epsilon g\left(x,\, y, \mu, \epsilon\right), \tag{5.2.6b}$$

where $\mu \in \mathbb{R}$ is a parameter, $0 < \epsilon \ll 1$, f, and g are smooth functions.

Let us now introduce the following definitions [10]:

Definition 5.3

A point $Q(x_m,\, y_m) \in M_0$ (critical set), is said to be a non-degenerate fold point if

$$f_x(x_m, y_m) = 0, f_{xx}(x_m, y_m) \neq 0, \text{ and } f_y(x_m, y_m) \neq 0. \tag{5.2.7}$$

Definition 5.4

The point $Q(x_m, y_m)$ is said to be a generic folded singularity or a canard point at $\mu = \mu_*$ if it is a non-degenerate fold point and

$$g(x_m, y_m, \mu_*, 0) = 0, \ g_x(x_m, y_m, \mu_*, 0) \neq 0 \text{ and } g_\mu(x_m, y_m, \mu_*, 0) \neq 0. \tag{5.2.8}$$

Definition 5.5

A canard of a slow–fast system is a trajectory if it follows an attracting slow manifold and stays within $\mathcal{O}(\epsilon)$ distance from the repelling slow manifold for a time $\mathcal{O}(1)$.

Definition 5.6

An orbit that lies in the intersection of the attracting and repelling slow manifolds is referred to as a maximal canard.

Let Q be a canard point. Using the transformation $X = x - x_m$, $Y = y - y_m$, $\lambda = \mu - \mu_*$, the generic folded singularity Q is shifted to the origin at $\lambda = 0$ and the system (5.2.6) can be derived to the following canonical form:

$$\frac{dX}{dt} = -Yh_1(X, Y, \lambda, \epsilon) + X^2 h_2(X, Y, \lambda, \epsilon) + \epsilon h_3(X, Y, \lambda, \epsilon), \tag{5.2.9a}$$

$$\frac{dY}{dt} = \epsilon \left[Xh_4(X, Y, \lambda, \epsilon) - \lambda h_5(X, Y, \lambda, \epsilon) + Yh_6(X, Y, \lambda, \epsilon) \right], \tag{5.2.9b}$$

where

$$h_3(X, Y, \lambda, \epsilon) = \mathcal{O}(X, Y, \lambda, \epsilon),$$

$$h_j(X, Y, \lambda, \epsilon) = 1 + \mathcal{O}(X, Y, \lambda, \epsilon), \ j = 1, 2, 4, 5.$$

One can then use the blow-up technique as introduced by Dumortier [26,15] to desingularize the folded singularity (canard point) as the origin of the system (5.2.9). We write the system (5.2.9) in the following form

$$\frac{dX}{dt} = -Yh_1(X, Y, \lambda, \epsilon) + X^2 h_2(X, Y, \lambda, \epsilon) + \epsilon h_3(X, Y, \lambda, \epsilon), \tag{5.2.10a}$$

$$\frac{dY}{dt} = \epsilon \left[Xh_4(X, Y, \lambda, \epsilon) - \lambda h_5(X, Y, \lambda, \epsilon) + Yh_6(X, Y, \lambda, \epsilon) \right],$$

(5.2.10b)

$$\frac{d\lambda}{dt} = 0,$$

(5.2.10c)

$$\frac{d\epsilon}{dt} = 0.$$

(5.2.10d)

We consider the blown-up space $\mathbb{S}^3 = \left\{ (\bar{X}, \bar{Y}, \bar{\lambda}, \bar{\epsilon}) \in \mathbb{R}^4 \mid \bar{X}^2 + \bar{Y}^2 + \bar{\lambda}^2 + \bar{\epsilon}^2 = 1 \right\}$ and define the blow-up map

$$\Psi : \mathbb{S}^3 \times [0, r_0] = B \to \mathbb{R}^4,$$

(5.2.11a)

$$\left(\bar{X}, \bar{Y}, \bar{\lambda}, \bar{\epsilon}, r \right) \to \left(r\bar{X}, r^2\bar{Y}, r\bar{\lambda}, r^2\bar{\epsilon} \right) = (X, Y, \lambda, \epsilon).$$

(5.2.11b)

The folded singularity $(0,0,0,0)$ is mapped to S^3 under this blow-up transformation and consequently, we need to investigate the dynamics on $\mathbb{S}^3$. To discuss the dynamics near the canard point, we define the following four coordinate charts for B.

$$K_1 : B_{\bar{Y}}^{\pm} \to \mathbb{R}^4 \text{ by } K_1\left(\bar{X}, \bar{Y}, \bar{\lambda}, \bar{\epsilon}, r \right) = \left(\bar{X}\bar{Y}^{-1/2}, \bar{\lambda}\bar{Y}^{-1/2}, \bar{\epsilon}\,\bar{Y}^{-1}, r\bar{Y}^{1/2} \right)$$

$$= (X_1, \lambda_1, \epsilon_1, r_1),$$

(5.2.12a)

$$K_2 : B_{\bar{\epsilon}}^{\pm} \to \mathbb{R}^4 \text{ by } K_2\left(\bar{X}, \bar{Y}, \bar{\lambda}, \bar{\epsilon}, r \right) = \left(\bar{X}\bar{\epsilon}^{-1/2}, \bar{Y}\bar{\epsilon}^{-1}, \bar{\lambda}\bar{\epsilon}^{-1/2}, r\bar{\epsilon}^{1/2} \right)$$

$$= (X_2, Y_2, \lambda_2, r_2),$$

(5.2.12b)

$$K_3 : B_{\bar{X}}^{\pm} \to \mathbb{R}^4 \text{ by } K_3\left(\bar{X}, \bar{Y}, \bar{\lambda}, \bar{\epsilon}, r \right) = \left(\bar{Y}\bar{X}^{-2}, \bar{\lambda}\bar{X}^{-1}, \bar{\epsilon}\bar{X}^{-2}, r\bar{X} \right)$$

$$= (Y_3, \lambda_3, \epsilon_3, r_3),$$

(5.2.12c)

$$K_4 : B_{\bar{\lambda}}^{\pm} \to \mathbb{R}^4 \text{ by } K_4\left(\bar{X}, \bar{Y}, \bar{\lambda}, \bar{\epsilon}, r \right) = \left(\bar{X}\bar{\lambda}^{-1}, \bar{Y}\bar{\lambda}^{-2}, \bar{\epsilon}\bar{\lambda}^{-2}, r\bar{\lambda} \right)$$

$$= (X_4, Y_4, \epsilon_4, r_4),$$

(5.2.12d)

where $B_{\bar{Y}}^{\pm} = B \cap \{\bar{Y} > 0\}$ and similarly, for $B_{\bar{\epsilon}}^+$, $B_{\bar{X}}^+$, and $B_{\bar{\lambda}}^+$. We then have to investigate the dynamics of the blown-up vector field in these charts, and connecting the results in different charts, the final results are derived by "blow down" [5,10]. Now, following the results derived in [5] and the

Equations (3.15) and (3.16) of [5], the expansions of singular Hopf bifurcation and maximal canard curves are given by

$$\lambda_H\left(\sqrt{\epsilon}\right) = -\frac{a_1 + a_5}{2}\epsilon + \mathcal{O}\left(\epsilon^{3/2}\right), \tag{5.2.13}$$

$$\lambda_c\left(\sqrt{\epsilon}\right) = -\left(\frac{a_1 + a_5}{2} + \frac{A}{8}\right)\epsilon + \mathcal{O}\left(\epsilon^{3/2}\right). \tag{5.2.14}$$

where

$$a_1 = \frac{\partial h_3}{\partial X}(0,0,0,0), \quad a_2 = \frac{\partial h_1}{\partial X}(0,0,0,0),$$

$$a_3 = \frac{\partial h_2}{\partial X}(0,0,0,0), \quad a_4 = \frac{\partial h_4}{\partial X}(0,0,0,0), \tag{5.2.15}$$

$$a_5 = h_6(0,0,0,0),$$

$$A = -a_2 + 3a_3 - 2a_4 - 2a_5. \tag{5.2.16}$$

Thus, based on [5], we have the following results on the existence of singular Hopf bifurcation and maximal canard.

Theorem 5.2.2

Let the origin be a generic folded singularity of the system (5.2.9). Then $\exists \epsilon_0 > 0$ and $\lambda_0 > 0$ such that for $0 < \epsilon < \epsilon_0$ and $|\lambda| < \lambda_0$, system (5.2.9) has an equilibrium point Q_0 in the neighborhood of the origin which converges to the folded singularity the origin as $(\epsilon, \lambda) \to (0,0)$. Moreover, $\exists$ a curve $\lambda = \lambda_H\left(\sqrt{\epsilon}\right)$ of singular Hopf bifurcation so that Q_0 is stable for $\lambda < \lambda_H\left(\sqrt{\epsilon}\right)$ and unstable for $\lambda = \lambda_H\left(\sqrt{\epsilon}\right)$. The expansion of $\lambda_H\left(\sqrt{\epsilon}\right)$ is given in Equation (5.2.13). The Hopf bifurcation is non-degenerate when $A \neq 0$. Hopf bifurcation is supercritical if $A < 0$, sub-critical if $A > 0$. (A is given in Equation 5.2.16).

Theorem 5.2.3

Let the origin be a generic folded singularity of the system (5.2.9). Then there exists a smooth function $\lambda_c\left(\sqrt{(\epsilon)}\right)$ such that for $\lambda = \lambda_c\left(\sqrt{\epsilon}\right)$, the system (5.2.9) admits a maximal canard (the expansion of the function λ_c is given in Equation 5.2.14).

Here, the Hopf bifurcation is termed as singular Hopf bifurcation in the sense that at the Hopf bifurcation threshold $\lambda = \lambda_H$, the eigenvalues of the Jacobian matrix become singular as $\epsilon \to 0$. Now, it happens in most of the cases that the canard cycle which emerges due to singular Hopf bifurcation at $\lambda = \lambda_H$ is of small amplitude and as λ tends to λ_c there is a rapid growth of the canard cycle and a further variation of λ leads to a relaxation oscillation. Such a phenomenon in literature is known as a "canard explosion." Canard explosion and relaxation oscillation are global phenomena for multiple time scale systems, as they characterize transitions between a small amplitude and large amplitude canard cycle. We refer to [5,10] for the detailed analysis of the phenomena of canard explosion and relaxation oscillation.

Definition 5.7

Canard explosion is a transition phenomenon from a small amplitude limit cycle to a large amplitude relaxation oscillation through a family of canard cycles and occurs upon variation of the control parameter within an exponentially small range.

Definition 5.8

A periodic solution Γ_ϵ of a slow–fast system is said to be a relaxation oscillation if it converges to a singular orbit Γ_0 consisting of alternating slow and fast segments of the slow–fast flow, forming a closed loop as the singular limit $\epsilon \to 0$ in the Hausdorff distance.

To investigate the existence of a relaxation oscillation, one can apply the entry-exit function as discussed below:

5.2.5 Entry–exit function

The entry–exit function plays an important role in obtaining the existence of periodic orbits that exhibit relaxation oscillations of a slow–fast system. We consider the following planar singularity perturbed vector field and restrict our attention to the positive quadrant only:

$$\frac{dx}{dt} = xF(x, y), \tag{5.2.17a}$$

$$\frac{dy}{dt} = \epsilon G(x, y), \tag{5.2.17b}$$

where $(x, y) \in \mathbb{R}_+^2$, $0 \leq \grave{o} \ll 1$ and F, G are C^∞ functions having the following properties:

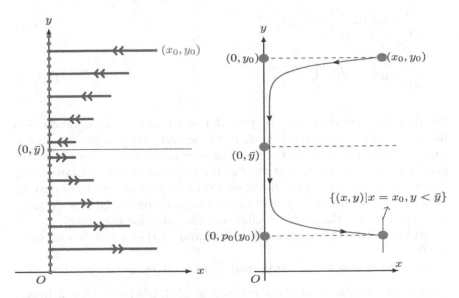

Figure 5.1 (Left) For the system (5.2.17), when $\epsilon = 0$, the y-axis consists of equilibria which are illustrated by solid circles. The y-axis is attracting when $y > \bar{y} > 0$ and is repelling when $y < \bar{y}$. (Right) A typical orbit of the system (5.2.17) for $0 < \epsilon \ll 1$, starting at (x_0, y_0) for $y_0 > \bar{y}$ and $x_0 > 0$, and exiting at $(x_0, p_0(y_0))$.

$$F(0,y) < 0, G(0,y) < 0 \text{ for } y > \bar{y} > 0 \text{ and } F(0,y) > 0 \text{ for } y < \bar{y} \quad (5.2.18)$$

The y-axis consists of equilibria attracting for $y > \bar{y}$ and repelling for $y < \bar{y}$ when $\epsilon = 0$. For $\epsilon > 0$, very small, a trajectory that starts at (x_0, y_0) with $x_0 > 0$ very small and $y_0 > \bar{y}$ is attracted towards the y-axis and then moves down gradually and then is repelled from the y-axis when it crosses the line $y = \bar{y}$ (see Figure 5.1). For $\epsilon > 0$, very small, the trajectory thus re-intersects the line $x = x_0$ at a point whose y-coordinate is $p_\epsilon(y_0)$ such that $\lim_{\epsilon \to 0} p_\epsilon(y_0) = p_0(y_0)$, where $p_0(y_0)$ is determined by

$$\int_{y_0}^{p_0(y_0)} \frac{F(0,y)}{G(0,y)} dy = 0. \quad (5.2.19)$$

The function $y_0 \to p_0(y_0)$ is called an entry-exit function.

5.3 APPLICATION IN A PREDATOR–PREY MODEL

A common framework to represent a two-species modified Leslie–Gower predator–prey model introduced by Leslie (1948, 1958) is to use a system of two ODEs:

$$\frac{du}{dT} = ru\left(1 - \frac{u}{K}\right) - vG(u),$$
(5.3.1a)

$$\frac{dv}{dT} = sv\left(1 - \frac{pv}{c+u}\right),$$
(5.3.1b)

the first equation describes the prey dynamics and the second equation models predator dynamics. Here u, v are the prey and predator biomass at time T, K is the carrying capacity of the prey population in the absence of predators, sp is the maximum value of the per capita reduction rate of v, c quantifies the amount to which the environment provides protection to the predator, s is the intrinsic growth rate of the predator and $G(u)$ is the functional response of the predator, satisfying the following properties
$G(0) = 0, G(u)$ is an increasing function, and $G(u) \to q$ as $u \to \infty$ for some $q > 0$.

For some examples of traditional functional response functions that meet these criteria, readers are referred to [27]. $p(x) = \frac{qu}{u+b}$ is a hyperbolic Holling type II functional response, where q is the capturing rate of prey by predator and b is the half-saturation constant. All the parameters r, K, b, c, p, and q are positive constants. Now, we introduce an Allee effect into the predator by introducing the term $H(v) = \frac{v}{v+m}$, called the weak Allee effect function, where m is the Allee effect constant. It is the probability that a female will meet and mate with at least one male during the reproductive phase. In general, to incorporate the Allee effect in the predator, the probability term $H(v)$ is multiplied with the birth rate s of the predator corresponding to the non-Allee growth equation of the predator. Due to this Allee effect function, the per capita growth rate of the predator is reduced from s to $\frac{v}{v+m}$. It is to mention that $H(v)$ satisfies the following basic properties (1) $H(0) = 0$, i.e., at a population size of zero, no mating takes place, (2) $\frac{dH(v)}{dv} > 0$, i.e., when population size grows, the likelihood of successful mating increases, and (3) $H(v) \to 1$ as $v \to \infty$, i.e., when the population is large, mating is almost guaranteed. The Allee effect is considered in the predator population because the predator population is more prone than their prey [28]. For this reason, we are curious to observe how the Allee effect in the predator population affects the dynamics of a slow–fast system. For further justification of such type of Allee effect function and more biological rationale, readers are referred to [28,29,30].

We now arrive at the following modified Leslie–Gower predator–prey model with Allee effect in predator population and Holling type II functional response, given by:

$$\frac{du}{dT} = ru\left(1 - \frac{u}{K}\right) - \frac{quv}{b+u}, \tag{5.3.2a}$$

$$\frac{dv}{dT} = sv\left(\frac{v}{v+m} - \frac{pv}{c+u}\right), \tag{5.3.2b}$$

subject to non–negative initial conditions $u(0) \geq 0, v(0) \geq 0$, and positive parameters r, K, e, q, b, d. The meaning of the variables and parameters involved in the model has already been discussed above. The predator equation in (5.3.2) indicates that the larger the value of m, the stronger the Allee effect, and the slower the predator population's per capita growth rate. Few real-life evidence from ecology where the predator population is affected by the Allee effect are the following (1) in the phytoplankton–zooplankton system, the zooplankton growth is known to be subject to the Allee effect [31], and (2) the spotted owls (*Strix occidentalis*) with mating limitation behavior due to its habitat loss [32]. We also consider that for model (5.3.2), $s \ll r$, i.e., the growth rate of the predator is very less than that of the prey.

To deal with the model (5.3.2) with a smaller number of parameters, we use the following rescaling transformations:

$$t = rT, \; x = \frac{u}{K}, \; y = \frac{pv}{K}, \tag{5.3.3}$$

we have

$$\frac{dx}{dt} = x(1-x) - \frac{\alpha xy}{\beta + x} = f(x, y, \mu), \tag{5.3.4a}$$

$$\frac{dy}{dt} = \epsilon y\left(\frac{y}{y+\gamma} - \frac{y}{\delta+x}\right) = \epsilon g(x, y, \mu), \tag{5.3.4b}$$

where x, y are the new dimensionless variables, $\mu = (\alpha, \beta, \gamma, \delta)$ with $\alpha = \frac{q}{pr}$, $\beta = \frac{b}{K}$, $\gamma = \frac{c}{K}$, $\delta = \frac{c}{K}$, and $\epsilon = \frac{s}{r}$ are the dimensionless parameters. All the parameters are positive and as $s \ll r$, we have $0 < \epsilon = s/r \ll 1$.

We have the following basic results for the system (5.3.4).

Lemma 5.3.1

The first quadrant $\mathbb{R}_+^2 = \{(x, y) \in \mathbb{R}^2 \mid x \geq 0, y \geq 0\}$ is invariant under the flow generated by the vector field $F = f\dfrac{\partial}{\partial x} + \epsilon g\dfrac{\partial}{\partial y}$.

Lemma 5.3.2

All the solutions of the model system (5.3.4) initiated from the interior of $\mathbb{R}^2_+$ are bounded.

5.3.1 Slow–fast analysis

With the time scaling $\tau = \epsilon t$, $0 < \epsilon \ll 1$ the system (5.3.4) transforms into the following topologically equivalent system:

$$\epsilon \frac{dx}{d\tau} = x(1-x) - \frac{\alpha xy}{\beta + x}, \tag{5.3.5a}$$

$$\frac{dy}{d\tau} = y\left(\frac{y}{y+\gamma} - \frac{y}{\delta + x}\right). \tag{5.3.5b}$$

The model system (5.3.4) or (5.3.5) is a standard form of a slow–fast system with t as the fast time scale and τ as the slow time scale, respectively. The variables x and y are referred to as fast and slow variables, respectively. In the singular limit $\epsilon \to 0$, the systems (5.3.4) and (5.3.5) transform into the following fast and slow subsystems.

$$\frac{dx}{dt} = x(1-x) - \frac{\alpha xy}{\beta + x}, \tag{5.3.6b}$$

$$\frac{dy}{dt} = 0, \tag{5.3.6b}$$

and

$$0 = x(1-x) - \frac{\alpha xy}{\beta + x}, \tag{5.3.7a}$$

$$\frac{dy}{d\tau} = y\left(\frac{y}{y+\gamma} - \frac{y}{\delta + x}\right). \tag{5.3.7b}$$

The slow flow corresponding to the slow subsystem (5.3.7) is constrained on the critical set M_0 given by

$$M_0 = \left\{(x, y) \in \mathbb{R}^2_+ \,\middle|\, f(x, y, \mu) = 0\right\}.$$

The critical set M_0 consists of two kinds of critical manifolds given by

$$M_{10} = \left\{(x, y) \in \mathbb{R}_+^2 \middle| x = 0\right\}, \tag{5.3.8a}$$

$$M_{20} = \left\{(x, y) \in \mathbb{R}_+^2 \middle| y = \phi(x) = \frac{(1-x)(\beta+x)}{\alpha}, \beta < 1, 0 < x < 1\right\} \tag{5.3.8b}$$

The function $\phi(x)$ has a local maximum at $x = x_m = \dfrac{1-\beta}{2}$ in the interior of $\mathbb{R}_+^2$ if $\beta < 1$. Henceforth, we will be assuming throughout the chapter the parametric condition $\beta < 1$ so that the critical manifold M_{20} is of parabolic shape, increases in $0 < x < x_m$ and decreases in $x_m < x < 1$. Consequently, the critical manifold M_{20} consists of two branches S_0^r and S_0^a where S_0^r is the branch from $P\left(0, \dfrac{\beta}{\alpha}\right)$ to $Q(x_m, y_m)$ (maximum point),

$x_m = \dfrac{1-\beta}{2}$, $y_m = \phi(x_m) = \dfrac{(1+\beta)^2}{4\alpha}$; S_0^a is the branch from Q to $R(1,0)$. Thus,

$$S_0^r = M_{20} \cap \left\{(x, y) \in \mathbb{R}_+^2 \middle| 0 < x < x_m\right\}, \tag{5.3.9}$$

$$S_0^a = M_{20} \cap \left\{(x, y) \in \mathbb{R}_+^2 \middle| x_m < x < 1\right\}. \tag{5.3.10}$$

Lemma 5.3.3

Consider $0 < \epsilon \ll 1$. Then

    i. M_{20} loses its normal hyperbolicity at P and Q.

    ii. The branches S_0^r and S_0^a are normally hyperbolic attracting and repelling.

Proof. (i) We have, $\dfrac{\partial f}{\partial x}\bigg|_{(x, \phi(x))} = \dfrac{x(1-\beta-2x)}{\beta+x}$. Therefore, $\dfrac{\partial f}{\partial x}$ has zero eigenvalue at P and Q and consequently M_{20} loses its normal hyperbolicity at P and Q. (ii) $\dfrac{\partial f}{\partial x}\bigg|_{(x, \phi(x))} > 0$, for $0 < x < x_m$ and $\dfrac{\partial f}{\partial x}\bigg|_{(x, \phi(x))} < 0$, for $x_m < x < 1$. Hence, the branches S_0^r and S_0^a are normally hyperbolic repelling and attracting, respectively.

Similarly, the normally hyperbolic repelling and attracting parts of the critical manifold M_{10}, denoted by S_0^{r+} and S_0^{a+} are given by

$$S_0^{r+} = M_{10} \cap \left\{(x, y) \in \mathbb{R}_+^2 \middle| 0 < y < \frac{\beta}{\alpha}\right\}, \tag{5.3.11}$$

$$S_0^{a+} = M_{10} \cap \left\{ (x, y) \in \mathbb{R}_+^2 \middle| y > \frac{\beta}{\alpha} \right\}. \tag{5.3.12}$$

The slow flow that evolves on the critical manifold M_{20} is given

$$\frac{dx}{d\tau} = \frac{\phi^2(x)\left(x^2 + (\alpha + \beta - 1)x + \alpha(\delta - \gamma) - \beta\right)}{\alpha\phi'(x)(x + \delta)\left(\phi(x) + \gamma\right)}, \tag{5.3.13}$$

and is not defined at the point Q. The points Q, P are known as the fold and transcritical points, respectively, as they correspond to fold and transcritical bifurcations for (5.3.6) considering y as a parameter. Now, for $0 < \epsilon \ll 1$, Fenichel's theorem tells us that, S_0^r and S_0^a can be perturbed to S_ϵ^r and S_ϵ^a which are within $\mathcal{O}(\epsilon)$ distance from S_0^r and S_0^a. The dynamics of the slow and fast subsystems (5.3.6) and (5.3.7), respectively, are shown in Figure 5.2.

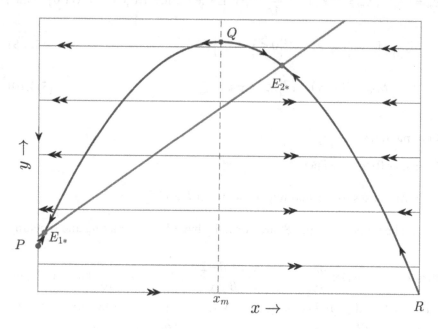

Figure 5.2 The dynamics of both the fast and slow subsystems (5.3.6) and (5.3.7), respectively, are presented. Solid circles on the straight line (predator nullcline) represent two possible interior equilibrium points. The generic transcritical point $P(0, \beta/\alpha)$ and generic fold point $Q(x_m, y_m)$ are the two non-hyperbolic points. The two branches of the critical manifold M_{20} (parabolic curve) are presented, of which S_0^r (from P to Q for $0 < x < x_m$) is normally hyperbolic repelling, and S_0^a (from Q to the point $R(1, 0)$ for $x_m < x < 1$) is normally hyperbolic attracting. Manifold M_{10} is along the positive y-axis. Double arrows on each of the horizontal lines indicate fast flow, whereas single arrows on the critical manifolds indicate slow flow.

5.3.2 Existence and linearized stability analysis

The system (5.3.4) has three equilibria on the co-ordinate axes, namely, the trivial equilibrium $E_0(0,0)$ and the boundary equilibria $E_1(0,\delta-\gamma)$, $E_2(1,0)$ where the boundary equilibrium E_1exists if $\delta > \gamma$. We have the following trivial results on the nature of the equilibria on the co-ordinate axes.

Lemma 5.3.4

i. The trivial equilibrium $E_0(0,0)$ is a saddle node.

ii. The boundary equilibrium $E_1(0,\delta-\gamma)$ is a hyperbolic stable node if $\delta > \gamma + \dfrac{\beta}{\alpha}$, a hyperbolic saddle if $\delta < \gamma + \dfrac{\beta}{\alpha}$ and a saddle node if $\delta = \dfrac{\beta}{\alpha} + \gamma$.

iii. The boundary equilibrium $E_2(1,0)$ is a saddle node.

The interior equilibria are the points of intersection of the non-trivial prey and predator nullclines given by

$$y = \frac{(\beta+x)(1-x)}{\alpha}, \; y = x + (\delta - \gamma).$$

Assuming $D = (\alpha+\beta-1)^2 - 4(\alpha(\delta-\gamma)-\beta)$, we consider the following parametric regions,

$$R_1 = \left\{ \mu = (\alpha, \beta, \gamma, \delta) \middle| D > 0, \alpha+\beta < 1, \delta > \gamma + \frac{\beta}{\alpha} \right\}, \tag{5.3.14a}$$

$$R_2 = \left\{ \mu = (\alpha, \beta, \gamma, \delta) \middle| D = 0, \alpha+\beta < 1 \right\}, \tag{5.3.14b}$$

$$R_3 = \left\{ \mu = (\alpha, \beta, \gamma, \delta) \middle| \{0 \leq \gamma - \delta < 1\} \cup \left\{ \gamma < \delta < \frac{\beta}{\alpha} + \gamma \right\}, \beta < 1 \right\} \tag{5.3.14c}$$

$$R_4 = \left\{ \mu = (\alpha, \beta, \gamma, \delta) \middle| \left\{ D > 0, \alpha+\beta > 1, \delta > \gamma + \frac{\beta}{\alpha} \right\} \cup \{D < 0\} \cup \{\gamma \geq \delta + 1\} \right\}. \tag{5.3.14d}$$

We now state the following results on the existence and stability of the interior equilibria of the system (5.3.4).

Lemma 5.3.5

i. If $\mu \in R_1$ then there exist two interior equilibrium points $E_{1^*}(x_{1^*}, y_{1^*})$ and $E_{2^*}(x_{2^*}, y_{2^*})$, where

$$x_{1^*} = \frac{1-(\alpha+\beta)-\sqrt{D}}{2}, \quad y_{1^*} = x_{1^*} + \delta - \gamma,$$

$$x_{2^*} = \frac{1-(\alpha+\beta)+\sqrt{D}}{2}, \quad y_{2^*} = x_{2^*} + \delta - \gamma.$$

The equilibrium E_{1^*} is a hyperbolic saddle and E_{2^*} is a stable equilibrium point if $x_{2^*} \geq x_m$. For $x_{2^*} < x_m$, the equilibrium E_{2^*} will be either stable or unstable, depending on whether Trace $J|_{E_{2^*}} <$ or > 0.

ii. If $\mu \in R_2$ then there exists only one interior equilibrium point $E_*(x_*, y_*)$ where $x_* = \dfrac{1-(\alpha+\beta)}{2}$, $y_* = x_* + \delta - \gamma$.

In this case, the non-trivial predator nullcline touches the non-trivial prey nullcline tangentially at the point E_*. The equilibrium E_* is a saddle node.

iii. If $\mu \in R_3$ then there exists only one interior equilibrium point $E_{2^*}(x_{2^*}, y_{2^*})$, stable if $x_{2^*} \geq x_m$ and for $x_{2^*} < x_m$, it will be either stable or unstable, depending on whether Trace $J|_{E_{2^*}} <$ or > 0.

iv. If $\mu \in R_4$ then the system (5.3.4) has no interior equilibrium in $\mathbb{R}_+^2$.

An example of the intersection of the non-trivial prey and predator nullclines is presented in Figure 5.3. A gradual increase in γ from small to large values indicates the number of interior equilibrium points of the system (5.3.4) varies from zero to two. Figure 5.3 also indicates that a saddle-node bifurcation will occur (for the predator nullcline), which will be described in Section 5.3.3.

5.3.3 Bifurcation scenario

The non-trivial prey and predator nullclines intersect the positive y-axis at the point $P\left(0, \dfrac{\beta}{\alpha}\right)$ and $E_1(0, \delta - \gamma)$ if $\delta > \gamma$ and consequently, based on the nature of the non-trivial nullclines we have that if E1 lies below the point P then there always exists a unique interior equilibrium point E_{2^*}, if E_1 lies above the point P then under certain parametric conditions (as mentioned in Lemma 5.3.5), there may exist zero, one, or two interior equilibrium points. Thus, we see that by varying the control parameter

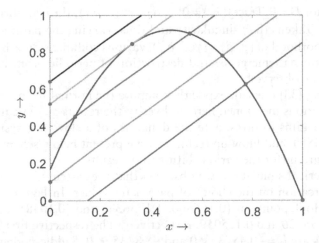

Figure 5.3 Relative positions of the nullclines of the system (5.3.4). The parabolic curve depicts the non-trivial prey nullcline and the non-trivial predator nullcline is a straight line passing through $E_1(0, \delta - \gamma)$. The figure depicts that the number of interior equilibrium points varies from zero to two for the parameter values $\alpha = 0.3$, $\beta = 0.05$, $\delta = 0.65$, and variable γ. Varying γ, five relative positions of predator nullclines are plotted sequentially from top to bottom lines for $\gamma = 0.01$ ($\mu \in R_4$, no interior equilibrium), $\gamma = 0.13125$ ($\mu \in R_2$, unique interior equilibrium), $\gamma = 0.3$ ($\mu \in R_1$, two interior equilibrium), $\gamma = 0.55$ ($\mu \in R_3$, unique interior equilibrium) and $\gamma = 0.8$ ($\mu \in R_3$, unique interior equilibrium), respectively. Equilibrium points are represented by circles.

δ it follows that for $\delta = \delta_{TC} = \gamma + \dfrac{\beta}{\alpha}$, the model system (5.3.4) undergoes a transcritical bifurcation as one interior equilibrium bifurcates from $E_1(0, \delta - \gamma)$ as γ passes through $\delta = \delta_{TC}$. Assuming the parametric conditions $\alpha + \beta < 1, \delta > \gamma + \dfrac{\beta}{\alpha}$, we have that for $D > 0$, there exist two interior equilibrium points E_{1*} and E_{2*} where E_{1*} is a hyperbolic saddle point; for

$$D = 0 \left(\gamma = \gamma_{SN} = \delta - \frac{\beta}{\alpha} - \frac{(\alpha + \beta - 1)^2}{4\alpha} \right)$$ the two equilibrium points E_{1*} and E_{2*}

coalesce at the degenerated saddle node equilibrium point $E_*(x_*, y_*)$ and for $D < 0$ there exists no interior equilibrium point. Thus, we have saddle node bifurcation of equilibria, i.e., the model system (5.3.4) undergoes a saddle node bifurcation as γ passes through $\gamma = \gamma_{SN}$. For $(\gamma, \delta) = (\gamma_{SN}, \delta_{TC})$, the model system (5.3.4) undergoes a saddle-node transcritical bifurcation topologically equivalent to co-dimension 2 cusp bifurcation as (γ, δ) passes through $(\gamma, \delta) = (\gamma_{SN}, \delta_{TC})$. Now, it may also happen that varying δ, there may take place Hopf bifurcation around E_{2*} for $\delta = \delta_H$ and will be studied in the next section in the realm of slow–fast analysis. We also have that for $D = 0$, $\text{TraceJ}(E_*) = 0$ $(\gamma, \delta) = (\gamma_{SN}, \delta_H)$, the equilibrium E_* is a Bogdanov–Takens (BT) singularity and thus, varying the parameter (γ, δ) in a neighborhood of $(\gamma, \delta) = (\gamma_{SN}, \delta_H)$, various codimension-2 BT bifurcation phenomena (emergence and destruction of periodic orbit, homoclinic orbit) will be observed.

Following [33] one can explicitly compute the normal forms of the various bifurcations mentioned here and verify the results analytically. But, as the chapter aims to investigate the dynamics of a slow–fast system in the realm of GSPT and blow-up technique, we present below schematic bifurcation diagrams for the various bifurcation results.

By numerical simulation, we observed that the system (5.3.4) undergoes Hopf bifurcation on the choice of parameter values. In Figure 5.4a, there are two Hopf points at $(0.156660, 1.173634)$ and $(0.104857, 1.129798)$ for $\gamma = 0.133026$ and 0.125059, respectively. The respective first Lyapunov coefficients are $l_1 = -0.45739 < 0$ and $-5.8235 < 0$. Saddle-node (LP) point and transcritical point (BP) are obtained for $\gamma = \gamma_{SN} = 0.125$ and $\gamma = 0.15$, respectively. Using the numerical bifurcation package MATCONT [34], we plot a two-parameter bifurcation diagram for the system (5.3.4) by choosing the bifurcation parameter γ and δ in Figure 5.4b with fixed parameters $\alpha = 0.4$, $\beta = 0.4$, $\epsilon = 0.1$. The Bogdanov-Takens (BT) bifurcation point is detected at $(\gamma_{BT}, \delta_{BT}) = (0.132789, 1.157789)$ with the normal form coefficients (a, b) of the BT bifurcation are $(-0.01131, 0.15908)$. Also, the system undergoes a Bautin or generalized Hopf (GH) bifurcation for $(\gamma_{GH}, \delta_{GH}) = (0.878646, 1.842204)$ with second Lyapunov coefficient $l_2 = -15.55193$.

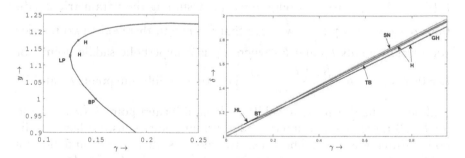

Figure 5.4   One and two-parameter bifurcation diagrams. (Left) One parameter bifurcation diagram concerning the bifurcation parameter γ. The other parameter values are $\alpha = 0.4$, $\beta = 0.4$, $\delta = 1.15$, $\epsilon = 0.1$. (Right) Two parameter bifurcation diagram in $\gamma - \delta$ plane with fixed parameters $\alpha = 0.4$, $\beta = 0.4$, $\epsilon = 0.1$. The H curve represents the Hopf bifurcation curve, the SN curve represents a saddle-node bifurcation, the HL curve stands for the homoclinic bifurcation curve and the TB curve is the transcritical bifurcation curve $\delta = \gamma + \beta/\alpha$.

5.3.4 Singular Hopf bifurcation and canard cycles

Here, we assume $\mu \in R_1 \cup R_3$ so that the existence of the interior equilibrium E_{2*} is ensured. We observe that for $\delta = \delta_* = \dfrac{(1+\beta)^2}{4\alpha} - \dfrac{1-\beta}{2} + \gamma$, the interior equilibrium E_{2*} coincides with the fold point $Q(x_m, y_m)$. i.e., $x_m = x_{2*}$, $y_m = y_{2*}$, and also, the following

$$f(x, y)\big|_{Q=E_{2*}} = 0, \quad g(x, y)\big|_{Q=E_{2*}} = 0,$$

$$\frac{\partial f(x, y)}{\partial x}\bigg|_{Q=E_{2*}} = 0, \quad \frac{\partial f(x, y)}{\partial y}\bigg|_{Q=E_{2*}} = -\alpha\frac{1-\beta}{1+\beta} < 0,$$

$$\frac{\partial g(x, y)}{\partial x}\bigg|_{Q=E_{2*}} = \frac{y_m^2}{(y_m+\gamma)^2}, \quad \frac{\partial g(x, y)}{\partial \delta}\bigg|_{Q=E_{2*}} = \frac{y_m^2}{(y_m+\gamma)^2},$$

$$\frac{\partial^2 f}{\partial x^2}\bigg|_{Q=E_{2*}} = \frac{2(\beta-1)}{\beta+1} < 0.$$

The singularity Q is then referred to as the generic folded singularity or the canard point. Using the transformation $X = x - x_m$, $Y = y - y_m$ and $\lambda = \delta - \delta_*$, the system (5.3.4) transforms into the following form

$$\frac{dX}{dt} = Y\left(a_{01} + a_{11}X + \mathcal{O}(X^2, XY, Y^2)\right) + X^2\left(\frac{a_{20}}{2} + \frac{a_{30}}{6}X + \mathcal{O}(X^2, Y^2)\right),$$

$$\text{(5.3.15a)}$$

$$\frac{dY}{dt} = \epsilon\left[X\left(b_{10} + \frac{b_{20}}{2}X + b_{11}Y\right) + \lambda\left(\frac{y_m^2}{(y_m+\gamma)^2} + \mathcal{O}(X, Y, \lambda)\right)\right.$$

$$\left. + Y\left(b_{01} + \frac{b_{02}}{2}Y + \mathcal{O}(X^2, XY, Y^2)\right),\right] \qquad \text{(5.3.15b)}$$

where

$$a_{01} = -\frac{\alpha(1-\beta)}{1+\beta}, \; a_{11} - \frac{4\alpha\beta^2}{(1+\beta)}, \; a_{20} = 2\frac{\beta-1}{1+\beta}, \; a_{30} = -\frac{24\beta}{(1+\beta)^2}, \; b_{10} = \frac{y_m^2}{(y_m+\gamma)^2},$$

$$b_{01} = -\frac{y_m^2}{(y_m+\gamma)^2}, \; b_{20} = -\frac{2y_m^2}{(y_m+\gamma)^3}, \; b_{02} = -\frac{2y_m(y_m+2\gamma)}{(y_m+\gamma)^3}, \; b_{11} = \frac{2y_m}{(y_m+\gamma)^2}.$$

To use the theory as developed in [5], we use the following rescaling

$$X = aX', \quad Y = bY', \quad t = ct'$$

where

$$a = -\frac{2b_{10}a_{01}}{a_{20}}\sqrt{-\frac{1}{a_{01}b_{10}}}, \quad b = \frac{2b_{10}}{a_{20}}, \quad c = \sqrt{-\frac{1}{a_{01}b_{10}}}.$$

The system (5.3.15) is then topologically equivalent to the following canonical form

$$\frac{dX'}{dt'} = -Y'h_1(X', Y') + X'^2 h_2(X', Y') + \epsilon h_3(X', Y') \qquad (5.3.16a)$$

$$\frac{dY'}{dt'} = \epsilon\left[X'h_4(X', Y') - \lambda' h_5(X', Y', \lambda') + Y'h_6(X', Y')\right], \qquad (5.3.16b)$$

where

$$h_1(X', Y') = 1 - bca_{11}X' + \mathcal{O}\left(X'^2, X'Y', Y'^2\right),$$

$$h_2(X', Y') = 1 + \frac{a^2c}{6}a_{30}X' + \mathcal{O}\left(X'^2, Y'^2\right),$$

$$h_3(X', Y') = 0, \quad h_4(X', Y') = 1 + \frac{a^2c}{2b}b_{20}X' + acb_{11}Y' + \mathcal{O}\left(X'^2, X'Y', Y'^2\right),$$

$$h_5(X', Y', \lambda') = 1 + \mathcal{O}(X', Y', \lambda'), \quad h_6(X', Y') = cb_{01} + \frac{bc}{2}b_{02}Y' + \mathcal{O}\left(X'^2, Y'^2\right),$$

$$\lambda' = -\frac{c\lambda}{b}\frac{y_m^2}{(y_m + \gamma)^2}.$$

Now, by Equations (5.3.12) and (5.3.13) of [5], we have

$$a_1 = \frac{\partial h_3}{\partial X'}(0,0) = 0, \qquad (5.3.17a)$$

$$a_2 = \frac{\partial h_1}{\partial X'}(0,0) = -bca_{11}, \qquad (5.3.17b)$$

$$a_3 = \frac{\partial h_2}{\partial X'}(0,0) = \frac{a^2 c}{6} a_{30},$$ (5.3.17c)

$$a_4 = \frac{\partial h_4}{\partial X'}(0,0) = \frac{a^2 c}{2b} b_{20},$$ (5.3.17d)

$$a_5 = h_6(0,0) = c b_{01},$$ (5.3.17e)

$$A = -a_2 + 3a_3 - 2a_4 - 2a_5 = bca_{11} + \frac{a^2 c}{2} a_{30} - \frac{a^2 c}{b} b_{20} - 2cb_{01}.$$ (5.3.18)

Hence, following Equations (5.3.15) and (5.3.16) of [5], the expansions of singular Hopf bifurcation and maximal canard curves are given by

$$\lambda'_H\left(\sqrt{\epsilon}\right) = -\frac{a_1 + a_5}{2}\epsilon + \mathcal{O}\left(\epsilon^{3/2}\right),$$

$$\lambda'_c\left(\sqrt{\epsilon}\right) = -\left(\frac{a_1 + a_5}{2} + \frac{A}{8}\right)\epsilon + \mathcal{O}\left(\epsilon^{3/2}\right).$$

In terms of original parameters, the singular Hopf and maximal canard curves can be written as

$$\delta_H\left(\sqrt{\epsilon}\right) = \delta_* + \frac{bb_{01}(y_m + \gamma)^2}{2y_m^2}\epsilon + \mathcal{O}\left(\epsilon^{3/2}\right),$$ (5.3.19)

$$\delta_c\left(\sqrt{\epsilon}\right) = \delta_* + \frac{b(y_m + \gamma)^2}{4y_m^2}\left(b_{01} + \frac{b}{2}a_{11} + \frac{a^2}{4}a_{30} - \frac{a^2}{2b}b_{20}\right)\epsilon + \mathcal{O}\left(\epsilon^{3/2}\right).$$ (5.3.20)

Hence, validating the results stated in Theorems 5.2.2 and 5.2.3 in Section 5.2, we have the existence of singular Hopf bifurcation and maximal canard for the system (5.3.4).

In Figure 5.5, we observe that when $\mu \in R_1$ and E_{2*} is a folded singularity (canard point). Varying δ near to δ_H we have several small amplitude canard cycles generated via subcritical Hopf bifurcation and as δ reaches δ_c the system admits a homoclinic orbit to the saddle point E_{1*}. Thus, we have the existence of canard cycles and a homoclinic orbit via the canard point. On the other hand, when $\mu \in R_3$ and E_{2*} is a folded singularity (canard point), varying δ near to δ_H, we have several small amplitude canard cycles generated via supercritical Hopf bifurcation and as δ reaches δ_c the rapid

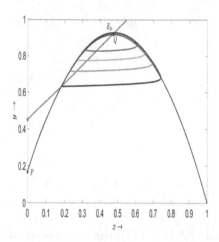 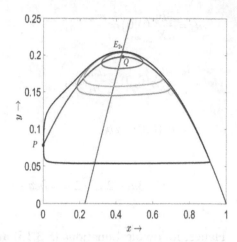

Figure 5.5 (Left) Canard cycles and the birth of homoclinic orbit via canard point for $\mu \in$ R_1. Four unstable periodic orbits are presented for four different values of δ, such as $\delta = 0.65$, 0.6499834, 0.649983132, 0.649983130261061 where the amplitude of the periodic orbits increases with the decrease of δ. For $\delta = 0.649983130261061$, the figure shows the birth of homoclinic orbit (the biggest periodic orbit) via the canard point connecting the saddle equilibrium point E_{1*}. The other parameter values are $\alpha = 0.30001$, $\beta = 0.05$, $\gamma = 0.2$, $\epsilon = 0.1$. (Right) The figure depicts the existence of the canard explosion phenomenon for $\mu \in R_3$. The stable periodic orbits are presented for different values of δ such as 0.181, 0.185, 0.1855 and 0.186 where the amplitude of the orbits increase with δ. The figure illustrates that the amplitude of the orbit drastically enlarged (canard explosion) when the bifurcation parameter δ is increased from $\delta = 0.1855$ (small amplitude cycle, the 3rd orbit away from Q) to $\delta = 0.186$ (big amplitude cycle called the relaxation oscillation). The other parameter values are $\alpha = 1.6$, $\beta = 0.125$, $\gamma = 0.41$, $\delta = 0.181$ and $\epsilon = 0.1$.

growth of the periodic orbits occurs showing the occurrence of a canard explosion. A further variation of the parameter leads to relaxation oscillation, whose existence employs the entry–exit function and Fenichel's theory will be shown in the next subsection.

5.3.5 Relaxation oscillation

Here, our target is to show the existence of relaxation oscillation for the system (5.3.4) for $0 < \epsilon \ll 1$ whenever $\mu \in R_3$ and $x_{2*} < x_m$ with the help of the entry-exit function, of which a preliminary idea has already been mentioned in Section 5.2.4. A relaxation oscillation for the system (5.3.4) is a periodic orbit Γ_ϵ which converges to a piece-wise smooth singular closed orbit Γ_0 consisting of slow fast segments as $\epsilon \to 0$ in the Hausdorff distance.

We know that the critical manifold M_{10} i.e., the y-axis is normally hyperbolic attracting for $y > \dfrac{\beta}{\alpha}$ and normally hyperbolic repelling for $y < \dfrac{\beta}{\alpha}$. We consider the system (5.3.4) and observe that for $\epsilon = 0$, the y axis consists of equilibria, attracting for $y > \dfrac{\beta}{\alpha}$ and repelling for $y < \dfrac{\beta}{\alpha}$. For $\epsilon > 0$, very small, a trajectory starting at (x_0, y_0), $x_0 > 0$, very small, $y_0 > \dfrac{\beta}{\alpha}$ gets attracted towards the y-axis and then drifts downward and then crosses the line $y = \dfrac{\beta}{\alpha}$ and gets repelled from the y-axis. Thus, for $\epsilon > 0$, very small, the trajectory re-intersects the line $x = x_0$ at $(x_0, p_\epsilon(y_0))$ such that

$$\lim_{\epsilon \to 0} p_\epsilon(y_0) = p_0(y_0)$$

where $p_0(y_0)$ is determined by

$$\int_{y_0}^{p_0(y_0)} \frac{1 - \dfrac{\alpha y}{\beta}}{y^2 \left(\dfrac{1}{y+\gamma} - \dfrac{1}{\delta} \right)} \, dy = 0. \tag{5.3.21}$$

Lemma 5.3.6

If $\gamma - \delta \geq 0$ or $\gamma - \delta < 0$ then there exists a unique y', where $0 < y' < \dfrac{\beta}{\alpha}$ or $\delta - \gamma < y' < \dfrac{\beta}{\alpha}$ such that

$$J(y') = \int_{y'}^{y_m} \frac{1 - \dfrac{\alpha y}{\beta}}{y^2 \left(\dfrac{1}{y+\gamma} - \dfrac{1}{\delta} \right)} \, dy = 0. \tag{5.3.22}$$

Proof. We have,

$$J(y) = \int_{y}^{y_m} \frac{1 - \dfrac{\alpha y}{\beta}}{y^2 \left(\dfrac{1}{y+\gamma} - \dfrac{1}{\delta} \right)} \, dy = -\frac{\delta}{\beta} \int_{y}^{y_m} \frac{(\beta - \alpha y)(y+\gamma)}{y^2 (y+\gamma-\delta)} \, dy$$

$$= -\frac{\delta}{\beta} \left[\frac{\alpha\gamma(\delta-\gamma) - \beta\delta}{(\gamma-\delta)^2} \int_{y}^{y_m} \frac{1}{y} \, dy + \frac{\beta\gamma}{\gamma-\delta} \int_{y}^{y_m} \frac{1}{y^2} \, dy + \frac{\delta(\alpha(\gamma-\delta)+\beta)}{(\gamma-\delta)^2} \int_{y}^{y_m} \frac{1}{y+\gamma-\delta} \, dy \right]$$

$$\to -\infty \text{ as } y \to 0^+, \ \gamma - \delta \geq 0 \text{ or as } y \to (\delta - \gamma)^+, \ \gamma - \delta < 0.$$

Further,

$$J'(y) = \frac{\delta}{\beta} \int_{y}^{y_m} \frac{(\beta - \alpha y)(y+\gamma)}{y^2(y+\gamma-\delta)} > 0,$$

either for $0 < y < \dfrac{\beta}{\alpha}$, $\gamma - \delta \geq 0$ or for $\delta - \gamma < y < \dfrac{\beta}{\alpha}$, $\gamma - \delta < 0$. Hence, $J(y)$ increases strictly for $0 < y < \dfrac{\beta}{\alpha}$, $\gamma - \delta \geq 0$ or for $\delta - \gamma < y < \dfrac{\beta}{\alpha}$, $\gamma - \delta < 0$.

We also have,

$$J\left(\frac{\beta}{\alpha}\right) = -\frac{\delta}{\beta}\int\limits_{\frac{\beta}{\alpha}}^{y_m} \frac{(\beta - \alpha y)(y + \gamma)}{y^2(y + \gamma - \delta)}\, dy > 0,$$

either   for $0 < y < \dfrac{\beta}{\alpha}$, $\gamma - \delta \geq 0$ or   for $\delta - \gamma < y < \dfrac{\beta}{\alpha}$, $\gamma - \delta < 0$.   Thus, it follows that there exists a unique y' where $0 < y' < \dfrac{\beta}{\alpha}$, $\gamma - \delta \geq 0$ or $\delta - \gamma < y' < \dfrac{\beta}{\alpha}$, $\gamma - \delta < 0$ such that $J(y') = 0$.

The critical manifolds M_{10} and M_{20} lose their normal hyperbolicity at $P\left(0, \dfrac{\beta}{\alpha}\right)$ and $Q(x_m, y_m)$. The point $Q(x_m, y_m)$ is a generic fold point for the system (5.3.4) and also a jump point as at this point the fast flow (5.3.6) is moved away from the critical manifold M_{20} and gets attracted toward S_0^{a+} the attracting branch of the critical manifold M_{10}. For the point P, we have,

$$\left.\frac{\partial f}{\partial x}\right|_P = 0 = \left.\frac{\partial f}{\partial y}\right|_P, \quad \left.\frac{\partial^2 f}{\partial x^2}\right|_P = 2\frac{(1-\beta)}{\beta} > 0, \; g\big|_P = \frac{\beta^2}{\alpha(\beta + \alpha\gamma)}\left(1 - \frac{\beta + \alpha\gamma}{\alpha\delta}\right) < 0,$$

and

$$\begin{vmatrix} \dfrac{\partial^2 f}{\partial x^2} & \dfrac{\partial^2 f}{\partial x \partial y} \\[2mm] \dfrac{\partial^2 f}{\partial x \partial y} & \dfrac{\partial^2 f}{\partial y^2} \end{vmatrix}_P = -\frac{\alpha^2}{\beta^2} < 0.$$

and hence, P is a generic transcritical point for the system (5.3.4). The point P is also a jump point, as at this point the fast flow (5.3.6) is moved away from the critical manifold M_{10}.

We now consider a singular slow–fast cycle Γ_0 as follows: From $S(0, y_m)$, follow the slow flow (5.3.7) down the y-axis to $T(0, y')$, follow the fast flow (5.3.6) to intersect the attracting branch S_0^a at $T'(x', y')$, follow the slow flow (5.3.7) along S_0^a to Q and then follow the fast flow (5.3.6) to the left of Q to $S(0, y_m)$. Consequently, we have a singular orbit Γ_0 consisting of slow and fast segments for which T, Q are jump points and T', S are drop points as at these points, the fast flow is moved toward the critical manifolds.

Theorem 5.3.7

Let $\mu \in R_3$, $x_{2^*} < x_m$ and N be a tubular neighborhood of Γ_0. Then for each fixed $0 < \epsilon \ll 1$, the system (5.3.4) has a unique relaxation oscillation $\Gamma_\epsilon \subset N$ which is strictly attracting with a characteristic multiplier bounded by $-\dfrac{K}{\epsilon}$ for some constant $K > 0$. Moreover, the cycle Γ_ϵ converges to Γ_0 in the Hausdorff distance as $\epsilon \to 0$.

Proof. Conditions stated in the theorem ensure that the system (5.3.4) has a unique interior equilibrium $E_{2^*}(x_{2^*}, y_{2^*})$ and the equilibrium lies to the left of the generic fold point Q. For $\epsilon > 0$ very small, following Fenichel's theorem S_0^a, S_0^{a+}, perturb to nearby slow manifolds S_ϵ^a and S_ϵ^{a+} and by Theorem (2.1) of [14], the slow manifolds S_ϵ^{a+} can be continued beyond the generic fold point Q and by Theorem (2.1) of [35], the slow manifold S_ϵ^{a+} can be continued beyond the generic transcritical point P. The slow manifold to S_ϵ^a (resp. S_ϵ^{a+}) lies close to S_0^a (resp. S_0^{a+}) until it arrives at the vicinity of the generic fold point Q (resp. generic transcritical point P).

We consider a small vertical section $\Delta = \left\{ (x_0, y) \mid y \in [y_m - \epsilon_0, \, y_m + \epsilon_0] \right\}$, $0 < \epsilon_0 \ll 1$. We know that for every point $(0, y_0)$, $y_0 \in [y_m - \epsilon_0, \, y_m + \epsilon_0]$ we can define $p_0(y_0)$ such that $0 < p_0(y_0) < \dfrac{\beta}{\alpha}$ for $\gamma - \delta \geq 0$ or $\delta - \gamma < p_0(y_0) < \dfrac{\beta}{\alpha}$ for $\gamma - \delta < 0$ by the result derived in Lemma 5.3.6.

We now follow tracking two trajectories $\Gamma_\epsilon^{1,2}$ starting on Δ at the points $\left(x_0, y^{1,2} \right)$. For $0 < \epsilon \ll 1$, it follows from Fenichel's theorem that $\Gamma_\epsilon^{1,2}$ get attracted toward the slow manifold S_ϵ^{a+} exponentially with a rate $\mathcal{O}\left(e^{-1/\epsilon}\right)$ and move downward slowly. Then by Theorem (2.1) of [14], $\Gamma_\epsilon^{1,2}$ pass by the generic transcritical point P contracting exponentially toward each other and leaving the repelling branch S_0^{r+} of the critical manifold M_{10} at the points $\left(0, p_0\left(y^{1,2}\right) \right)$ and then jump horizontally to $\left(x_0, p_\epsilon\left(y^{1,2}\right) \right)$ where $\lim\limits_{\epsilon \to 0} p_\epsilon\left(y^{1,2}\right) = p_0\left(y^{1,2}\right)$. The trajectories then follow two layers of the fast flow (5.3.6) and get attracted toward the slow manifold S_ϵ^a and pass the generic fold point Q contracting exponentially and thus, finally return to Δ.

Tracking the forward trajectories, we thus have a return map $\Pi : \Delta \to \Delta$ inducted by the flow of (5.3.4) for $0 < \epsilon \ll 1$. The return map Π is a contraction map as the trajectories contract toward each other with a rate $\mathcal{O}\left(e^{-1/\epsilon}\right)$ and by the contraction mapping theorem Π has a unique fixed point that is stable. This fixed point is the desired limit cycle Γ_ϵ which exists in a tubular neighborhood of the singular slow–fast cycle Γ_0 and as the contraction is exponential, the characteristic multiplier of Γ_ϵ is bounded above by $-\dfrac{K}{\epsilon}$ for some $K > 0$. Again, applying Fenichel's theorem, Theorem (2.1) of [35], and Theorem (2.1) of [14], we conclude that the periodic orbit Γ_ϵ converges to the singular orbit Γ_0 as $\epsilon \to 0$ in the Hausdorff distance.

For a geometrical description of the proof of Theorem 5.3.7, see Figure 5.6.

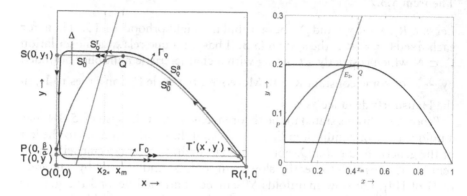

Figure 5.6   (Left) A schematic diagram of relaxation oscillations for the system (5.3.4) for $x_{2*} < x_m$: the singular orbit is Γ_0, and the relaxation oscillation orbit is presented by Γ_ϵ for $\epsilon > 0$. The prey and predator nullclines are depicted by the parabolic curve and the straight line. P and Q, the generic transcritical and generic fold points respectively are the two non-hyperbolic points on the parabolic curve depicted by solid circles. Γ_0 follows the following path: from S to T along y-axis downward, it follows the slow– flow (5.3.7); follows the fast–flow (5.3.6) along the horizontal line from T to T'; follows the slow–flow (5.3.7) from T' to Q along the attracting branch S_a^0, and finally, it follows the fast–flow (5.3.6) from Q to S along S_r^0. For $0 < \epsilon \ll 1$, by Fenichel's theorem, the submanifolds S_0^r and S_0^a can be perturbed to S_ϵ^r and S_ϵ^a which are within $0(\epsilon)$ distance from S_0^r and S_0^a. $\Delta = \left\{ \left(x_0, y\right) \mid y \in \left[y_m - \epsilon_0, y_m + \epsilon_0\right] \right\}$, $0 < \epsilon_0 \ll 1$ is a small vertical section. (Right) A numerical example of unique relaxation oscillation for the system (5.3.4), represented by a non-smooth orbit surrounding the unique interior equilibrium point generated for the parameter values $\alpha = 1.6$, $\beta = 0.125$, $\gamma = 0.41$, $\delta = 0.25$, and $\epsilon = 0.003$. The single arrows represent slow flow, whereas double arrows indicate fast flow.

Proposition 5.3.8

Let $\mu \in R_1 \cup R_2 \cup R_4$. Then the system (5.3.4) has no relaxation oscillation for $0 < \epsilon \ll 1$.

Proof. According to Hopf's theorem, the Poincaré index of a closed integral curve for a planar continuous vector field is 1, and consequently, a closed integral curve of a continuous planar vector field must enclose at least one equilibrium point [36]. Under the parametric condition $\mu \in R_1$, we have the existence of two interior equilibrium points E_{1*} and E_{2*} of which E_{1*} is a hyperbolic saddle and E_{2*} is either a stable or an unstable equilibrium point. The Poincaré indices of E_{1*} and E_{2*} are -1 and $+1$, respectively. Consequently, with the help of Hopf's theorem, it follows that there cannot be a periodic orbit enclosing only the equilibrium E_{1*} and also enclosing both the equilibrium points E_{1*} and E_{2*}. Therefore, we may have the possibility of having a periodic orbit enclosing E_{2*} only. In fact, we have shown in the previous subsection that due to singular Hopf bifurcation, there exists a canard cycle around the canard point E_{2*} for $0 < \epsilon \ll 1$ assuming the non-degeneracy condition $A \neq 0$. Now, if this periodic orbit is a relaxation oscillation for the system (5.3.4) for $0 < \epsilon \ll 1$ then the orbit should converge to a singular piece-wise smooth slow fast cycle Γ_0 constructed in Theorem 5.3.7 as $\epsilon \to 0$ in the Hausdorff distance. But we also know that the points Q, P are jump points, and correspondingly, if we construct such a singular piece-wise smooth closed orbit Γ_0 enclosing E_{2*} and consisting of attracting slow–fast segments, then its fast segment will be moved away from the critical manifold M_{10} at a point that lies below the transcritical point P and consequently, the singular piece-wise smooth closed orbit will also enclose the singularity E_{1*}. Hence, for $0 < \epsilon \ll 1$, the relaxation oscillation Γ_ϵ, if exists in the sense that $\Gamma_\epsilon \to \Gamma_0$ as $\epsilon \to 0$ then it should also enclose both the equilibrium points E_{1*} and E_{2*} which contradicts Hopf's theorem.

Let $\mu \in R_2$. Then the system (5.3.4) has a unique singular point E_*, which is a saddle-node singularity, having Poincaré index 0. Consequently, following Hopf's theorem has no periodic orbit and hence, no relaxation oscillation.

Let $\mu \in R_4$. Then the system (5.3.4) has no equilibrium point in the interior of $\mathbb{R}_+^2$ and, hence, no relaxation oscillation in $\mathbb{R}_+^2$.

5.3.6 Bi-stability

A phenomenon of bi-stability will occur for the system (5.3.4) whenever $\mu \in R_1$ and $x_{2*} > x_m$ (see Figure 5.7). Under these parametric restrictions, both the equilibria $E_1(0, \delta - \gamma)$ and E_{2*} are stable, whereas the equilibrium E_{1*} is a hyperbolic saddle. Thus, we have the basins of attraction for the equilibria E_{2*} and E_1 separated by the stable and unstable separatrices of the

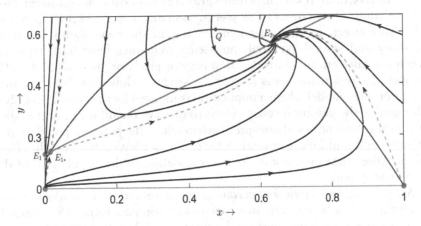

Figure 5.7 Phase portrait of the system (5.3.4) depicts the bistability scenario for $\mu \in R_1$ and $x_{2*} > x_m$. The parabolic curve and the straight line are the non-trivial prey and predator nullclines, respectively. The two attractor equilibria $E_1(0, \delta - \gamma)$ for $\delta > \gamma$ and E_{2*}, three saddle equilibria $(0, 0)$, $(1, 0)$ and E_{1*} are shown with circles. The stable and unstable manifolds of the saddle equilibria E_{1*} and $(1, 0)$ are shown by dashed curves. The attractive basins of the axial equilibrium point E_1 and interior equilibrium point E_{1*} are separated by the separatrix determined by the stable manifold of the saddle equilibrium point E_{1*} is drawn with a dashed curve. The parameter values used are given by $\alpha = 0.3$, $\beta = 0.05$, $\gamma = 0.05$, $\delta = 0.25$ and $\epsilon = 1 (\mu \in R_1)$.

saddle equilibrium E_{1*}. Geometrically, this signifies that if initially, the prey species lie to the left of the stable separatrix of E_{1*}, then at the long run, the prey species will die out otherwise both species will coexist.

A common phenomenon for the model system (5.3.4) is that the generalist predator species have the choice of alternate food source when its preferred prey is absent. This observation is reflected in the model system when $\mu \in R_4$, as in this case, there is no interior equilibrium for the system (5.3.4) but the axial equilibrium $E_1(0, \delta - \gamma)$ is a stable singularity. Thus, for $\mu \in R_4$, the prey species will die out in a long run, but the predator species will exist despite having the Allee effect on the predator species.

5.4 CONCLUSION

In this chapter, we have described briefly the significance of GSPT and blow-up technique to analyze multiple time scale dynamics and for illustration of the subsequent results, we have considered as an application a modified Leslie–Gower predator–prey model with Allee effect in predator population where the prey population reproduces much faster than the predator population. For instance, compared to the birth rate of hares and squirrels (prey), the death rate of coyotes and lynx (predators) is low. As a result of such consideration, a predator–prey system like (5.2.1) or (5.2.6) becomes a singularly perturbed problem.

Nowadays, many researchers from various fields of study have concentrated on the study of dynamics of slow–fast systems [1–6,21,22,18,37,7,8,20]. The Allee effect, on the other hand, is one of the most significant areas of study and several types of interesting dynamics have been reported when the Allee effect is introduced in prey or predator growth [20,27–29]. However, rare attention was paid to the study of slow–fast dynamics in a predator–prey model where a population is induced by the Allee effect [20]. Our goal of the present investigation is to observe the influence of the Allee effect on a slow–fast predator–prey system where the predator population is affected by a multiplicative weak Allee effect. We have addressed the effect of the Allee effect on the stability and oscillation of the populations of the slow–fast system.

After giving an overview of certain relevant analytical methodologies for analyzing a slow–fast system in a concise and compact form in Section 5.2, as a case study, we have considered a system (5.3.2) with prey and generalist predator population with Holling type II functional response in Section 5.3, where the growth of the generalist predator is logistic-type with prey-dependent environmental carrying capacity and induced by the Allee effect. To observe how the Allee effect constant and quantity of additional food sources in predator population affect the dynamics of the slow–fast system (5.3.2), we have considered their dimensionless versions namely γ and δ as controlling parameters throughout the chapter and obtained very rich and complicated dynamical phenomena.

The model system (5.3.2) has been analyzed with the tools from GSPT and the blow-up technique. The parameter space has been divided into four regions and consequently, the existence and local stability results of the various equilibria have been analyzed as the parameters are shifted from one region to another. We have specified the existence of various co-dimension 1 and 2 bifurcation results like saddle node bifurcation, transcritical bifurcation, saddle node-transcritical bifurcation, Hopf bifurcation, Bogdanov-Takens bifurcation etc., and substantiated the bifurcation results with the help of schematic bifurcation diagrams. Using the blow-up technique, we have analyzed the behavior of the system in the vicinity of a canard point and shown the existence of various rich phenomena like the emergence of a canard cycle due to singular Hopf bifurcation, canard explosion: transition from a small amplitude canard cycle to a homoclinic orbit; transition from a small amplitude canard cycle to a large amplitude relaxation oscillation. Within a constrained domain of δ, the amplitude of the canard cycles originating from singular Hopf bifurcation rises exponentially, giving birth to canard explosion. From the biological point of view, in response to the slight alteration in the parameter values, such a change of oscillations may serve as an early indicator of the impending regime shift.

We have also shown the existence of bi-stability when the parameters belong to the region R_1 and the co-existing equilibrium E_{2*} lies right to the fold point Q. In such a case, the basins of attraction for the co-existing equilibrium E_{2*} and the prey-free equilibrium E_1 are separated by the stable and unstable separatrices of the saddle point E_{1*}. We have also carried out sufficient numerical simulations to verify important analytical results derived in this chapter.

The stability property of the Hopf-bifurcating limit cycle stays unchanged due to the introduction of slow–fast time scale, but it causes a significant change in the size and shape of periodic orbits, and in certain parameter regimes, it causes a drastic change in the size of a periodic orbit. The structure of large amplitude relaxation oscillation for $0 < \epsilon \ll 1$ consists of two horizontal segments, one vertical segment, and another curvilinear segment like the associated segment of the slow manifold. From a biological viewpoint, the presence of the relaxation oscillation suggests that prey and predator may coexist in the system. When predator density falls below the lowest value of the critical curve, a prey breakout occurs in a relatively short period of time. When the prey density reaches a level sufficient to support predator reproduction, the predator density continues to expand slowly for a long time until it reaches a density greater than the maximum value of the critical curve. The prey density is then dropping significantly at a faster time scale. Because there is less food (prey), predator density gradually decreases over a faster time, and after a while, for lower prey population density causes predator density to decrease. In this way, the cycle continues, and prey and predator populations coexist. Also, the system experiences a transition from a small amplitude canard cycle to a homoclinic orbit and canard explosion, which indicate that a little change in parameter value can result in a significant change in system dynamics.

REFERENCES

1. FitzHugh, R. (1955). Mathematical models of threshold phenomena in the nerve membrane. *The bulletin of Mathematical Biophysics*, 17(4), 257–278.
2. Han, X., & Bi, Q. (2012). Slow passage through canard explosion and mixed-mode oscillations in the forced Van der Pol's equation. *Nonlinear Dynamics*, 68(1–2), 275–283.
3. Kristiansen, K. U. (2019). Geometric singular perturbation analysis of a dynamical target mediated drug disposition model. *Journal of Mathematical Biology*, 79(1), 187–222.
4. Krupa, M., Popovic, N., & Kopell, N. (2008). Mixed-mode oscillations in three time-scale systems: a prototypical example. *SIAM Journal on Applied Dynamical Systems*, 7(2), 361–420.
5. Krupa, M., & Szmolyan, P. (2001c). Relaxation oscillation and canard explosion. *Journal of Differential Equations*, 174(2), 312–368.
6. Li, J., Quan, T., & Zhang, W. (2018). Bifurcation and number of subharmonic solutions of a 4D non autonomous slow–fast system and its application. *Nonlinear Dynamics*, 92(2), 721–739.
7. Xia, Y., Zhang, Z., & Bi, Q. (2020). Relaxation oscillations and the mechanism in a periodically excited vector field with pitchfork–Hopf bifurcation. *Nonlinear Dynamics*, 101(1), 37–51.
8. Yaru, L., & Shenquan, L. (2020). Canard-induced mixed-mode oscillations and bifurcation analysis in a reduced 3D pyramidal cell model. *Nonlinear Dynamics*, 101(1), 531–567.
9. Guckenheimer, J. (2008). Singular Hopf bifurcation in systems with two slow variables. *SIAM Journal on Applied Dynamical Systems*, 7(4), 1355–1377.
10. Kuehn, C. (2015). *Multiple time scale dynamics* (Vol. 191). Springer, Cham.
11. Van der Pol, B. (1920). A theory of the amplitude of free and forced triode vibrations. *Radio Review*, 1, 701–710.
12. Fenichel, N. (1979). Geometric singular perturbation theory for ordinary differential equations. *Journal of Differential Equations*, 31(1), 53–98.
13. Fenichel, N., & Moser, J. (1971). Persistence and smoothness of invariant manifolds for flows. *Indiana University Mathematics Journal*, 21(3), 193–226.
14. Krupa, M., & Szmolyan, P. (2001a). Extending geometric singular perturbation theory to nonhyperbolic points—fold and canard points in two dimensions. *SIAM Journal on Mathematical Analysis*, 33(2), 286–314.
15. Dumortier, F. (1993). Techniques in the theory of local bifurcations: blow-up, normal forms, nilpotent bifurcations, singular perturbations. In Bifurcations and periodic orbits of vector fields, edited by Dana Schlomiuk (pp. 19–73). Springer, Dordrecht.
16. Dumortier, F., & Roussarie, R. (2001). Multiple canard cycles in generalized liénard equations. *Journal of Differential Equations*, 174(1), 1–29.
17. Rinaldi, S., & Muratori, S. (1992). Slow-fast limit cycles in predator-prey models. *Ecological Modelling*, 61(3–4), 287–308.
18. Wang, C., & Zhang, X. (2019). Canards, heteroclinic and homoclinic orbits for a slow-fast predator-prey model of generalized Holling type III. *Journal of Differential Equations*, 267(6), 3397–3441.

19. Ai, S., & Sadhu, S. (2020). The entry-exit theorem and relaxation oscillations in slow-fast planar systems. *Journal of Differential Equations, 268*(11), 7220–7249.

20. Zhao, L., & Shen, J. (2022). Canards and homoclinic orbits in a slow-fast modified May-Holling-Tanner predator-prey model with weak multiple Allee effect. *Discrete and Continuous Dynamical Systems-B, 27*(11), 6745–6769.

21. Saha, T., Pal, P. J., & Banerjee, M. (2021). Relaxation oscillation and canard explosion in a slow–fast predator–prey model with Beddington–DeAngelis functional response. *Nonlinear Dynamics, 103*(1), 1195–1217.

22. Saha, T., Pal, P. J., & Banerjee, M. (2022). Slow–fast analysis of a modified Leslie–Gower model with Holling type I functional response. *Nonlinear Dynamics, 108*, 4531–4555. https://doi.org/10.1007/s11071-022-07370-1

23. Grasman, J. (2012). *Asymptotic methods for relaxation oscillations and applications* (Vol. 63). Springer Science & Business Media, New York.

24. Verhulst, F. (2007). Singular perturbation methods for slow–fast dynamics. *Nonlinear Dynamics, 50*(4), 747–753.

25. Jones, C. K. (1995). Geometric singular perturbation theory. In *Dynamical systems* (pp. 44–118). Springer, Berlin, Heidelberg.

26. Dumortier, F. (1991). Local study of planar vector fields: singularities and their unfoldings. In Studies in mathematical physics, edited by H.W. Broer, F. Dumortier, S.J. van Strien, F. Takens (Vol. 2, pp. 161–241). Elsevier, Amsterdam.

27. Wang, J., Shi, J., & Wei, J. (2011). Predator–prey system with strong Allee effect in prey. *Journal of Mathematical Biology, 62*(3), 291–331.

28. Terry, A. J. (2015). Predator–prey models with component Allee effect for predator reproduction. *Journal of Mathematical Biology, 71*(6), 1325–1352.

29. Feng, P., & Kang, Y. (2015). Dynamics of a modified Leslie–Gower model with double Allee effects. *Nonlinear Dynamics, 80*(1), 1051–1062.

30. Zhou, S.-R., Liu, Y.-F., & Wang, G. (2005). The stability of predator–prey systems subject to the Allee effects. *Theoretical Population Biology, 67*(1), 23–31.

31. Kramer, A. M., Sarnelle, O., & Knapp, R. A. (2008). Allee effect limits colonization success of sexually reproducing zooplankton. *Ecology, 89*(10), 2760–2769.

32. Noon, B. R., & McKelvey, K. S. (1996). Management of the spotted owl: a case history in conservation biology. *Annual Review of Ecology and Systematics, 27*(1), 135–162.

33. Kuznetsov, Y. (1998). *Elements of applied bifurcation theory* (Vol. 112). Springer-Verlag, New York.

34. Dhooge, A., Govaerts, W., & Kuznetsov, Y. A. (2003). MATCONT: a MATLAB package for numerical bifurcation analysis of ODEs. *ACM Transactions on Mathematical Software (TOMS), 29*(2), 141–164.

35. Krupa, M., & Szmolyan, P. (2001b). Extending slow manifolds near transcritical and pitchfork singularities. *Nonlinearity, 14*(6), 1473.

36. Lawrence, P. (1991). *Differential equations and dynamical systems*. Springer-Verlag, New York.

37. Wang, Z., Zhang, Z., & Bi, Q. (2019). Relaxation oscillations in a nonsmooth oscillator with slow-varying external excitation. *International Journal of Bifurcation and Chaos, 29*(7), 1930019.

Chapter 6

Impact of periodic farming awareness campaign through media for crop pest control management
A mathematical study

Fahad Al Basir
Asansol Girls' College

Sagar Adhurya
Visva-Bharati University

Santanu Ray
Kerala University of Digital Sciences Innovation and Technology

CONTENTS

6.1 INTRODUCTION

The problem associated with pests has become evident worldwide as cultivation began. The total food supply of the world is being wasted due to the cause of pests in agriculture (Carvalho, 2006). On the other hand, major side-effects of synthetic pesticides on the environment, human health, and biodiversity are generating widespread concerns. Awareness of the farmers regarding the risk of synthetic pesticide use is thus one of the essential

DOI: 10.1201/9781003367420-6

factors. Using biopesticides to protect crops against pests requires indigenous knowledge about implementing such biological contents for pest management. Moreover, many botanical materials with potency against various insect pests have been identified (Walia et al., 2017). For example, the Palm Rhinoceros Beetle (*Oryctes rhinoceros*) is a major pest for coconut palm trees. This beetle kills the seedlings of young and old trees, leading to decreased yield. Research has shown that baculovirus (*Rhabdionvirus oryctes*) for *O. rhinoceros* has the potential for long-term control of the beetle through chronically infecting and killing the pest (Bedford, 1986).

In developing and underdeveloped countries, the mass media play a vital role in changing behaviours related to public health (Wallack and Montgomery, 1992). Awareness campaigns by radio, mobile, TV etc. related to farming can help in spreading farming knowledge among farmers about the risks of using pesticides on both human health and also the different related environmental hazards (Weinberger and Srinivasan, 2009, Fu and Akter, 2016, Satya Sai et al., 2019). Farmers' inadequate knowledge of pesticides, the influence of pesticide retailers, and lack of access to non-synthetic pest control methods are positively associated with pesticide overuse. In contrast, the propensity to overuse decreases with higher levels of education, training in Integrated Pest Management. Thus, farming awareness is necessary to reduce crop losses with minimal harmful side effects, which is a primary public concern (Bale et al., 2008, Birthal et al., 2015). Accessible pesticide information campaign helped communicate to the farmers about the severe risks pesticides have on human and environmental health and ways to minimise their adverse effects. Farmers learned about the use and dangers of pesticides mainly through verbal communication.

Mathematical modeling on farming awareness to control plant pests and diseases is limited. Al Basir et al. (2017) have shown that educating farmers through awareness campaigns results in better comprehensive development for *Jatropha curcas* plant mosaic disease management. Authors have described the participation of farming communities in *Jatropha* projects to protect the plant from the mosaic virus. In Al Basir et al. (2018a), the dynamics of the mosaic disease have been analysed in the presence of human intervention in the form of applying nutrients and insecticides depending on the level of population awareness about the disease. In a recent paper (Al Basir et al., 2019), a mathematical model has been proposed and analysed to control crop pests with farming awareness via radio, TV, and social media.

On the other hand, one of the most critical problems is the effect of time delays on the stability of the systems. Extensive research works are available based on delay-induced dynamical behaviours such as persistence, periodic oscillation, bifurcation, and chaos. A time delay for the time required by people to become aware is taken by Al Basir et al. (2018b) while controlling the mosaic disease with farming awareness. They have observed stable (supercritical) Hopf bifurcation taking delay as the bifurcation parameter.

In Al Basir et al. (2019), authors have considered a separate state variable, $A(t)$, for the level of awareness raised by the awareness campaign. But in this article, a mathematical model is formulated to protect crops through farming awareness, assuming that the rate of control measure/awareness activity is achieved through awareness activity which is proportional to the density of healthy pests. We consider that farmers require some time to become aware of eco-friendly control measures and their implementations.

The main aim of the present work is to study the impact of awareness programme and the time delay in pest management. To attain the goal, a mathematical model has been formulated (Section 6.2). The existence of equilibria (Section 6.3) and the stability analysis of equilibria have been investigated (Section 6.4) and the outcomes of the analytical results are clarified via numerical simulations (Section 6.5) with a discussion in Section 6.6.

Recently, mathematical models have investigated the impact of disease spread and control. In many studies, awareness campaigns are proportional to the number of infected individuals reported by a health organisation, which has been considered as a continuous process. Here, we assume that the awareness campaign is a discontinuous, periodic process. This awareness implementation feature can be appropriately modelled through impulsive differential equations, as it has the ability to capture periodic events (Liu and Chen, 2003, Yu et al., 2011, Li et al., 2015).

6.2 MATHEMATICAL MODEL FORMULATION

The following assumptions are made to formulate the mathematical model.

Let $S(t)$ be the susceptible pest population, $I(t)$ be the infected pest population and $A(t)$ is the level of awareness at any time t.

Logistic growth is assumed for susceptible pest growth rate r and carrying capacity K. Aware farmers use biocontrol for pest control. For example, they can use biopesticides to infect susceptible pests as infected pests are weak, so they are less harmful to crops (Al Basir et al., 2019).

Let λ be the infection rate of susceptible pests by biopesticides (generally viruses) because of the interactions and activity of aware farmers. Also, chemical pesticides are used (at a rate of γ) to speed up the control process but at a lower rate. Here, d is the natural mortality rate of pests and δ is the additional mortality rate of infected pests due to the awareness-induced infection.

Since the crop biomass is damaged by susceptible pests, their control measure on pests grows by seeing the density of susceptible pests. Level of awareness increases with the density of total pests in the field at a rate σ and due to the fading of memory level of awareness decreases at a rate μ.

Based on the above assumptions, the following mathematical model is obtained

$$\frac{dS}{dt} = rS\left(1 - \frac{S}{K}\right) - \lambda SA - \gamma AS,$$

$$\frac{dI}{dt} = \lambda SA - \gamma IA - (d + \delta)I, \tag{6.1}$$

$$\frac{dA}{dt} = \sigma(S + I) - \mu A,$$

with initial conditions:

$$S(0) > 0, \ I(0) > 0, \ A(0) > 0 \tag{6.2}$$

We now reformulate the model (6.1) to incorporate the effects of awareness programmes that are repeatedly performed at time instants separated by constant time intervals. Let us assume that through the action of mass media, the portion of level of awareness increases by a fixed amount ω every time in which the advertisement is made. Thus, the above model becomes:

$$\frac{dS}{dt} = rS\left(1 - \frac{S}{K}\right) - \lambda SA - \gamma AS, \text{ when } t \neq t_k$$

$$\frac{dI}{dt} = \lambda SA - \gamma IA - (d + \delta)I, \text{ when } t \neq t_k \tag{6.3}$$

$$\frac{dA}{dt} = \sigma(S + I) - \mu A, \text{ when } t \neq t_k$$

$$\Delta A = \omega A, \text{ at } t = t_k, \text{ where } k = 1, 2, 3, \ldots, n$$

Here $\Delta A = A(t^+) - A(t^-)$, where $A(t^-)$ is the level of awareness before an awareness campaign and $A(t^+)$ is the level of awareness after an awareness campaign.

First, we analyse the dynamics of the system without awareness campaign i.e., we analyse model (6.1). Then we analyse the impulsive model (6.3).

6.3 SYSTEM WITHOUT PERIODIC CAMPAIGN (WITH LOCAL AWARENESS ONLY I.E., MODEL 6.1)

In this section, we analyse the boundedness, equilibria, and their stability of the system (6.1) without impulses.

6.3.1 Boundedness

Theorem 6.1

All the solutions of (6.1), that are initiated from $\mathbb{R}^3_+$, will be confined in the region

$$\Gamma = \left\{ (S, I, A) \in \mathbb{R}^3_+ : 0 \le S + I \le K, \ 0 \le A \le \frac{\sigma K}{\mu} \right\} [[\text{Tab}]] \qquad (6.4)$$

Proof. From system (6.1) we have,

$$\frac{d(S+I)}{dt} \le r(S+I)\left(1 - \frac{S+I}{K}\right)$$

this imply that $\limsup_{t \to \infty} S + I = K$.
From the last equations of system (6.1), we obtain

$$\frac{dA}{dt} \le \sigma K - \mu A. [[\text{Tab}]] \qquad (6.5)$$

and thus $\limsup_{t \to \infty} A = \frac{\sigma K}{\mu}$

Finally, we get the region of attraction as

$$\Gamma = \left\{ (S, I, A) \in \mathbb{R}^3_+ : 0 \le S + I \le K, \ 0 \le A \le \frac{\sigma K}{\mu} \right\} [[\text{Tab}]] \qquad (6.6)$$

where all solutions are bounded.

6.4 EQUILIBRIA AND STABILITY

The model (6.1) has two equilibria namely the trivial equilibrium $E_1(0,0,0)$, and the coexistence equilibrium, $E^*(S^*, I^*, A^*)$, where,

$$A^* = \frac{rK\sigma}{\sigma\lambda + r\mu}, \ S^* = \frac{\mu A^*(d + \delta + \gamma A^*)}{\sigma(d + \delta + \gamma A^* + \lambda A^*)}, \ I^* = \frac{\lambda S^* A^*}{d + \delta + \gamma A^*}.$$

We see that both the equilibria always exist. Now we test the stable nature of the equilibria.

6.5 STABILITY ANALYSIS

Linearising the system (6.1) about any equilibrium point $E(S, I, A)$, We get the following system

$$\frac{dX}{dt} = PX(t)[[Tab]] \tag{6.7}$$

Here P is a 3×3 matrix, given as below:

$$P = \left[p_{ij} \right]$$

$$= \begin{bmatrix} r\left(1 - \dfrac{2S+I}{K}\right) - \lambda A - \gamma A & -\dfrac{rS}{K} & -\lambda S \\ \lambda A & -d(\mu + \delta) - \gamma A & \lambda I \\ \sigma & \sigma & -\mu \end{bmatrix} \cdot [[Tab]] \tag{6.8}$$

Actually P is the Jacobian matrix at the equilibrium points. The characteristic equation of the matrix P satisfy

$$H(\rho) = |\rho I - P| = 0$$

This gives,

$$H(\rho) = \rho^3 + L_1\rho^2 + L_2\rho + L_3 = 0.[[Tab]] \tag{6.9}$$

Here,

$$L_1 = -(p_{11} + p_{22} + p_{33}),$$

$$L_2 = p_{22}p_{33} - p_{23}p_{32} + p_{11}p_{33} - p_{13}p_{31} + p_{11}p_{22} - p_{21}p_{12},$$

$$L_3 = p_{11}(p_{22}p_{33} - p_{23}p_{32}) - p_{12}(p_{11}p_{33} - p_{13}p_{31}) + p_{13}(p_{11}p_{22} - p_{21}p_{12}).$$

The origin $E_0(0,0,0)$ is always unstable, in view of the fact that one of its eigenvalues, $r > 0$.

Now we study the possible occurrence of Hopf bifurcation. Hopf bifurcation can be seen when a characteristic equation has a pair of purely imaginary roots. Any of the parameters of the model may be a bifurcation parameter.

Theorem 6.2

The coexistence equilibrium point $E^*\left(S^*, I^*, A^*\right)$ is

i. stable if the following conditions are satisfied (as $L_1 > 0$, $L_2 > 0$)

$$L_3 > 0 \ and \ L_1 L_2 - L_3 > 0, [[\text{Tab}]] \tag{6.10}$$

ii. undergoes Hopf bifurcation around the endemic equilibrium point E^* whenever the critical parameter value $b = b^*$ contained in the following domain

$$\Gamma_{HB} = \left\{b \in R^+ : L_1\left(b^*\right)L_2\left(b^*\right) - L_3\left(b^*\right) = 0, \ with \ \dot{\sigma}_3 - \left(\dot{L}_1 L_2 + L_1 \dot{L}_2\right) \neq 0\right\} \tag{6.11}$$

$$.[[\text{Tab}]]$$

Proof. The first part follows from Routh-Hurwitz criterion. Thus we only provide the proof of part (ii) of the theorem.

Using the condition, $L_1 L_2 - L_3 = 0$, the characteristic Equation (6.9) becomes

$$\left(\rho^2 + L_2\right)\left(\rho + L_1\right) = 0, [[\text{Tab}]] \tag{6.12}$$

which has three roots namely $\rho_1 = +i\sqrt{L_2}$, $\rho_2 = -i\sqrt{L_2}$ and $\rho_3 = -L_1$. Therefore, a pair of purely imaginary eigenvalues exists for $L_1 L_2 - L_3 = 0$. Now we verify the transversality condition.

Differentiating the characteristic Equation (6.9) with respect to b, we have

$$\frac{d\rho}{db} = \frac{\rho^2 \dot{L}_1 + \rho \dot{L}_2 + \dot{L}_3}{3\rho^2 + 2\rho L_1 + L_2} \Big|_{\rho = i\sqrt{L_2}}$$

$$= \frac{\dot{L}_3 - \left(\dot{L}_1 L_2 + L_1 \dot{L}_2\right)}{2\left(L_1^2 + L_2\right)} + i\left[\frac{\sqrt{L_2}\left(L_1 \dot{L}_3 + L_2 \dot{L}_2 - L_1 \dot{L}_1 L_2\right)}{2L_2\left(L_1^2 + L_2\right)}\right].$$

Therefore,

$$\frac{dRe\rho}{db} \Big|_{b=b^*} = \frac{\dot{L}_3 - \left(\dot{L}_1 L_2 + L_1 \dot{L}_2\right)}{2\left(L_1^2 + L_2\right)} \neq 0 \Leftrightarrow \dot{L}_3 - \left(\dot{L}_1 L_2 + L_1 \dot{L}_2\right) \neq 0.[[\text{Tab}]] \tag{6.13}$$

and Hopf bifurcation occurs at $b = b^*$.

6.6 DYNAMICS OF THE SYSTEM WITH PERIODIC CAMPAIGN (WITH LOCAL AND GLOBAL PERIODIC CAMPAIGN)

In this section, we have analysed impulse periodic campaign-induced system (6.3) using impulsive differential equations for better understanding of the awareness effect considering a one-dimensional impulsive system.

$$\frac{dA}{dt} = \sigma(S + I) - \mu A, \text{ at } t \neq t_k$$

$$\Delta A = \omega A, \text{ for } t = t_k \text{ where } k = 1, 2, 3, \ldots, n, \; [[\text{Tab}]] \tag{6.14}$$

Now, taking the minimum level of awareness, the one-dimensional impulsive differential equation takes the form:

$$\frac{dA}{dt} = -\mu A, \text{ for } t \neq t_k$$

$$\Delta A = \omega A, \text{ for } t = t_k \text{ where } k = 1, 2, 3, \ldots, n, \; [[\text{Tab}]] \tag{6.15}$$

Let $\tau = t_{k+1} - t_k$ be the period of the campaign. The solution of the system (6.15) is,

$$A(t) = A\left(t_k^+\right) e^{-\mu(t - t_k)}, \text{ for } t_k < t \leq t_{k+1}. [[\text{Tab}]] \tag{6.16}$$

In presence of an impulsive effect, we have a recursion relation at the moments of impulse, given by

$$A\left(t_k^+\right) = A\left(t_k^-\right) + \omega.$$

Thus the level of awareness increase before and after the impulsive campaign is,

$$A\left(t_k^+\right) = \frac{\omega\left(1 - e^{-k\mu\tau}\right)}{1 - e^{-\mu\tau}}$$

and

$$A\left(t_{k+1}^-\right) = \frac{\omega\left(1 - e^{-k\mu\tau}\right) e^{-\mu\tau}}{1 - e^{-\mu\tau}}.$$

Thus the limiting value of the awareness level due to the awareness campaign before and after one cycle is as follows:

$$\lim_{k \to \infty} A\left(t_k^+\right) = \frac{\omega}{1 - e^{-\mu\tau}} \text{ and } \lim_{k \to \infty} A\left(t_{k+1}^-\right) = \frac{\omega e^{-\mu\tau}}{1 - e^{-\mu\tau}}$$

and

$$A\left(t_{k+1}^+\right) = \frac{\omega e^{-\mu\tau}}{1 - e^{-\mu\tau}} + \omega = \frac{\omega}{1 - e^{-\mu\tau}}.$$

Here we have found out that

$$A\left(t_k^+\right) - \frac{\omega}{1 - e^{-\mu\tau}} = \omega \frac{1 - e^{-k\mu\tau}}{1 - e^{-\mu\tau}} - \frac{\omega}{1 - e^{-\mu\tau}} = -\omega \frac{e^{-k\mu\tau}}{1 - e^{-\mu\tau}} < 0.$$

There does not exist a particular equilibrium point of an impulsive system, but equilibrium-like periodic orbits can be evaluated. Using the following analysis, we can show that there are two periodic orbits namely the disease-free periodic orbit and the endemic periodic orbit for the system (6.15) with impulses. Using the results from (Lakshmikantham et al., 1989, Yu et al., 2011), we have the following result.

Lemma 6.1

System (67.14) has a unique positive periodic solution $\tilde{A}(t)$ with period τ and given by

$$\tilde{A}(t) = \frac{\omega \exp\left(-\mu(t - t_k)\right)}{1 - \exp(-\mu\tau)}, \ t_k \leq t \leq t_{k+1}, \ \tilde{A}\left(0^+\right) = \frac{\mu}{1 - \exp(-\mu\tau)}.$$

On this basis, we have the following theorem.

6.6.1 Stability of periodic orbits

On the above basis, we study the stability of periodic orbits. We only focus on the disease-free periodic orbit deriving the following theorem.

Theorem 6.3

The pest-free periodic solution $\left(0, 0, \tilde{A}\right)$ of the system (6.15) is locally asymptotically stable if

$$\tau < \frac{1}{r} \int_0^\tau \cdot \left(\lambda\tilde{A} - \gamma\tilde{A}\right) d\tau. \tag{6.17}$$

Proof. Variational matrix at $\left(0,0,\tilde{A}\right)$ is given by,

$$
J_v = \begin{bmatrix} r - \lambda\tilde{A} - \gamma\tilde{A} & 0 & 0 \\ \lambda\tilde{A} & -d(\mu+\delta) - \gamma\tilde{A} & 0 \\ \sigma & \sigma & -\mu \end{bmatrix}. \tag{6.18}
$$

The monodromy matrix $\mathbb{D}$ of the variational matrix $J_v(t)$ is

$$
\mathbb{D}(\tau) = I\exp\left(\int_0^\tau J_v(t)dt\right) \tag{6.19}
$$

where I is the identity matrix. We thus have: $\mathbb{D}(\tau) = diag(\xi_1, \xi_2, \xi_3)$. Here, ξ_i, $i = 1,2,3$, are the Floquet multipliers and these are obtained using (6.19) as follows:

$$
\xi_1 = \exp\left[\int_0^\tau \left(r - \lambda\tilde{A} - \gamma\tilde{A}\right)d\tau\right], \quad \xi_2 =, \quad \xi_3 = \exp\left(-\mu\tau\right).
$$

Clearly $\xi_2 < 1$ and $\xi_3 < 1$. We only need to find the condition for which $\xi_1 < 1$. From above expression of ξ_1, if $\int_0^\tau \left(r - \lambda\tilde{A} - \gamma\tilde{A}\right)d\tau < 0$ holds, then $\xi_1 < 1$ will hold. Thus, according to Floquet theory, the periodic solution $\left(\tilde{E}\left(0,0,\tilde{A}(t)\right)\right)$ of the system (6.15) is locally asymptotically stable if the condition (6.17) holds.

6.7 NUMERICAL SIMULATION

In this section, numerical simulations are performed to investigate the dynamics of the system obtained in previous sections. Simulation results of the model without impulse are plotted in Figures 6.1–6.4.

Figure 6.1 is plotted for lower infection rate $\lambda = 0.15$. The coexisting equilibrium E^* exists in this case and is stable in nature. For some higher value of infection rate $\lambda = 0.25$, the equilibrium E^* becomes unstable (Figure 6.2). Solutions trajectories form a limit cycle in $S - I - A$ phase space. A Hopf bifurcation diagram is plotted in Figure 6.3 for λ. Another bifurcation diagram is plotted in Figure 6.4 for the control rate γ. In this case, the direction of Hopf bifurcation is reversed. We conclude that the impact of control γ is stabilising the coexisting equilibrium.

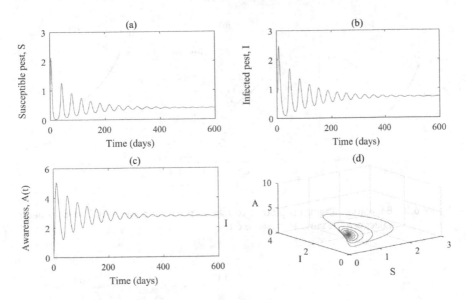

Figure 6.1   System populations are plotted (without impulses) with respect to time (days). The set of parameters is $r = 0.5, k = 10; \gamma = 0.01; \delta = 0.1, d = 0.1; \mu = 0.1;$ $a = 0.25$ and $\lambda = 0.15$.

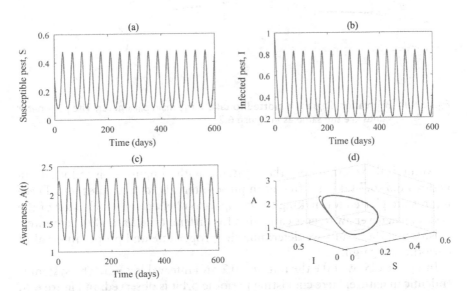

Figure 6.2   System populations are plotted (without impulses) with respect to time (days). The set of parameters is $r = 0.5, k = 10; \gamma = 0.01; \delta = 0.1, d = 0.1; \mu = 0.1;$ $a = 0.25$ and $\lambda = 0.25$.

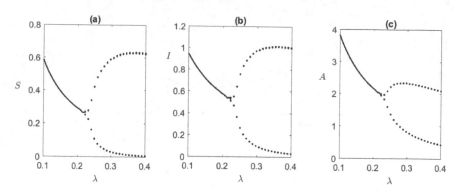

Figure 6.3   Bifurcation diagram is plotted to take λ the main parameter. The parameter values are the same as in Figure 6.1.

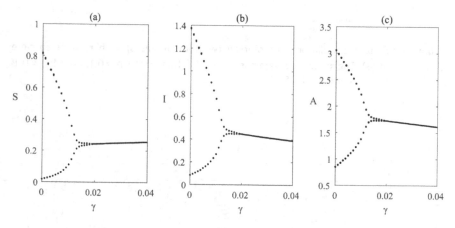

Figure 6.4   Bifurcation diagram is plotted to take γ the main parameter. The parameter values are the same as in Figure 6.2.

Numerical solution of the system with impulses is plotted in Figures 6.5–6.7 taking different impulse intervals (τ) and rates (ω). These figures are plotted for making the comparison between the behavior of the system whenever awareness campaign for different rates and different intervals. Using the results we determine the proper rate and proper interval of periodic campaign.

In Figure 6.5, we take the rate $\omega = 0.2$ and interval is $\tau = 5$. The system is endemic in nature. Here coexisting periodic orbit is observed. In Figure 6.6, we take $\tau = 2$ days and see that the pest-free periodic orbit exists and is stable (Theorem 6.3). We then decrease the rate of campaign $\omega = 0.1$, and see that the coexisting periodicity arises.

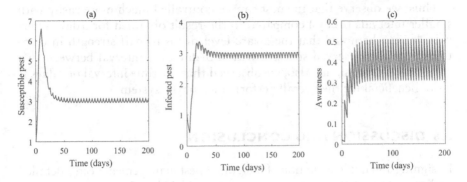

Figure 6.5   System populations are plotted for $\tau = 5$ days. Pest-free periodic orbit is obtained. Coexistence of periodic orbit is seen.

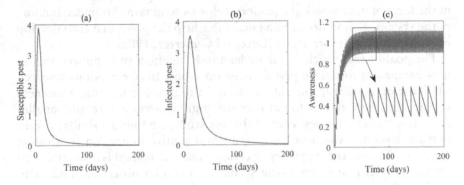

Figure 6.6   System populations are plotted for $\tau = 2$ days. Pest-free periodic orbit is obtained.

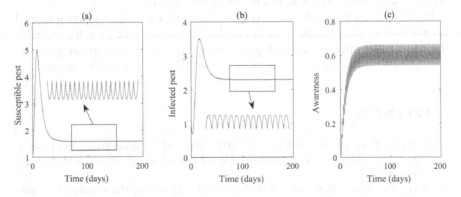

Figure 6.7   System populations are plotted for $\tau = 2$ days. Pest-free periodic orbit is obtained, $\omega = 0.1$.

Thus, we observe that the pest can be controlled much more easily with smaller intervals (2 days) compared to the result obtained for 5 days intervals. It is to be noted that the aware level drops to half strength in comparison to the observed strength with a 5-day time interval between two consecutive campaigns. Also, we observed that the time interval of 2 days is more beneficial for the overall performance of the system.

6.8 DISCUSSION AND CONCLUSION

Designing and implementation of integrated pest management need a detailed understanding of the agroecosystem, which includes plants, pests, natural pesticides, diseases, habitats, etc. The interactions between these components make the system so complex that it becomes difficult to get a fundamental understanding of these interactions. Model, the simplified picture of reality in the form of mathematical equations helps us to obtain this understanding of the system. This understanding may later help the policymaker to develop suitable management strategies (Getz and Gutierrez, 1982).

The goal of this article is the understand the effect of impulsive awareness campaigns to reduce pest damage on crops. In our previous research (Al Basir et al., 2019), the role of delay in organising farming awareness was investigated. It was found that the high awareness rate and smaller delay results in better pest control. In this study, we found a similar result, that the level of awareness has a significant poitive impact on controlling the pest. Whereas the frequency of conducting a campaign is also critical to obtain a better pest management scenario. A similar observation was also made in the case of Armyworm-maize model i.e., awareness level improves biomass yield (Daudi et al., 2021). The model also focused on the effect of the extent of biological and chemical control of the pest. Increasing the rate of biological control was found to destabilise the endemic equilibrium, whereas chemical control shows the opposite result.

In conclusion, our model suggests that frequent and higher awareness level is useful to control the pest. Although better control can be achieved with the selection of appropriate biological control measure. As a future perspective, the research should be done on the cost-effectiveness of this measure.

REFERENCES

Al Basir, F., Banerjee, A., and Ray, S., 2019. Role of farming awareness in crop pest management - A mathematical model. *Journal of Theoretical Biology*, 461, 59–67.

Al Basir, F., Blyuss, K.B., and Ray, S., 2018a. Modelling the effects of awareness-based interventions to control the mosaic disease of *Jatropha curcas*. *Ecological Complexity*, 36, 92–100.

Al Basir, F, Venturino, E., Ray, S., and Roy, P.K., 2018b. Impact of farming aware-ness and delay on the dynamics of mosaic disease in Jatropha curcas planta-tions. *Computational and Applied Mathematics*, 37 (5), 6108–6131.

Al Basir, F, Venturino, E., and Roy, P.K., 2017. Effects of awareness program for controlling mosaic disease in Jatropha curcas plantations. *Mathematical Methods in the Applied Sciences*, 40 (7), 2441–2453.

Bale, J., van Lenteren, J., and Bigler, F., 2008. Biological control and sustain-able food production. *Philosophical Transactions of the Royal Society B: Biological Sciences*, 363 (1492), 761–776.

Bedford, G.O. 1986. Biological control of the rhinoceros beetle (Oryctes rhi-noceros) in the South Pacific by Baculovirus. *Agriculture, Ecosystems & Environment*, 15, 141–147.

Birthal, P.S., Kumar, S., Negi, D.S., and Roy, D., 2015. The impacts of informa-tion on returns from farming: evidence from a nationally representative farm survey in India. *Agricultural Economics*, 46 (4), 549–561.

Carvalho F.P. 2006. Agriculture, pesticides, food security and food safety. *Environ-mental Science & Policy*, 9(7–8), 685–692.

Daudi, S., Luboobi, L., Kgosimore, M., and Kuznetsov, D., 2021. A fractional-order fall armyworm-maize biomass model with naturally beneficial insects and optimal farming awareness. *Results in Applied Mathematics*, 12, 100209.

Fu, X. and Akter, S., 2016. The impact of mobile phone technology on agricultural extension services delivery: Evidence from India. *The Journal of Development Studies*, 52 (11), 1561–1576.

Getz, W.M. and Gutierrez, A.P., 1982. A perspective on systems analysis in crop production and insect pest management. *Annual Review of Entomology*, 27 (1), 447–466.

Lakshmikantham, V., Bainov, D.D., and Simeonov, P.S., 1989. *Theory of Impulsive Differential Equations*. World Scientific, Singapore.

Li, X., Bohner, M., and Wang, C.-K., 2015. Impulsive differential equations: Periodic solutions and applications. *Automatica*, 52, 173–178.

Liu, X. and Chen, L., 2003. Complex dynamics of Holling type II Lotka–Volterra predator–prey system with impulsive perturbations on the predator. *Chaos, Solitons & Fractals*, 16 (2), 311–320.

Satya Sai, M., Revati, Gd., Ramya, R., Swaroop, A., Maheswari, E., and Kumar, M., 2019. Knowledge and perception of farmers regarding pesticide usage in a rural farming village, Southern India. *Indian Journal of Occupational and Environmental Medicine*, 23 (1), 32.

Walia, S., Saha, S., Tripathi, V. et al. 2017. Phytochemical biopesticides: some recent developments. *Phytochemistry Reviews*, 16, 989–1007.

Wallack, L. and Montgomery, K., 1992. Advertising for all by the year 2000: Public health implications for less developed countries. *Journal of Public Health Policy*, 13 (2), 204.

Weinberger, K. and Srinivasan, R., 2009. Farmers' management of cabbage and cauliflower pests in India and their approaches to crop protection. *Journal of Asia-Pacific Entomology*, 12 (4), 253–259.

Yu, H., Zhong, S., and Agarwal, R.P., 2011. Mathematics analysis and chaos in an ecological model with an impulsive control strategy. *Communications in Nonlinear Science and Numerical Simulation*, 16 (2), 776–786.

Chapter 7

Mathematical modeling and analysis of the impact of global warming on the dynamics of polar bear population

Alok Kumar Verma and Maitri Verma
Babasaheb Bhimrao Ambedkar University

CONTENTS

7.1 INTRODUCTION

The anthropogenic emission of greenhouse gases, mainly carbon dioxide (CO_2), in massive amounts is responsible for the rise in the average surface temperature of Earth. By year 2100, the surface temperature is predicted to exceed by 1.5°C above the mean surface temperature of 1850–1990 [1]. Among the trace gases, CO_2 gas is the most dominant one due to its contribution to anthropogenic global warming. It is estimated that enhanced carbon dioxide levels alone are responsible for approximately 66% of the warming observed between 1990 and 2020 [2]. Further, it is estimated that a doubling of CO_2 levels in the atmosphere may lead to a rise of 1.5°C–4.5°C in mean surface temperature [3]. The rising temperature caused by enhanced CO_2 levels may lead to a significant decline in the Arctic sea ice cover [4,5]. It is estimated that three square meters of Arctic sea ice are melted with each metric ton of CO_2 emissions [6]. From 1979 to 2012, the annual average

DOI: 10.1201/9781003367420-7

Arctic sea ice extent is found to decrease at a rate of 3.5%–4.1% per decade with the rate of declination of summer sea ice minimum lying in the range of 9.4%–13.6% per decade [1]. Experiments reveal that Arctic summer sea ice retreat has been unprecedented in the last three decades, and sea surface temperatures have indeed been abnormally high [1]. In comparison to the 1981–2010 average, the summer ice extent has decreased by around 13% per decade. Sea ice of the Arctic Ocean is shrinking in both incrassation and extent by melting faster than it refreezes in the winter [7]. The prediction of the current polar bear population is 26,000 with a 5% chance that it lies below 22,000 or above 31,000. It is estimated that there is 70% odd that the population of polar bears will see a decline of more than 33% in the next three generations [8]. Polar bears live in most Arctic areas which are covered by ice for most of the time in a year. The polar bear's primary habitat is sea ice. It is the largest carnivorous marine mammal because it spends many months of the year at sea ice. The Polar bear (pb) population relies on the sea ice of the Arctic Ocean praying on ringed seals (*Phoca hispida) and* bearded seals (*Eringnathus barbatus*), which make up the majority of their diet. The recent changes in the Arctic sea ice extent are found to negatively impact the population of polar bears [4,7–14]. Global warming is listed as the gravest menace to polar bears, mainly because the melting of its sea ice habitat degrades its ability to find enough food. Inadequate nutrition causes low fertility rates in adult females and low survival rates in cubs and juvenile bears [9]. Changes in sea ice affect pregnant female's ability to make suitable maternity dens. As the distance between the ice sheet and the coast increases, the female has to swim as long as possible to reach the appropriate den area on the ground [9,15]. Disease-causing bacteria and parasites will easily flourish in hot climates which affect polar bears population [4]. If no actions are taken to mitigate the climate changes, more than 30% of the global polar bear population is predicted be lost by the end of year 2050 [14].

Mathematical models are effective tools to comprehend the effect of the warming of our planet on the polar bear population. The effect of warming of the climate system on polar bear population is usually studied using numerical and empirical models [9–13,16–21]. However, studies which examine the qualitative properties of the dynamical model framed to investigate the long-term effect of temperature change on Arctic sea ice extent and the polar bear population are much needed. In recent years, several dynamical models are investigated to observe the dynamics of emission and mitigation of greenhouse gases [22–34]. Some mathematical models have investigated the impact of global warming induced by enhanced CO_2 levels on the melting of glaciers and sea level rise [35,36]. However, none of these studies explores the effect of the melting of glaciers on the population dynamics of polar bears. Further, these studies do not consider the temperature rise due to decrease in ice cover caused by global warming, i.e., the ice-albedo feedback. In this study, we have formulated a nonlinear mathematical model to examine the effect of temperature rise created by the enhanced CO_2 concentration on the Arctic sea ice cover and the polar bear population. It is considered that an enhanced level of CO_2 causes temperature rise which cause a reduction in sea ice cover, which further increases the temperature due to decrease in ice albedo.

7.2 MATHEMATICAL MODEL

To capture the effect of enhanced CO_2 levels and associated temperature rise on the dynamics of polar bear population, we have considered five dynamical variables, namely, $I_c(t)$ denoting the Arctic sea ice extent, $P(t)$ denoting the density of polar bear population, $X(t)$ denoting the concentration of CO_2 gas in the Earth's atmosphere, $N(t)$ denoting the human population and $T(t)$ denoting the mean surface temperature, at any time t. Let X_0 denotes the equilibrium CO_2 concentration in absence of human activities and T_0 be the equilibrium temperature when CO_2 level is $X = X_0$. The Arctic ice extent is assumed to grow at a constant rate q. The Arctic sea ice extent decreases due to ocean currents and winds [37], the natural declination rate of Arctic ice extent is taken as γI_c, where γ represents the natural declination rate coefficient of sea ice extent. The sea ice cover depletes due to increase in the temperature [38]. The depletion rate of sea ice extent due to rise in temperature is taken to be proportional to sea ice extent and temperature, i.e., $\beta I_c(T - T_0)$, where β denotes the melting rate coefficient of Arctic sea ice extent. Thus, the dynamics of the sea ice extent is governed by the following equation

$$\frac{dI_c}{dt} = q - \beta I_c (T - T_0) - \gamma I_c. \tag{7.1}$$

It is considered that the polar bear population is growing logistically having intrinsic growth rate s and a carrying capacity L. The melting of sea ice extent due to the enhanced temperatures causes reduction in the growth rate and carrying capacity of the polar bear population. The reduction in growth rate and carrying capacity polar bear population due to the increase in surface temperature are considered to be $\phi_1 P(T - T_0)$ and $\phi_2 P^2(T - T_0)$, respectively. Here, the constant ϕ_1 denotes the declination rate coefficient of the polar bear population due to the increased temperatures, and ϕ_2 denotes the reduction in carrying capacity of polar bear population due to increased temperatures. The availability of sea ice extent boosts the growth rate and carrying capacity of polar bear population [39,40]. Therefore, it is assumed that the growth rate and carrying capacity of polar bear population increase due to increase in sea ice extent at the rates $\psi_1 P I_c$ and $\psi_2 P^2 I_c$, respectively.

Here, the constants ψ_1 and ψ_2 denote the increase in growth rate and carrying capacity of polar bear population due to the increase in sea ice extent, respectively. Thus, the rate of change of polar bear population is given by

$$\frac{dP}{dt} = sP\left(1 - \frac{P}{L}\right) - \phi_1 P(T - T_0) - \phi_2 P^2(T - T_0) + \psi_1 PI_c + \psi_2 P^2 I_c. \qquad (7.2)$$

We assume that CO_2 is emitted from nature sources with a rate Q. The emissions of CO_2 gas from human activities depend on the population [41,42], thus the CO_2 emission rate from human activities is taken to be λN, where λ denotes the emission rate coefficient of CO_2 gas from anthropogenic sources. The removal rate of CO_2 gas is taken to be depending on the concentration of CO_2 gas (i.e., αX) [43]. Here, the constant α is the removal rate of CO_2 gas from the atmosphere. Thus, the dynamics of CO_2 in atmosphere is modeled as

$$\frac{dX}{dt} = Q - \alpha X + \lambda N. \qquad (7.3)$$

The human population is considered to grow logistically having a intrinsic growth rate r and a carrying capacity K. It is supposed that the population will decline due to the negative impacts of global warming [44–46]. The rate of decline in population due to global warming is taken to be $\theta(T - T_0)N$, where θ denotes the mortality rate because of the enhanced surface temperatures. Thus, the dynamics of human population is given by the differential equation

$$\frac{dN}{dt} = rN\left(1 - \frac{N}{K}\right) - \theta(T - T_0)N. \qquad (7.4)$$

The increase in surface temperature is considered to be proportional to the enhanced CO_2 concentration i.e., $\eta(X - X_0)$ [47]. Here, the constant η represents the growth rate of the mean surface temperature. The reduction in sea ice cover alters the albedo and the surface temperature of the Earth's surface, which contributes to the further warming of the Earth's atmosphere at a rate $\theta_1 \beta I_c(T - T_0)$, where θ_1 proportionality constant which represents the rise in surface temperature due to reduced ice-albedo. Thus, the rate of change in mean surface temperature is given by

$$\frac{dT}{dt} = \eta(X - X_0) - \eta_0(T - T_0) + \theta_1 \beta I_c(T - T_0). \qquad (7.5)$$

Here, η_0 denotes the declination rate coefficient of the surface temperature.

Thus, we arrive at the following model, capturing the dynamics of the problem:

$$\frac{dI_c}{dt} = q - \beta I_c (T - T_0) - \gamma I_c,$$

$$\frac{dP}{dt} = sP\left(1 - \frac{P}{L}\right) - \phi_1 P(T - T_0) - \phi_2 P^2(T - T_0) + \psi_1 PI_c + \psi_2 P^2 I_c,$$

$$\frac{dX}{dt} = Q - \alpha X + \lambda N, \qquad\qquad (7.6)$$

$$\frac{dN}{dt} = rN\left(1 - \frac{N}{K}\right) - \theta(T - T_0)N,$$

$$\frac{dT}{dt} = \eta(X - X_0) - \eta_0(T - T_0) + \theta_1 \beta I_c(T - T_0),$$

where,

$Q = \alpha X_0$, $I_c(0) > 0$, $P(0) \geq 0$, $X(0) \geq X_0$, $N(0) \geq 0$ and $T(0) \geq T_0$.

The descriptions of all the model parameters along with their unit are given in Table 7.1.

Lemma 7.1

The set

$$\Omega = \left\{ (I_c, P, X, N, T) \in R_+^5 : 0 < I_c \leq \frac{q}{\gamma}, 0 \leq P \leq P_m, X_0 \leq X \leq X_m, 0 \leq N \leq K, T_0 \leq T \leq T_m \right\}$$

where $X_m = X_0 + \dfrac{\lambda}{\alpha} K$, $P_m = \dfrac{L(s\gamma + \psi_1 q)}{s\gamma - \psi_2 qL}$, $T_m = T_0 + \dfrac{\eta \lambda \gamma K}{\alpha(\eta_0 \gamma - \theta_1 \beta q)}$,

constitutes the region of attraction for dynamical system (7.6) and it attracts all the solutions which start in R_+^5.

Thus, for bounded of model solutions, we assume that the following conditions always hold:

$$\frac{s}{L} - \psi_2 \frac{q}{\gamma} > 0 \text{ and } \eta_0 \gamma - \theta_1 \beta q > 0.$$

Table 7.1 Description and Units of Parameters of System (7.6)

Parameter	Description	Unit
q	The natural growth rate of sea ice extent	million square km per year
β	Melting rate coefficient of sea ice	per °C per year
X_0	Equilibrium CO_2 concentration in the absence of CO_2 emissions from human activities	ppm
T_0	Equilibrium surface temperature when $X = X_0$	°C
γ	Natural depletion rate coefficient of sea ice extent	per year
s	Intrinsic growth rate of polar bear population	per year
L	Carrying capacity of polar bear population	Pb
ϕ_1	Declination rate coefficient of the polar bears due to increase in temperature	per °C per year
ϕ_2	Declination rate coefficient of carrying capacity of polar bear population due to increase in temperature	per °C per pb per year
ψ_1	Growth rate coefficient of polar bear population due to increase in sea ice extent	per million square km per year
ψ_2	Growth rate coefficient of carrying capacity of polar bear population due to increase in sea ice extent	per million square km per pb per year
Q	Natural emission rate of CO_2 gas	ppm per year
λ	Emission rate coefficient of CO_2 from human activities	ppm per billion person per year
α	Removal rate coefficient of CO_2 gas	per year
r	Intrinsic growth rate of human population	per year
K	Carrying capacity of human population	billion person
θ	Mortality rate coefficient of human population due to increase in temperature	per °C per year
η	Growth rate coefficient of average surface temperature	°C per ppm per year
η_0	Depletion rate coefficient of average surface temperature	per year
θ_1	Proportionality constant represents the growth in the average surface temperature due to reduced ice-albedo	°C per million square km

7.3 EQUILIBRIUM STATES AND STABILITY ANALYSIS

7.3.1 Equilibrium states

The model (7.6) owns the following four non-negative equilibria:

i. $S_1 \left(\dfrac{q}{\gamma}, 0, \dfrac{Q}{\alpha}, 0, T_0 \right)$, always exists.

ii. $S_2\left(\dfrac{q}{\gamma}, \dfrac{L(s\gamma+\psi_1 q)}{s\gamma-\psi_2 qL}, \dfrac{Q}{\alpha}, 0, T_0\right)$, always exists.

iii. $S_3\left(I_{c_3}, 0, X_3, N_3, T_3\right)$, exists under the following condition:

$$\eta_0\left(\gamma+\dfrac{\beta r}{\theta}\right)-\theta_1\beta q > 0. \tag{7.7}$$

iv. $S^*\left(I_c^*, P^*, X^*, N^*, T^*\right)$, exists under the following conditions:

$$\eta_0\left(\gamma+\dfrac{\beta r}{\theta}\right)-\theta_1\beta q > 0,\ s-\phi_1\left(T^*-T_0\right)+\psi_1 I_c^* > 0. \tag{7.8}$$

The existence of the equilibrium points S_1 and S_2 are obvious. In the equilibrium S_3, the components I_{c3}, X_3, N_3 and T_3 satisfy the following equations:

$$q-\beta I_c\left(T-T_0\right)-\gamma\ I_c = 0, \tag{7.9}$$

$$Q+\lambda\ N-\alpha\ X = 0, \tag{7.10}$$

$$r\left(1-\dfrac{N}{K}\right)-\theta(T-T_0) = 0, \tag{7.11}$$

$$\eta(X-X_0)-\eta_0\left(T-T_0\right)+\theta_1\beta\ I_c\left(T-T_0\right) = 0. \tag{7.12}$$

From Equation (7.11), we get

$$T = T_0+\dfrac{r}{\theta}\left(1-\dfrac{N}{K}\right). \tag{7.13}$$

From Equation (7.10), we get

$$X = \dfrac{1}{\alpha}(Q+\lambda\ N) = X_0+\dfrac{\lambda N}{\alpha}. \tag{7.14}$$

From Equations (7.9) and (7.13), we get

$$I_c = \dfrac{q}{\beta(T-T_0)+\gamma}. \tag{7.15}$$

Using Equations (7.13)–(7.15) in (7.12), we get a quadratic equation in terms of N as given below

$$F(N) = \left[\gamma + \frac{\beta r}{\theta}\left(1 - \frac{N}{K}\right)\right]\left[\frac{\eta\lambda N}{\alpha} - \frac{\eta_0 r}{\theta}\left(1 - \frac{N}{K}\right)\right] + \frac{\theta_1\beta qr}{\theta}\left(1 - \frac{N}{K}\right) = 0. \quad (7.16)$$

From (7.16), note that

  i. $F(0) < 0$ provided the condition (7.7) holds.

  ii. $F(K) > 0$.

Thus, a unique positive root of (7.16), say N_3, exists in the interval $(0, K)$. Using this value of N_3 in Equations (7.13)–(7.15), we obtain the positive value of $T = T_3$, $X = X_3$ and $I_c = I_{c_3}$, respectively. Thus, S_3 exists under the condition (7.7).

The existence of S^* is similar to that of S_3 and can be obtained by solving the Equations (7.9)–(7.12) along with the following equation:

$$s\left(1 - \frac{P}{L}\right) - \phi_1(T - T_0) - \phi_2 P(T - T_0) + \psi_1 I_c + \psi_2 P I_c = 0 \quad (7.17)$$

Here it should be noted that the solution of system of Equations (7.7)–(7.12) and (7.17) gives $I_c^* = I_{c_3}$, $X^* = X_3$, $N^* = N_3$, $T^* = T_3$ and $P^* = \dfrac{s - \phi_1\left(T^* - T_0\right) + \psi_1 I_c^*}{\dfrac{s}{L} + \phi_2\left(T^* - T_0\right) - \psi_2 I_c^*}$. Thus, the equilibrium S^* exists under the conditions stated in (7.8).

7.3.2 Local stability analysis

In this subsection, we determine the local stability properties of equilibrium points S_1, S_2, S_3 and S^* of the system (7.6). The local stability analysis talks about the faith of the solution trajectories of the system (7.6) which initiated in a small neighborhood of the equilibrium state. The theorem stated below describes the local stability properties of the equilibria of system (7.6).

Theorem 7.1

  i. The boundary equilibrium states S_1 *and* S_2 are always unstable.

  ii. The boundary equilibrium S_3 is stable if the following conditions hold:

$$s - \phi_1 (T_3 - T_0) + \psi_1 I_{c3} < 0 \tag{7.18}$$

$$A_3 (A_1 A_2 - A_3) - A_1^2 A_4 > 0, \tag{7.19}$$

where, A_i $(i=1, 2, 3, 4)$ are defined in the proof.

iii. The equilibrium S^* is locally asymptotically stable under the condition (7.19).

Proof. The Jacobian matrix of system (7.6) can be written as

$$J = \begin{pmatrix} a_{11} & 0 & 0 & 0 & -\beta \, I_c \\ (\psi_1 + \psi_2 \, P) P & a_{22} & 0 & 0 & -(\phi_1 + \phi_2 \, P) P \\ 0 & 0 & -\alpha & \lambda & 0 \\ 0 & 0 & 0 & a_{44} & -\theta \, N \\ \theta_1 \beta (T - T_0) & 0 & \eta & 0 & -(\eta_0 - \theta_1 \beta \, I_c) \end{pmatrix}$$

where,

$$a_{11} = -\beta (T - T_0) - \gamma,$$

$$a_{22} = s\left(1 - \frac{2P}{L}\right) - \phi_1 (T - T_0) - 2\phi_2 P (T - T_0) + \psi_1 I_c + 2\psi_2 P \, I_c,$$

$$a_{44} = r\left(1 - \frac{2N}{K}\right) - \theta (T - T_0).$$

Let J_i be Jacobian matrix of system (7.6) evaluated at S_i for $i = 1, 2, 3$ and J^* be Jacobian matrix calculated at S^*.

i. The eigenvalues of J_1 are $-\gamma, \left(s + \psi_1 \dfrac{q}{\gamma}\right), -\alpha, r$ and $-\left(\eta_0 - \theta_1 \beta \dfrac{q}{\gamma}\right)$.

Since J_1 has positive eigenvalues, thus S_1 is always unstable.

Further, the eigenvalues of J_2 are $-\gamma, -\left(s + \psi_1 \dfrac{q}{\gamma}\right), -\alpha, r$ and $-\left(\eta_0 - \theta_1 \beta \dfrac{q}{\gamma}\right)$. Since r is a positive eigenvalue of matrix J_2, so S_2 is unstable.

ii. One eigenvalue of J_3 is found to be $\left(s - \phi_1 (T_3 - T_0) + \psi_1 I_{c3}\right)$ and the other eigenvalues are the roots of the characteristic equation:

$$\psi^4 + A_1 \psi^3 + A_2 \psi^2 + A_3 \psi + A_4 = 0, \tag{7.20}$$

where, $A_1 = \left(\beta(T_3 - T_0) + \gamma\right) + \alpha + \dfrac{rN_3}{K} + \left(\eta_0 - \theta_1\beta I_{c3}\right)$,

$A_2 = \left(\beta(T_3 - T_0) + \gamma\right)\left(\alpha + \dfrac{rN_3}{K} + \left(\eta_0 - \theta_1\beta I_{c3}\right)\right) + \alpha\dfrac{rN_3}{K}$

$\qquad + \left(\eta_0 - \theta_1\beta I_{c3}\right)\left(\alpha + \dfrac{rN_3}{K}\right) + \theta_1\beta^2 I_{c3}(T_3 - T_0)$,

$A_3 = \alpha\left(\beta(T_3 - T_0) + \gamma\right)\left(\dfrac{rN_3}{K} + \left(\eta_0 - \theta_1\beta I_{c3}\right)\right)$

$\qquad + \dfrac{rN_3}{K}\left(\eta_0 - \theta_1\beta I_{c3}\right)\left(\beta(T_3 - T_0) + \gamma + \alpha\right)$,

$\qquad + \eta\lambda\theta N_3 + \theta_1\beta^2 I_{c3}(T_3 - T_0)\left(\alpha + \dfrac{rN_3}{K}\right)$,

$A_4 = \alpha\dfrac{rN_3}{K}\left(\beta(T_3 - T_0) + \gamma\right)\left(\eta_0 - \theta_1\beta I_{c3}\right)$

$\qquad + \left(\beta(T_3 - T_0) + \gamma\right)\eta\lambda\theta N_3 + \theta_1\beta^2 I_{c3}(T_3 - T_0)\left(\alpha\dfrac{rN_3}{K}\right)$.

Here, all the $A_i{}'^s$ ($i = 1, 2, 3, 4$) are positive. Applying Routh-Hurwitz criterion, it can be concluded that all the roots of Equation (7.20) belong to the negative half of the complex plane if condition (7.19) holds. Thus S_3 is stable if the conditions (7.18) and (7.19) hold.

iii. One eigenvalue of matrix J^* is found to be $-\left(\dfrac{s}{L} + \phi_2(T^* - T_0) - \psi_2 I_c^*\right)P^*$,

which is negative, and the rest eigenvalues of J^* are the roots of Equation (7.20) which belongs to the negative half of complex plane under the condition (7.19).

Remark 7.1

It can be noted that if S^* exists, then $\left(s - \phi_1(T^* - T_0) + \psi_1 I_c^*\right) > 0$. Thus S_3 will be unstable whenever S^* exists.

7.3.3 Global stability of S^*

The characterization of global stability of S^* is obtained using Liapunov's direct method.

Theorem 7.2

A sufficient condition under which the equilibrium S^* is globally asymptotically stable inside Ω is given as:

$$\max\left\{\frac{\lambda^2\theta^2 K^2}{r^2\alpha}, \frac{\beta^2 q^2}{\gamma^2\left(\gamma+\beta\left(T^*-T_0\right)\right)}, \frac{1}{3}\frac{\left(\gamma+\beta\left(T^*-T_0\right)\right)\left(\phi_1+\phi_2 P_m\right)^2}{\left(\psi_1+\psi_2 P_m\right)^2}\right\}$$

$$<\frac{8}{75}\left(\eta_0-\theta_1\beta I_c^*\right)^2 \min\left\{\frac{\alpha}{\eta^2}, \frac{2}{3}\frac{\left(\gamma+\beta\left(T^*-T_0\right)\right)}{\theta_1^2\beta^2\left(T_m-T_0\right)^2}\right\}. \tag{7.21}$$

Proof. Define a scalar-valued positive definite function V on the region Ω as

$$V = \frac{1}{2}(I_c-I_c^*)^2 + m_1\left(P-P^*-P^*\ln\frac{P}{P^*}\right) + \frac{m_2}{2}(X-X^*)^2$$

$$+m_3\left(N-N^*-N^*\ln\frac{N}{N^*}\right) + \frac{m_4}{2}(T-T^*)^2,$$

where, m_1, m_2, m_3 *and* m_4 are positive constants to be chosen later on.
   The time derivative of the function 'V' is given as

$$\frac{dV}{dt} = -\left(\beta\left(T^*-T_0\right)+\gamma\right)(I_c-I_c^*)^2 - \beta I_c\left(I_c-I_c^*\right)\left(T-T^*\right)$$

$$-m_1\left(\frac{s}{L}+\phi_2\left(T^*-T_0\right)-\psi_2 I_c^*\right)(P-P^*)^2 - m_1\left(\phi_1+\phi_2 P\right)\left(P-P^*\right)\left(T-T^*\right)$$

$$+m_1\left(\psi_1+\psi_2 P\right)\left(P-P^*\right)\left(I_c-I_c^*\right) - \alpha m_2(X-X^*)^2$$

$$-m_3\frac{r}{K}(N-N^*)^2 + m_2\lambda\left(X-X^*\right)\left(N-N^*\right) - m_3\theta\left(N-N^*\right)\left(T-T^*\right)$$

$$+m_4\theta_1\beta\left(T-T_0\right)\left(T-T^*\right)\left(I_c-I_c^*\right)$$

$$-m_4\left(\eta_0-\theta_1\beta I_c^*\right)(T-T^*)^2 + m_4\eta\left(X-X^*\right)\left(T-T^*\right).$$

Now, choosing $m_2 = 1$, $\dfrac{dV}{dt}$ is negative definite inside Ω provided the inequalities stated below hold.

$$\frac{15\beta^2 I_{cm}^2}{4(\gamma+\beta(T^*-T_0))}\frac{1}{(\eta_0-\theta_1\beta I_c^*)} < m_4, \tag{7.22}$$

$$m_1(\phi_1 + \phi_2 P_m)^2 < \frac{2}{5}\left\{\frac{s}{L} + \phi_2(T^* - T_0) - \psi_2 I_c^*\right\}(\eta_0 - \theta_1\beta I_c^*)m_4, \qquad (7.23)$$

$$m_1(\psi_1 + \psi_2 P_m)^2 < \frac{2}{3}\left(\frac{s}{L} + \phi_2(T^* - T_0) - \psi_2 I_c^*\right)(\gamma + \beta(T^* - T_0)), \qquad (7.24)$$

$$\frac{\lambda^2 K}{r\alpha} < m_3, \qquad (7.25)$$

$$m_3 < \frac{2}{5\theta^2}\frac{r}{K}(\eta_0 - \theta_1\beta I_c^*)m_4, \qquad (7.26)$$

$$m_4 < \frac{2\alpha}{5\eta^2}(\eta_0 - \theta_1\beta I_c^*), \qquad (7.27)$$

$$m_4 < \frac{4(\gamma + \beta(T^* - T_0))}{15}\frac{(\eta_0 - \theta_1\beta I_c^*)}{\theta_1^2\beta^2(T_m - T_0)^2}. \qquad (7.28)$$

From inequalities (7.24) and (7.25), we choose

$$m_1 = \frac{1}{2}\frac{\left(\frac{s}{L} + \phi_2(T^* - T_0) - \psi_2 I_c^*\right)(\gamma + \beta(T^* - T_0))}{(\psi_1 + \psi_2 P_m)^2} \quad \text{and} \quad m_3 = \frac{3}{2}\frac{\lambda^2 K}{r\alpha}.$$ After

choosing m_1 and m_3, we can choose positive m_4 from inequalities (7.22), (7.23), (7.26)–(7.28) provided condition (7.21) holds.

Remark 7.2

It is noted that the condition (7.21) is easily holds for small values of parameters β, θ_1 and η, indicating the destabilizing effect of these parameters on the system's dynamics.

7.4 NUMERICAL SIMULATIONS

7.4.1 Parameters estimation

In this subsection, we have estimated the value of parameters of the proposed model system (7.6). The average Arctic sea ice extent data is taken for the period 1980–2020 [5]. The annual time series data for atmospheric CO_2 concentrations for the period 1980–2020 is obtained from the NOAA-ESRL data set [48]. Annual data for the human population and mean surface temperature for the period 1980–2020 are taken from [49] and [50], respectively. The parameters of the model are adjusted to fit the data. In 1980, the sea ice extent was 7.54 million square km [5], the average concentration of atmospheric CO_2 was 338.76 ppm (parts per million) [48], the world population was 4.432 in billion [49] and surface temperature was 14.18°C, thus we have taken $Ic(0) = 7.54$ million square km, $P(0)=28,000$ pb, $X(0)=338.76$ parts per million, $N(0) = 4.432$ billion and $T(0)=14.18$ degree Celsius. The pre-industrial concentration of CO_2 was 280 ppm [51], thus $X_0 = 280$ ppm. We take $r=0.032$ per year from [52]. The carrying capacity of humans is supposed to lie in a range of 8–16 billion [53]. We have taken the intrinsic growth rate of polar bear population $s=0.055$ per year [54]. By employing the least square method, the following estimated sets of parameter values are obtained:

$$q = 0.2, \beta = 0.031, T_0 = 13.5, \gamma = 0.02,$$

$$s = 0.055, L = 32000, \phi_1 = 0.001, \phi_2 = 0.0000006,$$

$$\psi_1 = 0.0005, \psi_2 = 0.0000001, Q = 5.6, X_0 = 280, \qquad (7.29)$$

$$\lambda = 0.596, \alpha = 0.02, r = 0.032, K = 12,$$

$$\theta = 0.002, \eta = 0.003, \eta_0 = 0.3, \theta_1 = 0.1.$$

Note that in the absence of data of polar bear population, the parameter values given in (7.29) provide only a rough estimate. Figure 7.1 demonstrates the actual data and model projection for the variables. This figure shows that actual data and model data are strongly correlated. For the parameter values given in (7.29), all the eigenvalues of J^* are found to belong to the left half of the complex plane, indicating the locally asymptotic stability of S^*. The global stability of the interior equilibrium S^* for $\beta=0.0031$ in $I_c - P - X$ and $I_c - P - T$ spaces are shown in Figure 7.2, keeping the rest parameter values same as in (7.29). It can be seen that the trajectories with different starts in $I_c - P - X$ and $I_c - P - T$ spaces converge to equilibrium values (I_c^*, P^*, X^*) and (I_c^*, P^*, T^*), respectively.

7.4.2 Effect of variations in parameters

In Figure 7.3, the time variations of sea ice extent and polar bear population for different values of melting rate coefficient β are plotted. It can be noted that the equilibrium levels of sea ice extent and polar bear population decrease as value of β increases. The first plot of Figure 7.4 depicts the contour plot of the equilibrium polar bear population as a function of ϕ_1 and ϕ_2, while the second plot of Figure 7.4 depicts the contour plot of the equilibrium polar bear population as function of ψ_1 and ψ_2. Here, the contour lines show the equilibrium polar bear population. From Figure 7.4a,

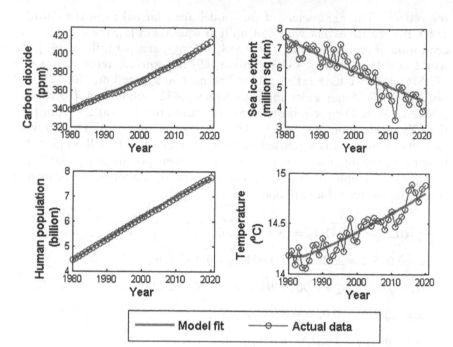

Figure 7.1 The actual data and model fit for the variables $I_c(t)$, $X(t)$, $N(t)$, $T(t)$.

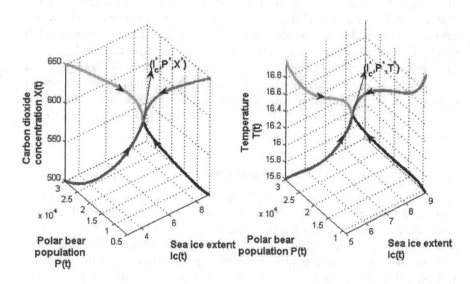

Figure 7.2 Global stability of S^* in (I_c, P, X) and (I_c, P, T) spaces at low melting rate coefficient $\beta = 0.0031$.

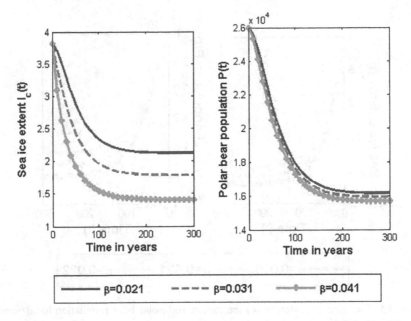

Figure 7.3 The time evolution of sea ice extent and polar bear population for different values of β.

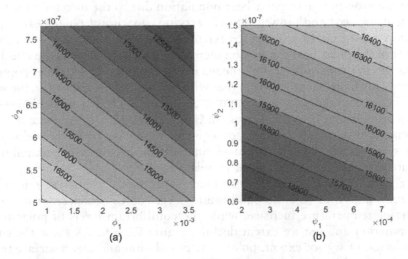

Figure 7.4 Contour plots of the equilibrium level of polar bear population as a function of (a) ϕ_1 and ϕ_2, (b) ψ_1 and ψ_2.

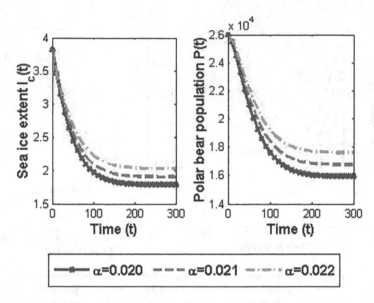

Figure 7.5 The time evolution of sea ice extent and polar bear population for different values of α.

we can see that as the parameters denoting the declination rate of polar bear population due to increase in temperature (i.e., ϕ_1) and the declination rate of carrying capacity of polar bear population due to increase in temperature (i.e., ϕ_2) increase, the equilibrium polar bear population decreases. Similarly, from Figure 7.4b, it can be observed that as the parameters denoting the growth rate of polar bear population due to the increase in sea ice extent (i.e., ψ_1) and growth rate of carrying capacity of polar bear population due to increase in sea ice extent (i.e., ψ_2) increase, the equilibrium level of the polar bear population increases. Figure 7.5 depicts the effect of changes in the parameter α on the sea ice extent and the polar bear population. This figure clearly depicts that with an increase in value of α, the sea-ice extent and the polar bear population increase. The reason behind these dynamics is that with the increase in CO_2 removal rate, the average surface temperature declines and as a consequence the reduction in the melting of ice sheet occurs. This in turn causes increase in the sea-ice extent and ultimately an increase in polar bear population.

Figure 7.6 depicts that with an increase in the CO_2 emission rate from human activities (λ), the equilibrium levels of CO_2 concentration and surface temperature increase, while the equilibrium levels of polar bear population and sea ice extent decline. Figure 7.7 and 7.8 show the time evolution of sea ice extent, polar bear population and mean surface temperature at different values of η and θ_1. From Figure 7.7, it can be seen that with an increase in the growth rate coefficient of mean surface temperature

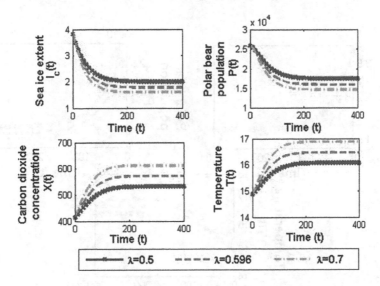

Figure 7.6 Time evolution of sea ice extent, polar bear population, concentration of carbon dioxide and average surface temperature for different values of λ.

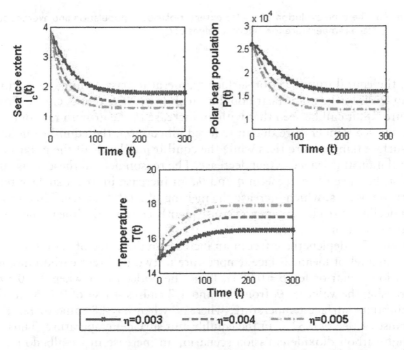

Figure 7.7 Time evolution of sea ice extent, polar bear population and average surface temperature for different values of η.

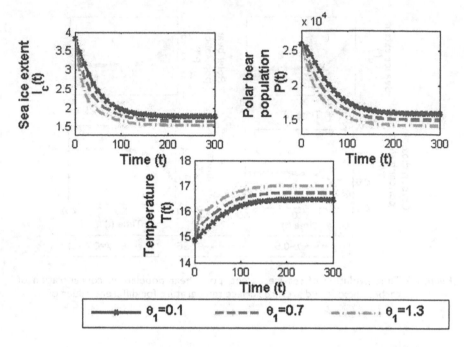

Figure 7.8 The time evolution of sea ice extent, polar bear population and average surface temperature for different values of θ_1.

(η), the equilibrium level of surface temperature rises while the equilibrium levels of the polar bear population and sea ice extent decrease. From Figure 7.8, it can be seen that with an increase in the growth rate of average surface temperature due to reduced albedo (θ_1), the equilibrium level of surface temperature rises while the equilibrium levels of the polar bear population and sea ice extent decrease. The reason behind these dynamics is that because of increase in η and θ_1, an increase in mean surface temperature occurs, which increase the melting of Arctic sea ice. This results in a decline of sea ice extent and consequently causes a decline in the polar bear population.

Figure 7.9 depicts the effect of an increase in the value of θ_1 on the equilibrium level of mean surface temperature in two different carbon dioxide emission scenarios for $\beta = 0.0031$. It can be noticed that when $\lambda = 0.596$, increasing the value of θ_1 from 0.1 and 1.7 causes a rise of 0.29°C in the equilibrium surface temperature; whereas, when $\lambda = 0.8$ same increase in θ_1 causes a rise of 0.32°C in the equilibrium surface temperature. Thus, in a high carbon dioxide emission scenario, an increase in ice-albedo effect causes more increase in the equilibrium level of surface temperature in comparison to that in low carbon dioxide emission scenario.

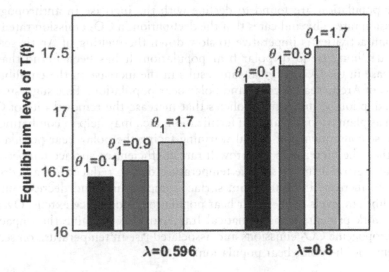

Figure 7.9 Effect of variation in the values of θ_l on the equilibrium levels of average surface temperature for different values of λ for $\beta=0.0031$.

7.5 CONCLUSIONS

The increase in global surface temperature is largely a consequence of excessive emissions of CO_2 and other trace gases because of human activities. The rise in mean surface temperature causes melting of the ice caps and glaciers. Sea ice melting has adversely affected the polar bear population due to the loss of habitat. This work presents a nonlinear differential equation model to investigate the impact of increase in CO_2 emissions on sea ice extent and the polar bear population. The proposed model captures the dynamic relationship between five variables namely; the sea ice extent, polar bear population, the atmospheric concentration of carbon dioxide, the human population and the mean surface temperature. The model possesses four non-negative equilibria; three boundary equilibria and an interior equilibrium. The stability properties of all the model equilibria are investigated. A sufficient criterion under which the positive equilibrium S^* is globally asymptotically stable is obtained by employing Lyapunov's direct method. It is found that the melting rate coefficient of sea ice extent and the growth rate coefficient of surface temperature due ice-albedo effect have a destabilizing effect on the system's dynamics. Employing the least squares method, the model parameters are estimated to fit the actual data of Arctic sea ice extent, atmospheric CO_2 concentration, human population and the average surface temperature. It is shown that an increase in CO_2 emission rate from human activities (*i.e.*, λ) results in an increase in the equilibrium surface temperature, while the equilibrium levels of sea ice extent and polar bear population are found to decline with the increase in anthropogenic emission rate. This indicates that the declination in CO_2 emission rate from human activities is imperative to slow down the melting of Arctic sea ice and declining trend of polar bear population. It has been found that an increase in the CO_2 removal rate results in the increase in the equilibrium levels of Arctic sea ice cover and polar bear population. This suggests that the adoption of mitigation policies that increase the removal rate of CO_2, such as plantation, ocean iron fertilization, etc., may help in combating the adverse consequence of global warming facing the polar bear population. Further, the increase in the growth rate coefficient of surface temperature (η) and growth in the surface temperature due to reduced ice-albedo (θ_1) lead to increase in equilibrium surface temperature, and decrease in the equilibrium levels of the polar bear population and sea ice extent. Overall, this work presents a mathematical framework to examine the impact of anthropogenic CO_2 emissions and associated rise in temperature on sea ice extent and the polar bear population.

ACKNOWLEDGEMENT

The first author (Alok Kumar Verma) thankfully acknowledges Council of Scientifc & Industrial Research (CSIR), New Delhi, India for financial support in form of senior research fellowship (09/961(0014)/2019-EMR-1).

REFERENCES

1. IPCC (2013). The physical science basis. In: Stocker, T.F., Qin, D., Plattner, G.K., Tignor, M., Allen, S.K., Boschung, J., Nauels, A., Xia, Y., Bex, V., & Midgley, P.M. (eds.) *Contribution of Working Group I to the Fifth Assessment Report of the Intergovernmental Panel on Climate Change.* Cambridge University Press, Cambridge.
2. WMO (2021). Greenhouse Gas Bulletin: World Meteorological Organization. https://public.wmo.int/en/media/press-release/greenhouse-gas-bulletin-another-year-another-record. Accessed 25 March 2022.
3. Schwartz, J. (2020). How Much Will the Planet Warm if Carbon Dioxide Levels Double? https://www.nytimes.com/news-features/climate-qa/how-much-will-earth-warm-if-carbon-dioxide-doubles. Accessed 25 March 2022.
4. Amstrup, S.C., Marcot, B.G., & Douglas, D.C. (2007). Forecasting the range-wide status of polar bears at selected times in the 21st century, administrative report, 123 pp., U.S. Geol. Surv., Alaska Sci. Cent., Anchorage, Alaska. http://www.usgs.gov/newsroom/special/polar/bear. Accessed 16 March 2022.
5. NASA (2020). Arctic sea ice minimum. https://climate.nasa.gov/sea-ice-extent.
6. Notz, D., & Stroeve, J. (2016). Observed Arctic sea-ice loss directly follows anthropogenic CO_2 emission. *Science*, 354, 747–750.

7. Whiteman, J.P., Harlow, H.J., Durner, G.M., Sprecher, R.A., Albeke, S.E., Regehr, E.V., Amstrup, S.C., & David, M.B. (2015).Summer declines in activity and body temperature offer polar bears limited energy savings. *Science*, 349, 295–298.
8. Pidcock, R. (2015). *Polar Bears and Climate Change: What Does the Science Say?* Carbon Brief Ltd. https://www.carbonbrief.org/polar-bears-and-climate-change-what-does-the-science-say
9. Derocher, A.E., Lunn, N.J., & Stirling, I. (2004). Polar bears in a warming climate. *Integrative and Comparative Biology*, 44, 163–176.
10. Oystein, W., Aars, J., & Erik, W.B. (2008). Effects of climate change on polar bears. *Science Progress*, 91(2), 151–173.
11. Christine, M.H., Caswell, H., Michael, C.R., Eric, V.R., Amstrup, S.C., & Stirling, I. (2010). Climate change threatens polar bear populations: A stochastic demographic analysis. *Ecological Society of America*, 91, 2883–2897.
12. Briggs, H., & Gill, V. (2020). Climate Change: Polar bears could be Lost by 2100, Science and Environment. https://www.bbc.co.uk/news/science-environment-53474445.
13. Fountain, H. (2020). Global Warming is Driving Polar Bears toward Extinction, Researchers Say. https://www.nytimes.com/2020/07/20/climate/polar-bear-extinction.
14. Bertrand, K. (2021). Polar Bears: How the Arctic's biodiversity has been impacted by climate change. https://arcticyouthnetwork.org. Accessed 16 March 2022.
15. Amstrup, S.C., DeWeaver, E.T., Douglas, D.C., Marcot, B.G., Durner, G.M., Bitz, C.M., & Bailey, D.A. (2010). Greenhouse gas mitigation can reduce sea-ice loss and increase polar bear persistence. *Nature*, 468, 955–958.
16. Ferguson, S.H., Taylor, M.K., & Messier, F. (2000). Influence of sea ice dynamics on habitat selection by polar bears. *Ecology*, 81, 761–772.
17. Regehr, E.V., Amstrup, S.C., & Stirling, I. (2006). Polar bear population status in the southern Beaufort Sea Open-File Report 2006-1337. U.S. Geological Survey, Reston, Virginia, USA.
18. Regehr, E.V., Hunter, C.M., Caswell, H., Amstrup, S.C., & Stirling, I. (2009). Survival and breeding of polar bears in the southern Beaufort Sea in relation to sea ice. *Journal of Animal Ecology*, 79, 117–127.
19. Rode, K.D., Amstrup, S.C., & Regehr, E.V. (2010). Reduced body size and cub recruitment in polar bears associated with sea ice decline. *Ecological Applications*, 20, 768–782
20. Ware, J.V., Rode, K.D., Bromaghin, J.F., Douglas, D.C., Wilson, R.R., Regehr, E.V., Amstrup, S.C., Durner, G.M., Pagano, A.M., Olson, J., & Robbins, C.T. (2017). Habitat degradation affects the summer activity of polar bears. *Oecologia*, 184, 87–99.
21. Johnson, A.C., & Derocher, A.E. (2020). Variation in habitat use of Beaufort Sea polar bears. *Polar Biology*, https://doi.org/10.1007/s00300-020-02705-3.
22. Tennakone, K. (1990). Stability of the biomass-carbon dioxide equilibrium in the atmosphere: Mathematical model. *Applied Mathematics and Computation*, 35, 125–130.
23. Caetano, M.A.L., Gherardi, D.F.M., & Yoneyama, T. (2011). An optimized policy for the reduction of CO_2 emission in the Brazilian Legal Amazon. *Ecological Modeling*, 222, 2835–2840.

24. Misra, A.K., & Verma, M. (2013). A mathematical model to study the dynamics of carbon dioxide gas in the atmosphere. *Applied Mathematics and Computation*, 219, 8595–8609.
25. Misra, A.K., & Verma, M. (2014). Modeling the impact of mitigation options on methane abatement from rice fields. *Adaptation Strategies for Global Change*, 19, 927–945.
26. Misra, A.K., & Verma, M. (2015). Impact of environmental education on mitigation of carbon dioxide emissions: A modelling study. *International Journal of Global Warming*, 7, 466–486.
27. Shukla, J.B., Chauhan, M.S., Sundar, S., & Naresh, R. (2015). Removal of carbon dioxide from the atmosphere to reduce global warming: A modeling study. *International Journal of Global Warming*, 7, 270–292.
28. Devi, S., & Gupta, N. (2018). Dynamics of carbon dioxide gas (CO_2): Effects of varying capability of plants to absorb CO_2. *Natural Resource Modeling*, 32, e12174.
29. Devi, S., & Gupta, N. (2020). Comparative study of the effects of different growths of vegetation biomass on CO_2 in crisp and fuzzy environments. *Natural Resource Modeling*, 33(3), e12263.
30. Devi, S., & Mishra, R.P. (2020). Preservation of the forestry biomass and control of increasing atmospheric CO_2 using concept of reserved forestry biomass. *International Journal of Applied and Computational Mathematics*, 6, 17.
31. Verma, M., Verma, A.K., & Misra, A.K. (2021). Mathematical modeling and optimal control of carbon dioxide emissions from energy sector. *Environment, Development and Sustainability*, 23, 13919–13944.
32. Verma, M., & Verma, A.K. (2021). Effect of plantation of genetically modified trees on the control of atmospheric carbon dioxide: A modeling study. *Natural Resource Modeling*, 34, e12300.
33. Misra, A.K., & Jha, A. (2021). Modeling the effect of population pressure on the dynamics of carbon dioxide gas. *Journal of Applied Mathematics*, 67(2), 1–18.
34. Mandal, S., Islam, M.S., Biswas, M.H.A., & Akter, S. (2021). Modeling the optimal mitigation of potential impact of climate change on coastal ecosystems. *Heliyon*, 7, e07401.
35. Shukla, J.B., Verma, M., & Misra, A.K. (2017). Effect of global warming on sea level rise: A modeling study. *Ecological complexity*, 32, 99–110.
36. Shukla, J.B., Arora, M.S., Verma, M., Misra, A.K., & Takeuchi, Y. (2021). The impact of sea level rise due to global warming on the coastal population dynamics: A modeling study. *Earth Systems and Environment*, 5, 909–926.
37. Moon, T.A., Overeem, I., Druckenmiller, M., Holland, M., Huntington, H., Kling, G., Lovecraft, AL., Miller, G., Scambos, T., Schadel, C., Schuur, E.A.G., Trochim, E., Wiese, F., Williams, D., & Wong, G. (2019). The expanding footprint of rapid Arctic change. *Earth's Future*, 7, 212–218.
38. EPA (2021).Climate Change Indicators: Arctic Sea Ice. https://www.epa.gov/climate-indicators/climate-change-indicators-arctic-sea-ice. Accessed 16 March 2022.
39. Stirling, I., & Parkinson, C.L. (2006). Possible effects of climate warming on selected populations of polar bears (*Ursus maritimus*) in the Canadian Arctic. *Arctic*, 59, 261–275.

40. Regehr, E.V., Laidre, K.L., Akcakaya, H.R., Amstrup, S.C., Atwood, T.C., Lunn, N.J., Obbard, M., Stern, H., Thiemann, G.W., & Oystein, W. (2016). Conservation status of polar bears (*Ursus maritimus*) in relation to projected sea ice declines. *Biology Letters* 12, 0556. http://dx.doi.org/10.1098/rsbl.2016.0556.

41. Newell, N.D., & Marcus, L. (1987). Carbon dioxide and people. *Palaios*, 2, 101–103.

42. Onozaki, K. (2009). Population is a critical factor for global carbon dioxide increase. *Journal of Health Sciences*, 55, 125–127.

43. Nikol'skii, M.S. (2010). A controlled model of carbon circulation between the atmosphere and the ocean. *Computational Mathematics and Modeling*, 21, 414–424.

44. McMichael, A.J., Woodruff, R.E., & Hales, S. (2006). Climate change and human health: Present and future risks. *Lancet*, 367, 859–869.

45. Kurane, I. (2010).The effect of global warming on infectious diseases. *Osong Public Health Res Perspect*, 1, 4–9.

46. Casper, J.K. (2010). *Greenhouse Gases: Worldwide Impacts*. Facts on File, Inc., New York.

47. Boucher, O., Haigh, J., Hauglustaine, D et al. (2001). Radioactive forcing of climate change. In: Joos, S.F., Srinivasan, J. (eds) *Climate Change 2001: The Scientific Basis, Contribution of Working Group I to the Third Assessment Report of the Intergovernmental Panel on Climate Change*. Cambridge University Press, Cambridge.

48. NOAA (2021a) Trends in atmospheric carbon dioxide, Mauna Loa, CO_2 annual mean data. https://gml.noaa.gov/ccgg/trends/data/html. Accessed 25 June 2021.

49. World Bank (2021) Population, total. https://data.worldbank.org/indicator/sp.pop.totl. Accessed 25 June 2021.

50. NOAA (2021b) Global surface temperature anomalies. https://www.ncdc.noaa.gov/monitoringreferences/faq/anomalies.phpanomalies. Accessed 25 June 2021.

51. IPCC (2001). The carbon cycle and atmospheric carbon dioxide. In: Houghton, J.T., Ding, Y., Griggs, D.J., Noguer, M., Vander Linden, P.J., Dai, X., Maskell, K., Johnson, C.A. (eds) *Climate Change 2001: The Scientific Basis. Contribution of Working Group I to the Third Assessment Report of the Intergovernmental Panel on Climate Change*. Cambridge University Press, Cambridge, New York.

52. Verma, M., & Misra, A.K. (2018). Optimal control of anthropogenic carbon dioxide emissions through technological options: A modeling study. *Computational and Applied Mathematics*, 37, 605–626.

53. UNEP (2012). *One Planet, How Many People? A Review of Earths Carrying Capacity*. UNEP Global Environmental Alert Service.A discussion paper for the year of RIO+20, Rio de Janeiro.

54. Regehr, E.V., Wilson, R.R., Rode, K.D., Runge, M.C., & Stern, H.L. (2017). Harvesting wildlife affected by climate change: A modeling and management approach for polar bears. *Journal of Applied Ecology*, 54, 1534–1543.

Chapter 8

Rainfall-runoff modeling using SWAT model

A case study of middle Godavari basin, Telangana State, India

Aadhi Naresh, Harish Gupta, and Mudavath Gopal Naik
Osmania University

Sandeep Hamsa
Water and Land Management Training and Research Institute

Manne Mohan Raju
Government of TS

Dinesh C. S. Bisht
Jaypee Institute of Information Technology

CONTENTS

DOI: 10.1201/9781003367420-8

8.1 INTRODUCTION

Rainfall-runoff relation of any watershed is dependent on many factors. Among these factors, climate, land use/land cover and soils are prominent ones. Hence, these parameters play a defining role in any hydrological modeling. The results from the model will help find the condition of the watershed and the impact of geological, and physical factors on a watershed. The assessment of physical and geological factors is much needed to accurately predict the runoff, and proper watershed management is very much needed. The major principle behind land management includes changes in land use/ land cover entails corresponding variations in watershed conditions and hydrologic response units. Watershed response in the form of runoff depth and peak flow can be used as good indicators of the watershed condition and forecast the difficulties associated with land cover change.

Many parts of the world have been experiencing changes in river flow patterns for a few decades and these changes may have been for a few decades, and these changes may come from various causes, namely land-use change, climate, agricultural practices, and population change. Hence, these changes affect the hydrological cycle by changing the rainfall intensities, actual evapotranspiration, and runoff (Booij et al., 2019). Excess precipitation in a watershed after meeting all the requirements such as infiltration, and evaporation accumulates as runoff and at last collected at the outlet of the watershed. The quantity of runoff generated at the outlet of a stream is influenced by many factors such as aggregation of the different climatic, physiologic, and as well as geologic conditions of the watershed. In India, during the monsoon, the quantity of runoff in streams is very high, and in some areas due to insufficient carrying capacity of the streams, overflow from the streams creates flooding. During the non-monsoon, there is less flow or no flow. Due to it, the irrigation dependents could not provide proper irrigation scheduling in time and it creates a drought-like situation. Various studies have presented that there is a growing number of flood events (Rogger et al., 2016; Hirabayashi et al., 2013) and drought events (Samaniego et al., 2018; Dai, 2013) around the globe, attributed due to the effect of land-use change or climate change. Numerous studies have revealed that because of land-use change and agricultural increase affect the entire process of the hydrological cycle majorly such as interception, infiltration, and evapotranspiration (ET), resulting in a change of surface and subsurface flows (Gashaw et al., 2018; Jaksa & Sridhar, 2015; Ahiablame et al., 2017; Welde & Gebremariam, 2017; Wang et al., 2014; Niraula et al., 2015; Seong & Sridhar, 2017; Sridhar and Wedin, 2009; Sridhar & Anderson, 2017).

In the same way, because of climate change, the entire hydrological system gets disturbed and affects the spatial and temporal distribution of water resources in an area (Paul et al., 2018; Ghosh et al., 2012; Sridhar et al., 2013; Jin & Sridhar, 2012). Over the global scale of climate, India has experienced a significant increase in maximum rainfall intensity, and spatial variability has been observed over the last 50 years (Vinnarasi & Dhanya, 2016; Bisht et al., 2017a, b). The changes in both rainfall and temperature reveal significanft variations in future Indian climate i.e., may experience wetter and warmer climate in the future. Due to the change in rainfall pattern and intensity, there will be a huge variation in the hydrology of regions, including droughts, especially in the tropical semi-arid regions of peninsular India, where water availability shows a major role in agricultural and economic development (Kumar et al., 2018; Shrestha et al., 2017; Pervez & Henebry, 2015; Bisht et al., 2019). Recently, studies estimated that the changes in streamflow were resulted mainly due to the land use/land cover or climate variability (Anand et al., 2018; Marhaento et al., 2017; Zhang et al., 2016; Aich et al., 2015; Nie et al., 2011; Wang et al. 2014b; Wang et al., 2013; Pokhrel et al., 2018; Sridhar et al., 2019). From these studies, it is found that the study findings are useful for considering the different causes of hydrological variations and as well as developing adaptive responses. But, the results obtained from these studies are somewhat different because climate uncertainties and land-use changes may differ from place to place due to their geographical variations, requiring further study at regional scales.

This study focuses mainly on rainfall-runoff modeling in the middle Godavari basin at the Central Water Commission (CWC) gauging station of Mancherial, located in Telangana State, India. The Godavari River in this basin includes from its confluence with Manjira to its confluence with Pranahita. The middle Godavari basin is an interstate sub-basin between Telangana state and Maharashtra. This middle Godavari basin majorly covers all agricultural land (88.5%), and the cultivation under these areas depends on river water and rainfall. Generally, the basin has large undulating plains divided by low, flat-topped hills ranges. This basin's primary soil types are black soils, red soils, lateritic soils, alluvium, mixed soils, and saline and alkaline soils. Due to the uneven and erratic rainfall distribution, the basin of 13 districts is prone to drought in a total of 42 districts (Central Water Commission, 1987). Proper handling of water resources especially with water deficiency is very much difficult to manage. Hydrological models are generally categorized into two groups such as lumped and distributed type. Due to the connection of local variations and geographic information system (GIS) in distributed models their execution in large watersheds is time-consuming (Ahmadi et al., 2019). There are various hydrological models available for rainfall-runoff modeling and can be selected for doing analysis based on the availability of input data. Among them, the few are variable infiltration capacity model (VIC), TOPMODEL, Hydrologiska Byråns Vattenavdelning (HBV), MIKESHE and soil and water assessment tool (SWAT) model have been used widely throughout the world (Devia et al., 2015). Soil and Water Assessment Tool (SWAT) can quantify the flow accurately by using the different input components of the hydrological cycle such as rainfall, evapotranspiration, temperature, relative humidity, solar radiation etc., at the watershed outlet for the historical and future period. Because of these different variations and heterogeneity of hydrologic processes in space and time, the unavailability of data due to a lack of monitoring stations or absence of measurements during extreme rainfall-runoff modeling is still an age-old problem for particular watersheds and time periods. For optimal and accurate model results requires homogeneous data and reliable data preparation (Alizadeh et al., 2017; Arefinia et al., 2020). Hence, keeping the above facts and data available, an attempt is made to study the rainfall-runoff modeling for the middle Godavari basin using the Soil and Water Assessment Tool (SWAT).

8.2 STUDY AREA

The Godavari River basin (Figure 8.1) is the largest river in peninsular India and it is called as 'Dakshina Ganga' of India. The basin area accounts for 9.50% of the total geographical area of the country, and it is spread

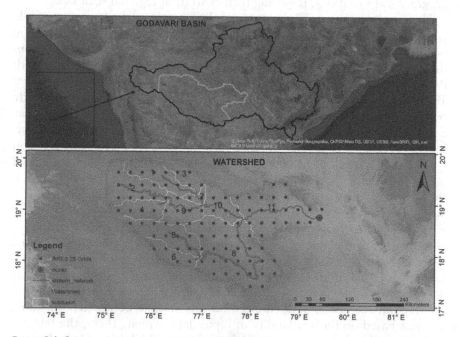

Figure 8.1 Stream Network of Middle Godavari River Basin with gauge station at Mancherial and IMD rainfall observation locations.

over an extent of 73° 24′–83° 4′ E, and 16° 19′–22° 34′ N. It starts in the Sahyadris ranges, at the height of 1,067 m above m.s.l (mean sea level) near Trimbakeshwar in the Nashik district of Maharashtra and it flows from the Western to the Eastern Ghats through the Deccan Plateau. The main river divides the states of Telangana and Maharashtra, as well as Telangana and Chhattisgarh. It flows through the four states, Maharashtra Telangana, Chhattisgarh, and Andhra Pradesh, and finally, falls into the Bay of Bengal. The river has a total length of 1,465 km. The entire Godavari basin has been divided into 12 sub-basins by the Krishna-Godavari commission. The Godavari basin has a tropical climate, with evaporation losses ranging from 1,800 to 2,440 mm over different areas of the basin (CWC, 1999). The Godavari basin receives an average rainfall of 1,132 mm, out of which the monsoon season only contributes almost 84% of total rainfall. During the monsoon season, rainfall is huge in this basin and contributes to huge runoff through the river, which creates floods in the parts of Andhra Pradesh, Maharashtra, Chhattisgarh, and Orissa.

The middle basin of the Godavari River (Figure 8.1) covers an area of 17,205 sq. km, from its confluence with the Manjira to its confluence with the Pranahita. The Mancherial gauge discharge observation site (18°50′9″N 79°26′41 ″E) was chosen to examine the stream flow behaviour of this region. Under the middle Godavari basin, there are five main water resource projects: Pochampad dam, Kadam dam, Sripada Yellampalli barrage, Sundilla barrage and Annaram barrage. Sundilla barrage and Annaram barrage were part of the Kaleshwaram project.

8.3 MATERIALS AND METHODS

8.3.1 SWAT model inputs

The Soil Water Assessment Tool (Arnold et al. 2012a, b, 1998) is a physically-based semi-distributed hydrological model which has been developed by the Agriculture Research Service (ARS) of the United States Department of Agriculture (USDA). This tool can mimic the rainfall-runoff process as well as stream water quality (Neitsch et al., 2011; Young et al., 1989; Taylor et al., 2016; Thilakarathne et al., 2018; Sehgal et al., 2018).

SWAT has the ability to perform the simulation of different hydrological parameters of a watershed with the help of water balance mathematical formulations daily, and monthly (Neitsch et al., 2002, 2009) time steps. One of the methods available for runoff estimation in SWAT model is Soil Conservation Service (SCS) Curve Number (SCS-CN), and the other calculations such as Potential Evapotranspiration (PET) predicted using any of the following methods: Penman-Monteith, Priestley-Taylor, or Hargreaves methods. This study used the SCS-CN approach for continuous runoff simulation (Williams et al., 2012).

For runoff simulation in SWAT, the major basic input data required are spatial data, including a digital elevation model (DEM), land use, and soil raster maps. The temporal dataset involves hydrological data (measured stream flow) and climate data (precipitation, temperature, solar radiation, relative humidity, and wind speed).

8.3.1.1 Temporal data

The only high-resolution gridded daily rainfall and temperature (0.25×0.25) available at Indian Meteorological Department (IMD, 2014) for the period 1972–2013 were used as the input for the SWAT model. The watershed covers 103 grid points within its boundary, and for which rainfall and temperature were assigned in the SWAT weather database to set up the model. The daily observed streamflow discharge gauge data collected from Central Water Commission (CWC) for the period 1972–2013 at the Mancherial station were considered for calibration, as shown in Table 8.1.

8.3.1.2 Spatial data

This study includes a digital elevation model (DEM) of 30×30 m resolution based Shuttle Radar Topography Mission (SRTM) was used to show the topography pattern. The DEM dataset was downloaded from the USGS website (SRTM, 2006). The soil map for the respective study area was extracted from Food and Agriculture Organization (FAO) webpage (LSMSD, 2012). The soil characteristics (Table 8.2), such as soil class, soil texture, and others, were used from the FAO soil database (Figure 8.2). The detailed properties of soil, such as soil texture, soil depth, bulk density, hydraulic conductivity, and organic carbon content, were included in the

Table 8.1 Input Data Sources and Their Details

Data Type	Scale/Resolution	Source
DEM	30×30 m	SRTM
LULC	1×1 km	Waterbase
Soil	10 ×10 km	Waterbase
Weather	0.25°×0.25°	Indian Meteorological Department (IMD)
Observed discharge	Daily	Central Water Commission (CWC)

Table 8.2 Distribution of Soil Types for the Study Area

Global Soil Code	Soil Texture
BV12-3b-3696	Clay Loam
LC13-2a-3813	Sandy-clay loam
VC43-3ab-3861	Clay

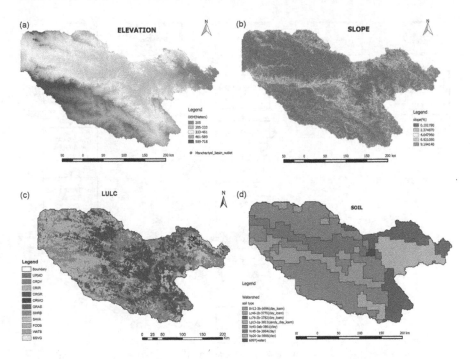

Figure 8.2 Thematic maps generated used as inputs for the model: (a) Elevation map of 30 m SRTM DEM (b) Slope Map (c) Land use/Land cover Map (d) Soil Map.

soil database of the world (FAO). The land-use data were extracted from global land-use data (Waterbase, 2013). The study area has been categorized into 11 land-use/land-cover classes (Figure 8.2). The elevation and slope variation for the watershed is generated using SRTM DEM (30 m) in the QGIS tool, and these both have been divided into five types of classes, as shown in Figure 8.2.

The major parameters which influence the output result due to the input variability (Holvoet et al., 2005) for the SWAT model were identified using the sensitivity analysis. In this study, 18 parameters from various sources of published literature were used for the sensitivity analysis. The SWAT model result obtained has been used further for calibration, validation, and identification of sensitive parameters using SWAT-CUP. The performance of the model is evaluated using the two statistical parameters as the Nash-Sutcliffe Efficiency (NSE) coefficient and coefficient of determination (R^2).

8.4 METHODOLOGY

This study would illustrate a strategy for influencing the rainfall variability on streamflow. In this approach 42 years of gridded rainfall and

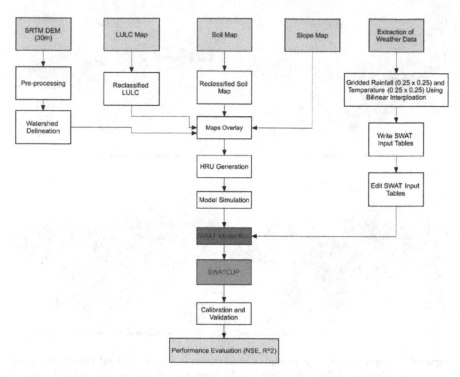

Figure 8.3 Schematic representation of flowchart of the proposed methodology.

temperature data were used to study the variability effect on stream flow and its trends concerning rainfall and temperature. Here, the model formulation included 42 years of streamflow data, and it is divided into two parts; calibration and validation. The first part consists of 21 years (1972–1993) of data used for calibration, and the remaining 21 years (1994–2013) of data used for validation. Similarly, DEM, Global land use/land cover, and soil data (2012) data were incorporated into the model formulation as one of the inputs. Finally, simulations of stream flow have been carried out using a semi-distributed hydrological model such as SWAT. Moreover, the next step includes calibration and validation which have been performed using SWAT-CUP. The formulation of the rainfall-runoff modeling flow chart was used in this study, shown in Figure 8.3.

8.4.1 SWAT model description

SWAT is a physically-based continuous period, semi-distributed hydrological river basin model. It was developed to study the effects of the hydrological process and non-point source pollution in river basins. The process of SWAT modeling for any watershed or basin includes spatial distribution, which divides the area into many sub-watersheds. These sub-watersheds are further divided into hydrologic response units (HRUs) based on land use/land cover (LULC), shape, and soil attribute values.

The SWAT model drives mainly on the water balance approach and is the only driving factor behind the process involved. The SWAT model used the water balance approach to compute the runoff and peak flows (Arnold et al., 1998) can be expressed as:

$$SW_t = SW_0 + \sum_{i=1}^{t} \left(R_{\text{day}, i} - Q_{\text{surf}, i} - E_{a,i} - W_{\text{seep},i} - Q_{\text{gw},i} \right) \tag{8.1}$$

where SW_0 represents its initial soil water content and SW_t shows the soil water content on day t. All other units are considered in millimeters, and the time (t) is measured in days. All forms of water loss have been subtracted using Equation (8.1) on any of the day i from precipitation of that particular day (R_{day}), as well as surface runoff ($Q_{\text{surf}, i}$), evapotranspiration ($E_{a, i}$), loss to vadose zone ($W_{\text{seep}, i}$), and return flow ($Q_{\text{gw}, i}$) (Neitsch et al., 2011). Equation (8.1) can be modified as required variables of interest. In this study, to compute the runoff, the required Equation (8.2) has been derived from the USDA soil conservation service runoff curve number (CN) method (USDA, 1972) as presented:

$$Q_{\text{surf}} = \frac{\left(R_{\text{day}} - I_a \right)^2}{\left(R_{\text{day}} - I_a + S \right)} \tag{8.2}$$

where Q_{surf} is the accumulated depth of rainfall excess, it will act as runoff. R_{day} is the rainfall depth of a particular day, and I_a is the initial abstraction which includes infiltration, interception, and surface storage. S is the retention parameter calculated from the Curve Number (CN).

8.4.2 Sequential uncertainty fitting (SUFI)-2 algorithm description

In this algorithm, checking the uncertainties associated with the model results are quantified by a measure known as the P-factor. P-factor is the percentage of measured data related by the 95% prediction uncertainty (95PPU). Another factor available for verifying the strength of a calibration/uncertainty is the R-factor, which is represented as the average thickness of the 95PPU band divided by the standard deviation of the observed data. Hence, SUFI-2 tries to find the relation between observed vs. predicted with the less possible uncertainty band. Theoretically, the value of the P-factor ranges between 0% and 100%, whereas the R-factor ranges between 0 and ∞. The true combination of P and R values of 1 and zero indicates a perfect simulation that matches with measured data.

8.5 PERFORMANCE EVALUATION CRITERIA

8.5.1 Coefficient of determination (R^2)

The coefficient of determination (R^2) illustrates the proportion of the total variance. Its range varies from 0.0 to 1.0, where better agreement is indicated by a higher value. R^2 lies between 0 and 1, and 0.5 is considered satisfactory (Van Liew et al., 2003).

$$R^2 = \left[\frac{\sum_{i=1}^{N}\left(O_i - \bar{O}\right)\times\left(P_i - \bar{P}\right)}{\left[\sum_{i=1}^{N}\left(O_i - \bar{O}\right)^2\right]^{0.5} \times \left[\sum_{i=1}^{N}\left(P_i - \bar{P}\right)^2\right]^{0.5}} \right]^2 \tag{8.3}$$

where O corresponds to observed runoff and P to simulated runoff. $\bar{O}$ and $\bar{P}$ represent the mean values of observed and simulated runoff.

8.6 RESULTS AND DISCUSSION

8.6.1 Land use/land cover (LULLC) statistics and trend patterns for precipitation, temperature

The surface flow, type of erosion, and quantity and quality of sediment transported through the water discharge mainly depend on the current land use and land cover in the basin (Jordan et al., 2005; Valentin et al., 2008). In this study, global land use (Waterbase) and a modified version of Anderson Level-1 type were considered (Table 8.3). The area of extent clipped for the watershed is carried out in QGIS and labeled according to the global land use and Anderson Level-1 (Figure 8.2c). The statistics of the LULC for the watershed as shown in Table 8.3. As per the statistics, the watershed covers a maximum of 30.14% of dryland and cropland pasture (CRDY). This highest percentage of CRDY class has been observed in the western part of the watershed.

The long-term variation of temperature and precipitation in the period 1969–2013 for the middle Godavari river basin was considered for the Mann-Kendall trend test. Additionally, annual stream flows for the period 1969–2013 were also investigated. The average annual rainfall and stream flow is the average daily rainfall and stream flow of that particular year, respectively. The annual variability of average maximum temperature, minimum temperature, average rainfall, and average stream flow is shown in Figure 8.4. As per the Kendall Test statistics report shown in Table 8.4, the minimum temperature of TS value, and 0.323 indicates that there was no trend observed, for the maximum temperature of TS 2.33 indicates that there is an increasing trend detected. Whereas the rainfall of TS −0.274

Table 8.3 Distribution of Landuse for the Study Area Based on Global Land-Use and Modified to Anderson Level-I

Modified Land-Use Classes of Anderson level _I	SWAT_LU Classification	Land-Use Type	% of the Watershed
Urban and built-up land	URMD	Residential of Medium Density	0.08
Agricultural land	CRDY	Dryland and Cropland Pasture	30.14
	CRIR	Irrigated Cropland and Pasture	28.74
	CRGR	Mosaic Cropland/grassland	28.75
	CRWO	Cropland-Woodland Mosaic	0.79
Range land	GRAS	Grassland	0.08
	SHRB	Shrubland	4.78
	SAVA	Savana	5.58
Forest	FODB	Deciduous broad-leaf forest	0.06
Water	WATR	Waterbodies	0.93
Barren land	BSVG	Barren or sparsely vegetated lands, including bare land and sand/stone mining activities	0.07

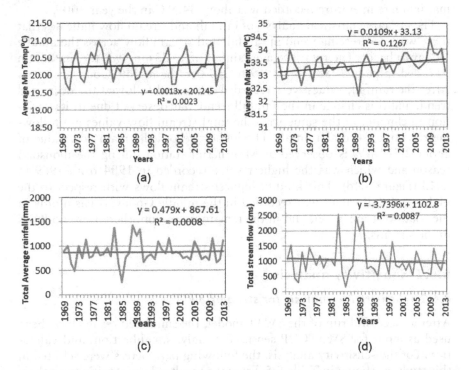

Figure 8.4 The Mann-Kendall trend test for minimum and maximum temperature, rainfall, and stream flow for the middle Godavari River basin during the period 1969–2013.

Table 8.4 MK Test for Minimum Temperature, Maximum Temperature, Rainfall, and Streamflow

S. No	Parameter	Time	Kendal Test Statistics (TS)	Nature of Trend
1	Minimum Temperature (°C)	1969–2013	0.323	No trend
2	Maximum Temperature (°C)	1969–2013	2.33	Increasing trend
3	Rainfall (mm)	1969–2013	−0.274	Decreasing trend
4	Streamflow (cumec)	1969–2013	−3.521	Decreasing trend

indicates a decreasing trend and the stream flow of TS value of −3.521 pointed out as decreasing trend. The Kendall test statistics values for the individual parameters are presented in Table 8.4.

The average minimum temperature continuously increase the up to the year 1979 and was recorded at a maximum of 21°C in the year 1979. Meanwhile, the average minimum temperature afterward decreases and increases, i.e. no trend has been observed (Figure 8.4a). The reason for the average minimum temperature values recorded (Figure 8.4a) is due to the southwest monsoon (June–September). On the contrary, the average maximum temperature plot is following the increasing trend and the average maximum temperature recorded was about 34.5°C in the year 2009.

The primary statistical analysis of rainfall and stream flow indicates that there was a decreasing trend in rainfall and stream flow across the middle Godavari basin. The total average rainfall of 1,433 mm was the maximum recorded value in 1988 from 1969 to 2013 in the middle Godavari basin. After the rainfall, values were recorded lower and followed the decreased trend, and this change in the rainfall trend was observed due to its variation in climate. In the same way, the total stream flow values are following the same trend as rainfall. The highest recorded stream flow value of about 2,510 m³/s is observed at Mancherial station during the monsoon season and which was the highest value recorded in 1984 from 1969 to 2013 (Figure 8.4d). This kind of highest stream flows with respect to the highest rainfall values was recorded in the middle Godavari basin only in the southwest monsoon, and in the remaining season, there were no flows in the streams.

8.6.2 Sensitivity analysis

8.6.2.1 Sensitivity analysis for streamflow

After a successful run of the SWAT model, the simulated results have been used as input for SWATCUP sensitivity analysis, calibration, and validation. For the sensitivity analysis, the following parameters were selected in this study, as shown in Table 8.5. From the results of the sensitivity analysis of stream flow, it is indicated that the parameter R_SOL_Z.sol is the most

Table 8.5 Optimized Sensitivity Parameters For Stream Flow of Middle Godavari Basin

Parameter Name	t-Stat	P-Value	Rank
R_SOL_Z(..).sol	0.2143	0.8955	1
R_RCHRG_DP.gw	0.1889	0.8811	2
R_SOL_AWC(..).sol	0.1824	0.8850	3
R_ EPCO.hru	0.1687	0.8935	4
RSLSUBBSN.hru	−0.1569	0.9009	5
R_CN2.mgt	0.1544	0.9024	6
V_GW_DELAY.gw	-0.1379	0.9127	7
R_CH_K2.rte	-0.1325	0.9161	8
R_OV_N.hru	−0.1176	0.9254	9
R_SOL_K(..).sol	−0.1073	0.9319	10
V_GWQMN.gw	0.1043	0.9338	11
RESCO.hru	−0.0900	0.9427	12
R GW_REVAP.gw	0.0861	0.9453	13
V ALPHA_BF.gw	−0.0572	0.9636	14
R_CH_N2.rte	−0.0550	0.9649	15
R_CANMX.hru	0.0525	0.9665	16
R_SURLAG.bsn	−0.0462	0.9706	17
R_SOL BD(..).sol	0.0188	0.9879	18

sensitive parameter. Subsequently, others in the next order are least sensitive according to its rank. The t-stat and P-values with best-fitted values are shown in Table 8.5. This study analysis results that the soil parameter of first-order rank has shown more impact on flow in the stream.

8.6.3 Calibration and validation of streamflow

To model the effective stream flow between the observed and simulated, it is important to calibrate the model for the selected baseline period perfectly. For model calibration and validation, the baseline periods such as 1972 to 1992 and 1993 to 2013, respectively, were used for stream flow records at the Mancherial gauge station. During the calibration, different sets of sensitive parameters were tested using performance indicators such as R^2. After several tests and iterations, the best-fitted values (Table 8.5) concerning the high R^2 value of 0.8336 during calibration and 0.6229 at validation are shown in Figure 8.5. Based on the results obtained, the model performed very well during the calibration and satisfactory during the validation. Overall, this model analysis indicates that it can capture the watershed behavior and flow at the Mancherial gauge station during the spatio-temporal variations of rainfall, temperature, and LULC for the middle Godavari basin.

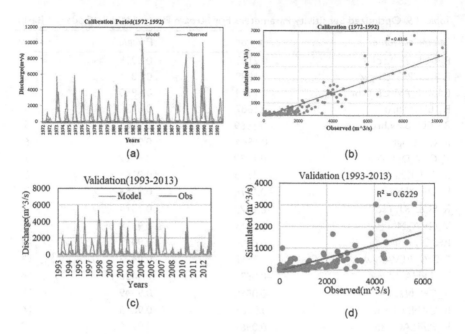

Figure 8.5 Observed vs simulated annual stream flow hydrograph during (a) calibration (1972–1992) and (b) validation (1993–2013).

8.7 CONCLUSIONS

This study highlighted the rainfall-runoff process behavior at the Mancherial gauge station for 42 years under the land use/land cover and climate effects. Trend analysis with the help Mann-Kendall test estimated that there was an increasing trend in rainfall till 1988, and after that, it decreased up to 2013. And in the case of stream flow also there was an increasing trend observed up to 1982, and after that, it decreased up to 2013. These stream flow variations across the middle Godavari basin, sub-basins, and mainstream were common. It is observed from the results that, there was considerable variation in stream flow with respect to measured flow. During the calibration period (1972–1992), the coefficient of determination was observed as 0.833, while in the case of the validation period (1993–2013) is 0.622. However, this stream flow variation is sometimes higher because of the existence of more dams and its release of flow during the monsoon if it reaches maximum flood levels. Although the Godavari River reaches maximum flow during the monsoon and this water-filled into the dams up to its maximum water level, and the remaining water gets released into the ocean through major streams. The extra and stored water in dams is diverted to agricultural activities during the on-season and off-season, respectively.

ACKNOWLEDGMENTS

The authors are duly thankful to the Central Water Commission (CWC), Hyderabad, and the Indian Meteorological Department (IMD) for providing data.

REFERENCES

Ahiablame, L., Sinha, T., Paul, M., Ji, J. H., & Rajib, A. (2017). Streamflow response to potential land use and climate changes in the James River watershed, Upper Midwest United States. *Journal of Hydrology: Regional Studies*, 14(August), 150–166. https://doi.org/10.1016/j.ejrh.2017.11.004

Ahmadi, M., Motamedvaziri, B., Ahmadi, H., Moeini, A., & Zehtabiyan, G. R. (2019). Assessment of climate change impact on surface runoff, statistical downscaling and hydrological modeling. *Physics and Chemistry of the Earth*, 114(August). https://doi.org/10.1016/j.pce.2019.09.002

Aich, V., Liersch, S., Vetter, T., Andersson, J. C. M., Müller, E. N., & Hattermann, F. F. (2015). Climate or land use? - Attribution of changes in river flooding in the Sahel zone. *Water*, 7(6), 2796–2820. https://doi.org/10.3390/w7062796

Alizadeh, M. J., Kavianpour, M. R., Kisi, O., & Nourani, V. (2017). A new approach for simulating and forecasting the rainfall-runoff process within the next two months. *Journal of Hydrology*, 548, 588–597. https://doi.org/10.1016/j.jhydrol.2017.03.032

Anand, J., Gosain, A. K., & Khosa, R. (2018). Prediction of land use changes based on land change modeler and attribution of changes in the water balance of Ganga basin to land use change using the SWAT model. *Science of the Total Environment*, 644, 503–519. https://doi.org/10.1016/j.scitotenv.2018.07.017

Arefinia, A., Bozorg-Haddad, O., Oliazadeh, A., & Loáiciga, H. A. (2020). Reservoir water quality simulation with data mining models. *Environmental Monitoring and Assessment*, 192(7). https://doi.org/10.1007/s10661-020-08454-4

Arnold, J. G., Kiniry, J., Srinivasan, R., Williams, J., Haney, E., & Neitsch, S. 2012a. Soil and water assessment tool, input/output documentation version 2012, 654. Rep. No. TR-439. College Station, TX: Texas Water Resources Institute.

Arnold, J. G., Moriasi, D. N., Gassman, P. W., Abbaspour, K. C., White, M. J., Griensven, V., & Liew, V. 2012b. SWAT: Model use, calibration, and validation. *Transactions of the ASABE 55* (4): 1491–1508. https://doi.org/10.13031/2013.42256.

Arnold, J. G., Srinivasan, R., Muttiah, R. S., & Williams, J. R. (1998). Large area hydrologic modeling and assessment part I: model development. *Journal of the American Water Resources Association*. 34(1), 73–89.

Bisht, D. S., Chatterjee, C., Raghuwanshi, N. S., & Sridhar, V. (2017a). An analysis of precipitation climatology over Indian urban agglomeration. *Theoretical and Applied Climatology*, 133(1–2), 421–436. https://doi.org/10.1007/s00704-017-2200-z

Bisht, D. S., Chatterjee, C., Raghuwanshi, N. S., & Sridhar, V. (2017b). Spatio-temporal trends of rainfall across Indian river basins. *Theoretical and Applied Climatology*, 132(1–2), 419–436. https://doi.org/10.1007/s00704-017-2095-8

Bisht, D. S., Sridhar, V., Mishra, A., Chatterjee, C., & Raghuwanshi, N. S. (2019). Drought characterization over India under projected climate scenario. *International Journal of Climatology*, 39(4), 1889–1911. https://doi.org/10.1002/joc.5922

Booij, M. J., Bouaziz, L., Hegnauer, M., & Kwadijk, J. C. J. (2019). Effect of hydrological model structures on extreme high discharge simulations. In *27th IUGG General Assembly*. Montreal: IGAC.

Central Water Commission. 1987. *Flood Atlas of India*. R. K. Puram, New Delhi: Publication Division of Central Water Commission.

Central Water Commission. 1999. *Statistical Profile of Godavari Basin. Publication Division of Central Water Commission*. R. K. Puram, New Delhi, 242.

Dai, A. (2013). Increasing drought under global warming in observations and models. *Nature Climate Change*, 3(1), 52–58. https://doi.org/10.1038/nclimate1633

Devia, G. K., Ganasri, B. P., & Dwarakish, G. S. (2015). A review on hydrological models. *Aquatic Procedia*, 4(Icwrcoe), 1001–1007. https://doi.org/10.1016/j.aqpro.2015.02.126

Gashaw, T., Tulu, T., Argaw, M., & Worqlul, A. W. (2018). Modeling the hydrological impacts of land use/land cover changes in the Andassa watershed, Blue Nile Basin, Ethiopia. *Science of the Total Environment*, 619–620, 1394–1408. https://doi.org/10.1016/j.scitotenv.2017.11.191

Ghosh, S., Das, D., Kao, S. C., & Ganguly, A. R. (2012). Lack of uniform trends but increasing spatial variability in observed Indian rainfall extremes. *Nature Climate Change*, 2(2), 86–91. https://doi.org/10.1038/nclimate1327

Hirabayashi, Y., Mahendran, R., Koirala, S., Konoshima, L., Yamazaki, D., Watanabe, S., Kim, H., & Kanae, S. (2013). Global flood risk under climate change. *Nature Climate Change*, 3(9), 816–821. https://doi.org/10.1038/nclimate1911

Holvoet, K., van Griensven, A., Seuntjens, P., & Vanrolleghem, P. A. (2005). Sensitivity analysis for hydrology and pesticide supply towards the river in SWAT. *Physics and Chemistry of the Earth*, 30(8–10), 518–526. https://doi.org/10.1016/j.pce.2005.07.006

http://www.waterbase.org/download_data.html

IMD (India Meteorological Department). (2014). India Meteorological Department (IMD) gridded temperature data set for the period 1953–2013. Accessed December 22, 2017. https://mausam.imd.gov.in/.

Jaksa, W. T., & Sridhar, V. (2015). Effect of irrigation in simulating long-term evapotranspiration climatology in a human-dominated river basin system. *Agricultural and Forest Meteorology*, 200, 109–118. https://doi.org/10.1016/j.agrformet.2014.09.008

Jin, X., & Sridhar, V. (2012). Impacts of climate change on hydrology and water resources in the Boise and Spokane river basins. *Journal of the American Water Resources Association*, 48(2), 197–220. https://doi.org/10.1111/j.1752-1688.2011.00605.x

Jordan, G., Van Rompaey, A., Szilassi, P., Csillag, G., Mannaerts, C., & Woldai, T. (2005). Historical land use changes and their impact on sediment fluxes in the Balaton basin (Hungary). *Agriculture, Ecosystems and Environment*, 108(2), 119–133. https://doi.org/10.1016/j.agee.2005.01.013

Kumar, N., Tischbein, B., Beg, M. K., & Bogardi, J. J. (2018). Spatio-temporal analysis of irrigation infrastructure development and long-term changes in irrigated areas in Upper Kharun catchment, Chhattisgarh, India. *Agricultural Water Management, 197*, 158–169. https://doi.org/10.1016/j.agwat.2017.11.022

LSMSD (Legacy Soil Maps and Soils Databases). (2012). "Legacy soil maps and soils databases (LSMSD), soil survey, FAO soils portal." Accessed February 13, 2018. https://www.fao.org/soils-portal/soil-survey/soil -maps-and-databases/en/.

Marhaento, H., Booij, M. J., Rientjes, T. H. M., & Hoekstra, A. Y. (2017). Attribution of changes in the water balance of a tropical catchment to land use change using the SWAT model. *Hydrological Processes, 31*(11), 2029–2040. https://doi.org/10.1002/hyp.11167

Neitsch, S. L., Arnold, J. G., Kiniry, J. R. & Williams, J. R. (2009). Soil and water assessment tool theoretical documentation version 2009, 618. Technical Rep. No. 406. College Station, TX: Texas Water Resources Institute.

Neitsch, S. L., Arnold, J. G., Kiniry, J. R. & Williams, J. R. (2011). *Soil and Water Assessment Tool; Theoretical Documentation: Version 2009*. College Station, TX: Texas Water Resource Institute.

Neitsch, S. L., Arnold, J. G., Kiniry, J. R., Williams J. R, & King, K. W. (2002). *Soil and Water Assessment Tool Theoretical Documentation: Version 2000*. Temple, TX: Texas A&M Universty, Blackland Research and Extension Center.

Nie, W., Yuan, Y., Kepner, W., Nash, M. S., Jackson, M., & Erickson, C. (2011). Assessing impacts of Landuse and Landcover changes on hydrology for the upper San Pedro watershed. *Journal of Hydrology, 407*(1–4), 105–114. https://doi.org/10.1016/j.jhydrol.2011.07.012

Niraula, R., Meixner, T., & Norman, L. M. (2015). Determining the importance of model calibration for forecasting absolute/relative changes in streamflow from LULC and climate changes. *Journal of Hydrology, 522*, 439–451. https://doi.org/10.1016/j.jhydrol.2015.01.007

Paul, S., Ghosh, S., Mathew, M., Devanand, A., Karmakar, S., & Niyogi, D. (2018). Increased spatial variability and intensification of extreme monsoon rainfall due to urbanization. *Scientific Reports, 8*(1), 1–10. https://doi.org/10.1038/s41598-018-22322-9

Pervez, M. S., & Henebry, G. M. (2015). Assessing the impacts of climate and land use and land cover change on the freshwater availability in the Brahmaputra River basin. *Journal of Hydrology: Regional Studies, 3*, 285–311. https://doi.org/10.1016/j.ejrh.2014.09.003

Pokhrel, Y., Burbano, M., Roush, J., Kang, H., Sridhar, V., & Hyndman, D. W. (2018). A review of the integrated effects of changing climate, land use, and dams on Mekong river hydrology. *Water, 10*(3), 1–25. https://doi.org/10.3390/w10030266

Rogger, M., Agnoletti, M., Alaoui, A., Bathurst, J. C., Bodner, G., Borga, M., Chaplot, V., Gallart, F., Glatzel, G., Hall, J., Holden, J., Holko, L., Horn, R., Kiss, A., Quinton, J. N., Leitinger, G., Lennartz, B., Parajka, J., Peth, S., ... & Blöschl, G. (2016). Land use change impacts on floods at the catchment scale: Challenges and opportunities for future research. *Water Resources Research. 53*, 5209–5219. https://doi.org/10.1002/2017WR020723.Received

Samaniego, L., Thober, S., Kumar, R., Wanders, N., Rakovec, O., Pan, M., Zink, M., Sheffield, J., Wood, E. F., & Marx, A. (2018). Anthropogenic warming exacerbates European soil moisture droughts. *Nature Climate Change, 8*(5), 421–426. https://doi.org/10.1038/s41558-018-0138-5

Sehgal, V., Sridhar, V., Juran, L., & Ogejo, J. A. (2018). Integrating climate forecasts with the Soil and Water Assessment Tool (SWAT) for high-resolution hydrologic simulations and forecasts in the Southeastern U.S. *Sustainability, 10*(9). https://doi.org/10.3390/su10093079

Seong, C., & Sridhar, V. (2017). Hydroclimatic variability and change in the Chesapeake Bay watershed. *Journal of Water and Climate Change, 8*(2), 254–273. https://doi.org/10.2166/wcc.2016.008

Shrestha, S., Shrestha, M., & Shrestha, P. K. (2017). Evaluation of the SWAT model performance for simulating river discharge in the Himalayan and tropical basins of Asia. *Hydrology Research, 49*(3), 846–860. https://doi.org/10.2166/nh.2017.189

SRTM (Shuttle Radar Topography Mission). (2006). Shuttle radar topography mission (SRTM) data subset, digital elevation data set. Accessed March 6, 2018. https://srtm.csi.cgiar.org/srtmdata/.

Sridhar, V., and D. A. Wedin. (2009). Hydrological behavior of grasslands of the Sandhills: Water and energy balance assessment from measurements, treatments and modeling. *Ecohydrology 2*(2), 195–212. https:// doi.org/10.1002/eco.61.

Sridhar, V., & Anderson, K. A. (2017). Human-induced modifications to land surface fluxes and their implications on water management under past and future climate change conditions. *Agricultural and Forest Meteorology, 234–235*, 66–79. https://doi.org/10.1016/j.agrformet.2016.12.009

Sridhar, V., Jin, X., & Jaksa, W. T. A. (2013). Explaining the hydroclimatic variability and change in the Salmon River basin. *Climate Dynamics, 40*(7–8), 1921–1937. https://doi.org/10.1007/s00382-012-1467-0

Sridhar, V., Kang, H., & Ali, S. A. (2019). Human-induced alterations to land use and climate and their responses for hydrology and water management in the Mekong River Basin. *Water, 11*(6). https://doi.org/10.3390/w11061307

Sridhar, W. (2017). Hydrological behaviour of grasslands of the Sandhills of Nebraska: Water and energy-balance assessment from measurements, treatments, and modelling. *Ecohydrology, 130*(February), 126–130. https://doi.org/10.1002/eco

Taylor, S. D., He, Y., & Hiscock, K. M. (2016). Modelling the impacts of agricultural management practices on river water quality in Eastern England. *Journal of Environmental Management, 180*, 147–163. https://doi.org/10.1016/j.jenvman.2016.05.002

Thilakarathne, M., Sridhar, V., & Karthikeyan, R. (2018). Spatially explicit pollutant load-integrated in-stream *E. coli* concentration modeling in a mixed land-use catchment. *Water Research, 144*, 87–103. https://doi.org/10.1016/j.watres.2018.07.021

USDA-SCS. (1972). *Section 4: Hydrology. In National Engineering Handbook (NEH).* Washington, DC: USDA Soil Conservation Service.

Valentin, C., Agus, F., Alamban, R., Boosaner, A., Bricquet, J. P., Chaplot, V., de Guzman, T., de Rouw, A., Janeau, J. L., Orange, D., Phachomphonh, K., Phai, D. D., Podwojewski, P., Ribolzi, O., Silvera, N., Subagyono, K., Thiébaux, J. P., Toan, T. D., & Vadari, T. (2008). Runoff and sediment losses from 27 upland catchments in Southeast Asia: Impact of rapid land use changes and conservation practices. *Agriculture, Ecosystems and Environment, 128*(4), 225–238. https://doi.org/10.1016/j.agee.2008.06.004

Van Liew, M. W., and J. Garbrecht. (2003). Hydrologic simulation of the Little Washita River Experimental Watershed using SWAT. *Journal of American Water Resources Association, 39*(2), 413–426.

Vinnarasi, R., & Dhanya, C. T. (2016). Journal of Geophysical Research. *Nature, 175*(4449), 238. https://doi.org/10.1038/175238c0

Wang, B., Yim, S. Y., Lee, J. Y., Liu, J. & Ha, K. J. (2014a). Future change of Asian-Australian monsoon under RCP 4.5 anthropogenic warming scenario. *Climate Dynamic.* 42 (1–2): 83–100. https://doi.org/10.1007/s00382-013-1769-x.

Wang, W., Shao, Q., Yang, T., Peng, S., Xing, W., Sun, F., & Luo, Y. (2013). Quantitative assessment of the impact of climate variability and human activities on runoff changes: A case study in four catchments of the Haihe River Basin China. *Hydrological Processes, 27*(8), 1158–1174. https://doi.org/10.1002/hyp

Welde, K., & Gebremariam, B. (2017). Effect of land use land cover dynamics on hydrological response of watershed: Case study of Tekeze Dam watershed, northern Ethiopia. *International Soil and Water Conservation Research, 5*(1), 1–16. https://doi.org/10.1016/j.iswcr.2017.03.002

Williams, J. R., Kannan, N., Wang, X., Santhi, C., & Arnold, J. G. (2012). Evolution of the SCS runoff curve number method and its application to continuous runoff simulation. *Journal of Hydrologic Engineering, 17*(11), 1221–1229. https://doi.org/10.1061/(asce)he.1943-5584.0000529

Young, R. A., Onstad, C., Bosch, D., & Anderson, W. (1989). AGNPS: A nonpoint-source pollution model for evaluating agricultural watersheds. *Journal of Soil and Water Conservation.* 44 (2), 168–173.

Zhang, L., Nan, Z., Yu, W., & Ge, Y. (2016). Hydrological responses to land-use change scenarios under constant and changed climatic conditions. *Environmental Management, 57*(2), 412–431. https://doi.org/10.1007/s00267-015-0620-z

Chapter 9

Mach number impact on Richtmyer-Meshkov instability in shock-refrigerant-22 bubble interaction

Satyvir Singh

Nanyang Technological University

Mukesh Kumar Awasthi

Babasaheb Bhimrao Ambedkar University

CONTENTS

9.1 INTRODUCTION

A corrugated interface forms between two different fluids when they are exposed to a shock wave, according to the Richtmyer-Meshkov (RM) instability (Richtmyer, 1960; Meshkov, 1969). The interface deforms continually following the shock passage caused by baroclinic and pressure perturbation phenomena. Due to the formation of enormous small-scale vortices and subsequent increase in secondary instabilities, turbulent mixing results. The RM instability occurs in a wide range of man-made and natural occurrences when impulsively generated flows occur (Brouillette, 2002; Zhou, 2017a, b). This instability causes mixing between the inner fuel and the outer shell material in inertial confinement fusion (ICF), limiting final compression and lowering energy gain (Lindl et al., 1992). The RM instability can be utilized in air-breathing supersonic and hypersonic engines to promote fuel and oxidizer mixing (Yang and Zukoski, 1993). As a result, understanding the evolution of RM instability is critical.

DOI: 10.1201/9781003367420-9

203

The analysis of shock-bubble interaction is one of the key research topics for shedding light on the complicated flow structure of RM instability. Much research has been undertaken to examine the advancement of RM instability in the process of shock-bubble interaction of various configurations, containing various gases, from experimental, numerical, and theoretical perspectives. Various experimental results and a greater understanding of shock-bubble interaction were accomplished by Haas and Sturtevant (1987), Jacobs (1992), Layes et al. (2009), Ranjan et al. (2007), Zhai et al. (2011), as well as many others, using progressively improved experimental methodologies, after the important research work of Markstein (1957).

Numerous outstanding numerical approaches have been developed to provide a clearer picture of the intricate shock-bubble interactions by Rudinger and Somers (1960), Quirk and Karni (1996), Zabusky and Zeng (1998), Bagabir and Drikakis (2001), Giordano and Burtschell (2006), Niederhaus et al. (2008), and many more. On the other hand, Picone and Boris (1988), Yang et al. (1988), Samtaney and Zabusky (1994), and Li et al. (2019) developed distinct circulation models based on theoretical predictions of the velocity circulation to be used in various conditions.

In the recent development of shock-bubble interaction, the Atwood number effect on the growth of RM instability induced by the interaction of a shock wave with a square bubble was explored numerically by Singh (2020). The dynamics of a shock-heavy cylindrical bubble interaction in both diatomic and polyatomic gases were studied numerically by Singh and Battiato (2021). Subsequently, Singh et al. (2021) examined the bulk viscosity effects on the flow field of a shock-light cylindrical bubble interaction. Further, this study was extended to the interaction of a shock wave and light helium square bubble by Singh (2021a). Recent numerical simulations of the RM instability of the heavy SF_6 square bubble in diatomic and polyatomic gases were published by Singh and Battiato (2022a).

The shock Mach number is a significant regulating parameter for the exploration of flow-mixing events in RM instability studies. Bagabir and Drikakis used a numerical model to examine the impacts of shock Mach numbers on the shock-light bubble interaction. When Zhu et al. (2018) used numerical simulations to examine the impact of different incoming shock Mach numbers on the flow fields of a shock-SF_6 bubble interaction, they discovered that the gas bubble deforms differently depending on the shock Mach number. Singh (2021b) recently looked at the impact of the shock Mach number on the development of the RM instability brought on by the shock-helium square bubble interaction. Singh and Battiato (2022b) expanded on this research to examine the impact of shock Mach number on the convergent RM instability in a square heavy gas cylinder.

In this study, we present the computational modeling of RM instability in shock-refrigerant 22 (R_{22}) bubble interaction and examined the flow characteristics under the shock Mach number effects (M_s=1.15, 1.27, 1.5, and 1.9).

For this purpose, a two-dimensional unsteady system of compressible multi-component Euler flows is simulated with a high-order modal discontinuous Galerkin approach. Impacts of the shock Mach number on wave evolution, shock-focusing, jet formation, bubble deformation, vorticity production, and enstrophy evolution are highlighted. We organize this chapter as follows: Section 5.2 outlines the numerical method, validation of solver, and computational setup. Section 5.3 discusses the Mach number effects on the shock-R_{22} bubble interaction in depth. Finally, in Section 5.4, some concluding remarks are made, along with suggestions for further research on this subject.

9.2 NUMERICAL METHOD AND VALIDATION

9.2.1 Governing equations

In this work, a two-dimensional system of compressible Euler equations for an ideal binary gas mixture is adopted, which can be written in the conservative form (Marquina and Mulet, 2003) as

$$\frac{\partial U}{\partial t} + \frac{\partial}{\partial x}F(U) + \frac{\partial}{\partial y}G(U) = 0, \tag{9.1}$$

With

$$U = \begin{bmatrix} \rho \\ \rho u \\ \rho v \\ \rho E \\ \rho \phi \end{bmatrix}, \quad F(U) = \begin{bmatrix} \rho u \\ \rho u^2 + p \\ \rho uv \\ (\rho E + p)u \\ \rho u\phi \end{bmatrix}, \quad G(U) = \begin{bmatrix} \rho v \\ \rho uv \\ \rho v^2 + p \\ (\rho E + p)v \\ \rho v\phi \end{bmatrix}.$$

In the above expression, ρ denotes the density, u and v represent the velocity components in x- and y- directions, respectively, E signifies the total energy, ϕ indicates the mass fraction, and p refers to the static pressure determined by the ideal gas law

$$p = (\gamma_{mix} - 1)\left(\rho E - \frac{1}{2}(u^2 + v^2)\right), \tag{9.2}$$

where γ_{mix} represents the specific heat ratio of a gas mixture, which is evaluated from the following mathematical expression (Marquina and Mulet, 2003) as

$$\gamma_{mix} = \frac{C_{p1}\phi + C_{p2}(1-\phi)}{C_{v1}\phi + C_{v2}(1-\phi)}. \tag{9.3}$$

Here, subscripts 1 and 2 are meant for bubble and ambient gas, respectively. C_p and C_v represents the specific heat coefficients at constant pressure and volume, respectively.

A high-order explicit modal discontinuous Galerkin method is applied for solving the two-dimensional, unstable system of compressible Euler equations for multi-component gas flows (Singh, 2022). The considered computational domain is partitioned into non-overlapping uniform rectangular meshes. For both the volume and surface integrations, the Gauss-Legendre quadrature rule is used. Using the HLLC technique for two-component flows, the numerical fluxes at the elemental interfaces are estimated in this work (Abgrall and Karni, 2001). The finite element space solutions are approximated using the scaled Legendre polynomial expansion with third-order precision. An explicit third-order precise SSP Runge-Kutta scheme is applied to conduct the time discretization (Singh, 2018; Singh et al., 2022).

9.2.2 Validation study

An excellent comparison between the experimental results of Ding et al. (2018) and the current results are shown in Figure 9.1 to demonstrate the validation of the employed modal DG solver. In this validation, a weak moving shock with the strength of $M_s=1.20$ interacts with a stationary cylindrical SF_6 gas bubble, which is enclosed by air. The numerical findings demonstrate that the square bubble is continuously distorted by shock waves that are propagating and being reflected, and they also demonstrate the identical beginning condition, resolution, shock wave evolution, and thickness of the diffusion layer. The experimental and current results exhibit good qualitative agreement, demonstrating the dependability of the current numerical solver.

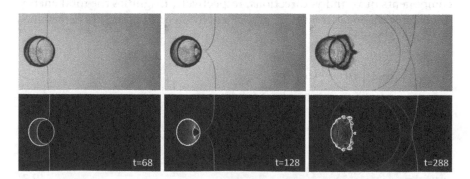

Figure 9.1   Comparison of numerical schlieren images between the experimental results (upper) and present results (bottom) at different time instants.

9.2.3 Computational setup

Figure 9.2 displays the computational setup of the shock-bubble interaction problem. A rectangular with dimensions of $170 \times 90 \, mm^2$ is chosen for the computational domain. A stationary cylindrical bubble collides with an incident shock (IS) wave driving in the $x-$ direction from left to right sides. For simulations, four shock Mach numbers with $M_s = 1.15$, 1.27, 1.50, and 1.90 are selected, which represent the strength of propagating IS wave. From the left-hand edge of the computational domain, the shock wave is initially positioned at $x = 30 \, mm$. The radius of gas bubble is considered as $r = 25 \, mm$, while the center of gas bubble is fixed at $(x_c, y_c) = (65, 45) \, mm$. Around the gas bubble, the initial pressure and temperature are taken to be $P_0 = 101,325$ Pa, and $T_0 = 273$ K, respectively. Nitrogen gas surrounds the heavy refrigerant 22 (R_{22}) gas that fills the cylinder bubble. The left boundary is designated as the inflow, and the upper, bottom and right borders serve as the outflow. We use the ambient state on the right-hand side of the incident shock wave to start the numerical simulation. The Rankine-Hugoniot conditions are used to determine primitive variables at the shock wave's left side.

9.2.4 Mesh refinement analysis

To precisely represent the interface flow structure, a mesh refinement analysis is executed by considering one test case study on a shock-bubble interaction with $M_s = 1.27$. In this regard, the domain is partitioned by five distinct rectangular meshes with 200×100, 400×200, 800×400, 1200×600, and 1400×700, respectively. To compare the influence of mesh resolutions on the computational outcomes, these five different meshes were tested. The density profiles along the center-line of the bubble interface are shown in Figure 9.3 with five different mesh resolutions. The outcomes demonstrate that the density dissipations decrease as mesh resolution rises. Therefore, the computational mesh with grid points 1200×600 is considered in the following numerical tests to ensure numerical accuracy.

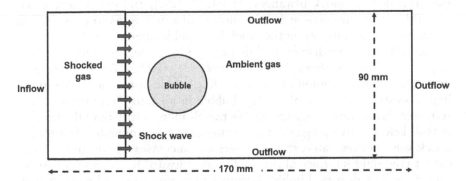

Figure 9.2   Computational setup of shock-bubble interaction problem.

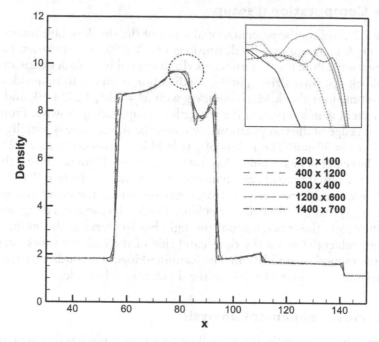

Figure 9.3   Grid refinement study with five different grid points: density profiles along the centreline of the computed bubble.

9.3 RESULTS AND DISCUSSION

This section analyses how the shock-R_{22} bubble interaction flow characteristics are affected by the shock Mach number. The effects of an initial interface disruption on the gas bubble shape, wave evolutions, vorticity generation, and time evolution of enstrophy are highlighted. Four different incoming shock Mach numbers, including one light shock case ($M_s=1.15$), two intermediate shock instances ($M_s=1.27$, 1.50), and one strong shock case ($M_s=1.90$), are chosen for the numerical simulations to examine the impact of Mach numbers on the shock-R_{22} bubble interaction.

Figure 9.4 uses the density, and density gradient contours and shows the impact of incoming shock Mach numbers ($M_s=1.15$, 1.27, 1.5, and 1.9) on the time flow evolution of the shock-R_{22} bubble interaction. Due to the high acoustic impedance of the R_{22} bubble in all four circumstances, the refracted shock wave inside the bubble travels more slowly than the IS wave outside before converging near the downstream pole. Immediately after the shock wave concentrates, the local pressure increases significantly. Then, close to the right interface of the bubble, an outward R_{22} jet appears in the centerline. The distorted bubble becomes more complex and more vorticities appear on the bubble surface as well as the outer jet when the reflected transmitted shock wave impinges on it once more. It can be seen that the shock and bubble interact more strongly when the shock Mach number is large. Additionally, the R_{22} bubble distortion grows and becomes more noticeable as the shock Mach number rises. It should be observed that, as shown in Figure 9.4, the situation of $M_s = 1.27$ has the longest outward R_{22} jet structure among the four examples, necessitating more in-depth research. The transmitted shock waves that are reflected when $M_s = 1.50$ and 1.90 are stronger than those that are reflected when $M_s = 1.15$ and 1.27. As a result, the size of the bubbles substantially shrinks, and the resulting wave patterns become more complicated. The strength and size of the

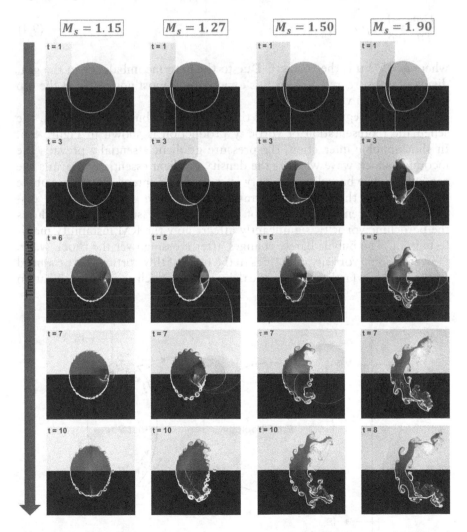

Figure 9.4   Shock Mach number effect on the time evolution of density contour (upper) and numerical schlieren image (bottom) in shock-R_{22} bubble interaction.

wrapped vortices also substantially increase with incoming shock Mach numbers. The baroclinic vorticity deposition produces these vortices, which are discernible at the bubble-to-surrounding-gas border.

An essential process in transport phenomena like turbulent mixing generation is the baroclinic vorticity production term. Due to the imbalance of the density and pressure gradients caused by the passage of the incoming shock wave, this production term is formed in the flow fields and initially dispersed on the bubble interface during the interaction process. The vorticity transport equation can then be written using the baroclinic vorticity production term as follows:

$$\frac{D\omega}{Dt} \equiv \frac{1}{\rho^2} \nabla\rho \times \nabla p, \tag{9.4}$$

where $\omega = \nabla \times \mathbf{u}$ is the vorticity. Due to the extreme imbalance in the gradients of pressure and density, this expression is most noticeable at the top and bottom of a vertical bubble.

During the beginning phase of the shock-heavy bubble interaction, the schematic representation of vorticity production is shown in Figure 9.5. In shock-bubble interaction, the pressure gradient essentially prevails the incoming shock wave whereas the density gradient essentially prevails the gas bubble. The baroclinic vorticity is created and spread along the bubble interface when the shock wave travels through the gas bubble. The discrepancy between the gradients of density and pressure, which produces the baroclinic vorticity, significantly affects how the RM instability manifests itself. The bubble barely changes after crossing over the shock wave. However, some vorticity quantities in the form of tiny vortices are produced at the bubble's farthest left vertical interface, which is also the location

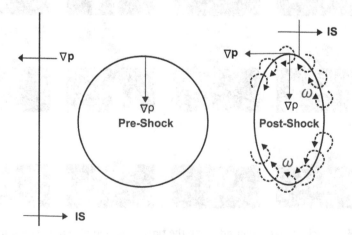

Figure 9.5 Schematic diagram of the vorticity production in shock-heavy bubble interaction.

where the gradients of pressure and density are perfectly aligned. These vortices become the primary characteristic of the shock-bubble interaction flows in later stages.

Figure 9.6 demonstrates the impact of shock Mach number on the contour of vorticity in the shock-R_{22} bubble interaction at $t = 8$. The vorticity is typically zero everywhere at the beginning of an encounter. The baroclinic vorticity is largely localized at the bubble interface where the R22 gas and ambient gas discontinuity arises when the incoming shock wave travels through the bubble. One may notice that the density and pressure gradients are orthogonal at the top and bottom of the bubble, where the vorticity is greatest. At the interface along the bubble's axis, when the density and pressure gradients are collinear, it is zero. Positive and negative vorticity is created in significant amounts on the bubble's top and bottom interfaces. The upper and lower surfaces of the bubble may be seen to have significant amounts of positive and negative vorticity, respectively. The distorted shock wave also generates a little amount of positive and negative vorticity on the upper and lower interfaces during aftershock focusing, with the primary negative (positive) vorticity occurring in the outward jet head on the lower-half (top-half) plane. For the various shock Mach values during the interaction process in shock-bubble interaction, a sizable disparity in vorticity production is seen. A minor amount of vorticity is created nearby the rolled-up vortices on the bubble interface for $M_s = 1.15$. For high Mach values, these rolled-up vortices are more noticeable. In summation, when rolled-up vortices form at large shock Mach numbers, vorticity creation plays a significant role.

To further understand how Mach numbers affect the interaction process, one of the fundamental fields in the shock-bubble interaction—enstrophy—is discussed here. The time evolution of enstrophy is evaluated as

$$\mathrm{Ens}(t) = \frac{1}{2} \int_{\Omega} \omega^2 \, dx \, dy. \tag{9.5}$$

where Ω denotes the considered computed domain.

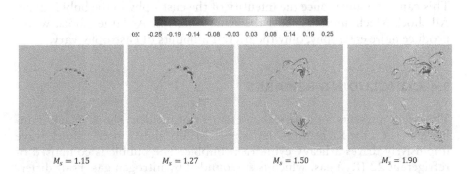

Figure 9.6 Shock Mach number effect on the contours of vorticity distribution in shock-R_{22} bubble interaction at time instant.

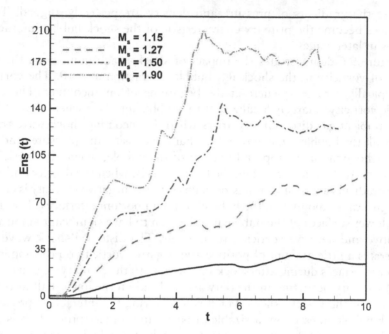

Figure 9.7   Shock Mach number effect on the time evolution of enstrophy in shock-R$_{22}$ bubble interaction.

Figure 9.7 shows how the interaction process of the shock wave and the R$_{22}$ bubble is affected by the incoming shock Mach number on the time evolution of enstrophy. Before the shock wave compasses the upstream pole of the bubble interface, the value of enstrophy is zero. The creation of baroclinic vorticity causes an increase throughout the crossing of the shock wave. The severity of the enstrophy grows as waves from the reflected shock and incoming shock press on the bubble contact. As a result, the increased vorticities encourage the mixing of gases inside and outside the gas bubble, speeding up the transfer and consumption of vorticity energy. This can gradually reduce the intensity of the enstrophy in the bubble zone. All shock Mach numbers show the same behavior. As larger shock waves produce more enstrophy, only the overall amounts of enstrophy vary.

9.4 CONCLUDING REMARKS

This study presents a numerical assessment of incident shock Mach number effects on the Richtmyer-Meshkov (RM) instability after a shock wave impulsively drives a heavy cylindrical bubble. The bubble is composed of refrigerant 22 (R$_{22}$) gas, which is surrounded by nitrogen gas. Four different shock Mach numbers are considered: $M_s = 1.15$, 1.27, 1.50, and 1.90.

To produce high-resolution numerical simulations of the RM instability, we employ a two-dimensional explicit modal discontinuous Galerkin scheme. The numerical outcomes show that the growth of RM instability in a heavy bubble is significantly influenced by the incident shock Mach number. The shock Mach number has a considerable impact on flow morphology, leading to complex wave patterns, shock focusing, the formation of jets, bubble deformation, and the development of vortices. Due to the bubble's converging structure, there is a localized area of high pressure, which results in the generation of an outward jet. As the shock Mach number rises, it is found that the bubble deforms differently. Physical phenomena including the creation of vortices and the evolution of enstrophy are then used to explore the impacts of shock Mach numbers.

REFERENCES

Abgrall, R., & Karni, S. (2001). Computations of compressible multifluids. *Journal of Computational Physics, 169*, 594–623.

Bagabir, A., & Drikakis, D. (2001). Mach number effects on shock-bubble interaction. *Shock Waves, 11*, 209–218

Brouillette, M. (2002). The Richtmyer-Meshkov instability. *Annual Review of Fluid Mechanics, 34*, 445–468.

Ding, J., Liang, Y., Chen, M., Zhai, Z., Si, T., & Luo, X. (2018). Interaction of planar shock wave with three-dimensional heavy cylindrical bubble. *Physics of Fluids, 30*, 106109.

Giordano, J., & Burtschell, Y. (2006). Richtmyer-Meshkov instability induced by shock-bubble interaction: Numerical and analytical studies with experimental validation. *Physics Fluids, 18*(3), 036102.

Haas J.F., & B. Sturtevant, B. (1987). Interaction of weak shock waves with cylindrical and spherical gas inhomogeneities. *Journal of Fluid Mechanics, 181*, 41.

Jacobs, J.W. (1992). Shock-induced mixing of a light-gas cylinder. *Journal of Fluid Mechanics, 234*, 629–649.

Layes, G., Jourdan, G., & Houas, L. (2009). Experimental study on a plane shock wave accelerating a gas bubble. *Physics of Fluids, 21*, 074102.

Li, D., Wang, W., & Guan, B. (2019). On the circulation prediction of shock-accelerated elliptical heavy gas cylinders. *Physics Fluids, 31*, 056104.

Lindl, J.D., McCrory, R.L., & Campbell, E.M. (1992). Progress toward ignition and burn propagation in inertial confinement fusion. *Physics Today, 45*, 32.

Markstein, G.H. (1957). A shock-tube study of flame front-pressure wave interaction. *Symposium on Combustion, 6*(1), 387–398.

Marquina, A., & Mulet, P. (2003). A flux-split algorithm applied to conservative models for multicomponent compressible flows. *Journal of Computational Fluids, 185*, 120–138.

Meshkov, E.E. (1969). Instability of the interface of two gases accelerated by a shock wave. *Fluid Dynamics, 4*(5), 101–104.

Niederhaus, J.J., Greenough, J. A., Oakley, J.G., Ranjan, D., Anderson, M.H., & Bonazza, R. (2008). A computational parameter study for the three-dimensional shock-bubble interaction. *Journal of Fluid Mechanics*, *594*, 85.

Picone, J.M., & Boris, J.P. (1988). Vorticity generation by shock propagation through bubbles in a gas. *Journal of Fluid Mechanics*, *189*, 23–51.

Quirk J.J., & Karni, S. (1996). On the dynamics of a shock–bubble interaction. *Journal of Fluid Mechanics*, *318*, 129–163.

Ranjan, D., Niederhaus, J.H.J., Motl, B., Anderson, M.H., Oakley, J., & Bonazza, R. (2007). Experimental investigation of primary and secondary features in high-Mach-number shock-bubble interaction. *Physical Review Letters*, *98*, 024502.

Richtmyer, R.D. (1960). Taylor instability in shock acceleration of compressible fluids. *Communications on Pure and Applied Mathematics*, *13*, 297.

Rudinger, G., & Somers, L.M. (1960). Behaviour of small regions of different gases carried in accelerated gas flows. *Journal of Fluid Mechanics*, *7*(2), 161–176.

Samtaney, R., & Zabusky, N.J. (1994). Circulation deposition on shock-accelerated planar and curved density-stratified interfaces: Models and scaling laws. *Journal of Fluid Mechanics*, *269*, 45–78.

Singh, S. (2018). Development of a 3D discontinuous Galerkin method for the second-order Boltzmann-Curtiss based hydrodynamic models of diatomic and polyatomic gases. (Doctoral dissertation, Gyeongsang National University South Korea. Department of Mechanical and Aerospace Engineering.)

Singh, S. (2020). Role of Atwood number on flow morphology of a planar shock-accelerated square bubble: A numerical study. *Physics of Fluids*, *32*, 126112.

Singh, S. (2021a). Numerical investigation of thermal non-equilibrium effects of diatomic and polyatomic gases on the shock-accelerated square light bubble using a mixed-type modal discontinuous Galerkin method. *International Journal of Heat and Mass Transfer*, *169*, 121708.

Singh, S. (2021b). Contribution of Mach number on the evolution of Richtmyer-Meshkov instability induced by a shock-accelerated square light bubble. *Physical Review Fluids*, *6*, 044001.

Singh, S. (2022). An explicit modal discontinuous galerkin approach for compressible multicomponent flows: application to shock-bubble interaction. In *Computing and Simulation for Engineers*, edited by Ziya Uddin, Mukesh Kumar Awasthi, Rishi Asthana, & Mangey Ram. CRC Press, Boca Raton, FL.

Singh, S., & Battiato, M. (2021). Behavior of a shock-accelerated heavy cylindrical bubble under nonequilibrium conditions of diatomic and polyatomic gases. *Physical Review Fluids*, *6*, 044001.

Singh, S., & Battiato, M. (2022a). Numerical simulations of Richtmyer-Meshkov instability of SF6 square bubble in diatomic and polyatomic gases. *Computers and Fluids*, *242*, 105502.

Singh, S., & Battiato, M. (2022b). Numerical investigation of shock Mach number effects on convergent Richtmyer-Meshkov instability in a heavy square bubble. *Physica D: Nonlinear Phenomena*, (under review).

Singh, S., Battiato, M. & Myong, R.S. (2021). Impact of bulk viscosity on flow morphology of shock-accelerated cylindrical light bubble in diatomic and polyatomic gases. *Physics of Fluids*, *33*, 066103.

Singh, S., Karchani, A., Chourushi, T., & Myong, R.S. (2022). A three-dimensional modal discontinuous Galerkin method for second-order Boltzmann-Curtiss constitutive models of rarefied and microscale gas flows. *Journal of Computational Physics, 457,* 111052.

Yang, J., Kubota, T., & Zukoski, E.E. (1988). A model for characterization of a vortex pair formed by shock passage over a light-gas inhomogeneity. *Journal of Fluid Mechanics, 258,* 217–244.

Yang, T.K., & Zukoski, E. (1993). Applications of shock-induced mixing to supersonic combustion. *AIAA Journal, 31,* 854–862.

Zabusky, N.J., & Zeng, S.M. (1998). Shock cavity implosion morphologies and vortical projectile generation in axisymmetric shock-spherical fast/slow bubble interactions. *Journal of Fluid Mechanics, 362,* 327–346.

Zhai, Z., Si, T., Luo, X., & Yang, J. (2011). On the evolution of spherical gas interfaces accelerated by a planar shock wave. *Physics Fluids, 23,* 084104.

Zhou, Y. (2017a). Rayleigh-Taylor and Richtmyer-Meshkov instability induced flow, turbulence, and mixing. I. *Physics Reports, 720,* 1–136.

Zhou, Y. (2017b). Rayleigh-Taylor and Richtmyer-Meshkov instability induced flow, turbulence, and mixing. II. *Physics Reports, 723,* 1–60.

Zhu, Y., Yang, Z., Pan, Z., Zhang, P., & Pan J. (2018). Numerical investigation of shock-SF6 bubble interaction with different Mach numbers. *Computers & Fluids, 177,* 78.

Chapter 10

One-dimensional weak shock wave in generalized magnetogasdynamics

Akmal Husain

University of Petroleum and Energy Studies

Vijai Krishna Singh

Institute of Engineering & Technology

Syed Aftab Haider

Shia P. G. College

CONTENTS

10.1 INTRODUCTION

In scientific fields as well as in many processes related to our environment, we can see examples of differential equations quite easily. Most of the abstract and physical processes occurring around us are non-linear in nature and in general mathematically can be modelled by nonlinear partial differential equations. In the context of theoretical and experimental concerns, such nonlinear mathematical models play a capacious role in the fields of space science research, oceanography, continuum mechanics, etc. Many complex physical phenomena abound in the areas of astrophysics, plasma physics, gas dynamics, etc. are mathematically modelled as quasi-linear hyperbolic systems of nonlinear PDEs. Because of the many physical and natural intricacies involved, finding exact solutions for systems of this type of non-linear hyperbolic equations is a challenging and difficult task, and only in some special cases, we can explicitly determine the solutions.

DOI: 10.1201/9781003367420-10

Shock wave phenomenon in continuum mechanics is a dynamic process in which discontinuous waves propagate and this propagation occurs due to the sudden release of the enormous amount of energy in a very short period; e.g. streamer discharges, lighting propagation (natural spark discharge), transient luminous events etc. It is generally observed that the process of shock is caused by high temperature in which ionization of gases takes place, so the influence of the magnetic field becomes very significant. It is a well-known fact that the magnetic field extends throughout the universe, so the study of the effect of magnetic field on non-linear wave propagation phenomena and resultant characteristic flow fields in various dynamical mediums becomes very important as it includes many applications in the fields of hypersonic aerodynamics, atmospheric science, nuclear science, astrophysics etc.

Keeping in view the importance of the magnetic field in various phenomena encountered in the field of gas dynamics, the study of wave propagation under the influence of a magnetic field has immense importance for theoretical and experimental aspects. It has been intensively studied by many researchers to better understand the effects of magnetic fields on the dynamics of shock waves [See; Bazer and Ericson (1959), Bazer and Fleischman (1959), Kanwal (1960), Murray (1961), Greenspan (1962), Helliwell (1963), Ferraro and Plumpton (1966), Gordan and Helliwell (1967), Friedel (1967) and reference cited therein]. In keeping with other authors (Bazer and Ericson, 1959; Bazer and Fleischman, 1959; Kanwal, 1960; Murray, 1961; Greenspan, 1962; Helliwell, 1963; Ferraro and Plumpton, 1966; Gordan and Helliwell, 1967; Friedel, 1967), Ray (1971) considered the explosion problem in an ideal magnetogasdynamic medium of constant density and pressure and constitute a closed form set of non-similarity solutions. Bertram (1973) investigated the downstream state problem from a fixed magnetogasdynamic shock and obtained an analytical solution consisting of a single strength parameter for the downstream state and fixed parameters of the upstream state. Rosenau and Frankenthal (1976) discussed the global features of the self-similar solution for the equatorial propagation of axisymmetric MHD shock waves permeated by an azimuthal magnetic field. Sharma (1981) reported the impact of the transverse magnetic field on the growth and decay behavior of weak discontinuities and discusses the various possibilities for the evolution of waveforms. Sharma et al. (1987) used asymptotic analysis technique and examine the effect of axial magnetic field on the growth and decay behavior of a sawtooth profile. They reported that the magnetic field greatly affected the decay behavior of weak waves in cylindrically symmetric flow in comparison of identical flow. Nath (1989) obtained a similar solution for a theoretical model of MHD shock waves with cylindrical geometry including monochromatic radiation. Radha and Sharma (1995) presented a unified linear and non-linear analysis for small amplitude discontinuities in a perfectly conducting and radiating medium and observe that in contrast to linear analysis, non-linear solution shows wave distortion and consequent shock formation. By using the theory of Majda-Rosales for resonant interaction, Ali and Hunter (1998) described resonant triads of magnetohydrodynamic waves. Taylor and Cargill (2001) presented a general theory of self-similar expansion waves in MHD flows and, obtained an analytical solution for flow-field orientation and slow and fast wave modes. Lock and Mestel (2008) demonstrated a finite annular z-pinch solution to the equation of ideal plasma and ideal magnetogasdynamic plasma. They found that because of the choice of initial conditions (due to flux freezing) numerical solutions of the full ideal MGD equation are somewhat resistive. Genot (2009) solved the system of anisotropic MHD equations describing shock transitions and obtained an explicit expression for the density compression ratio for shock configuration and arbitrary upstream parameters.

In all above aforementioned studies, most of the investigations concerned the propagation dynamics of shock waves in a medium supplemented with perfect gas while generally the problems associated with shock propagation occur under high pressure or low temperature therefore the non-ideal medium model explains the non-linear wave problem better than the ideal gas medium model. With the increasing strength of shock waves, non-ideal gas effects become very prominent and play a vital role in theoretical and experimental investigations. Over the past few decades, several investigations have been devoted to discuss the propagation dynamics of shock waves in real fluids under the action of the magnetic field. Singh et al. (2011a) formulated the shock propagation problem in a non-ideal medium under the influence of a transverse magnetic field and, using the self-similar method, concluded that as compared to isothermal flow, the impact of magnetic field on the flow variables is more significant in adiabatic flow. Singh et al. (2014) obtained the characteristic solution for the problem of leading characteristic in a non-ideal media influenced by a magnetic field and noticed that despite initial strength, all compression waves terminates to a shock wave for identical and cylindrical geometry. By using the asymptotic approach method Nath et al. (2017) derived a growth equation which characterizes the disturbances in a high-frequency domain and also assessed how the magnetic field and parameter of non-idealness affected the wave phenomenon in the form of a sawtooth profile. The convergence of the cylindrical shock problem at the axis of symmetry in a non-ideal magnetogasdynamic flow has been investigated by Chauhan et al. (2018) by applying the perturbation-series technique method. For a one-dimensional non-ideal quiescent relaxing media, Shah and Singh (2019) obtained the evolution equations for weak and strong shock in non-planar geometry by using a systematic approximation method. Singh and Arora (2021a) determined the first and second-order approximate solutions to the problem of propagation of cylindrical shock in a magnetic field. Their result shows that a significant enhancement of magnetic field causes a decrease in pressure and density variable while an increase in magnetic pressure and fluid velocity.

Nath and Devi (2021) applied the lie group invariance method to obtain a similar solution of shock waves in a rotating axisymmetric non-ideal gas in a generalized magnetic field under the assumption that the azimuthal and axial velocities and magnetic field to be varying according to a power law. Singh and Arora (2021b) investigated the evolution and steepening behaviour of shocks in a simplified van der Waals radiating gas and studied the imperative effect of real gases on the flow system. With the help of modified shock-jump conditions, Avramenko et al. (2022) represented an analytical analysis of shock dynamics in a non-ideal medium which adequately describes the dynamics of shock propagation in real gaseous media.

Since the gas-dynamical problems are modelled by a system of quasi-linear partial differential equations, so the main complexity for finding a solution to such systems is the existence of discontinuity in the solution. In recent years, analytical solutions of gas-dynamic flow equations in various gas dynamic regimes have been a subject of discussion among investigators and by using different analytical perspectives, many precise solutions have been given by several authors [See; Oliveri and Speciale (2005), Singh et al. (2010), Ram et al. (2013), Husain et al. (2020, 2021) and reference cited therein]. Within the framework of these references, the precise explicit solution was given for strong discontinuities however for weak shock waves within the context of magnetic field no such studies have been performed. Singh et al. (2011b) obtained a particular solution of weak discontinuity waves in ideal gasdynamic flow via a straightforward technique proposed by Murata (2006) and examine that the class of the flow includes the instantaneous constant energy for a particular region. Later, Chaudhary and Singh (2018) extended the work of Singh et al. (2011b) and reported the closed-form solution of compressible quasi-one dimensional weak shock problem in a non-ideal adiabatic gas flow. Nath (2021) pointed out the problem of propagation of ionizing cylindrical shock in a rotational axisymmetric real gas flow in the presence of a magnetic field and provided first and second-order approximate analytical solutions by using perturbation techniques. Husain et al. (2022) considered converging weak shock problem in a mixture of non-ideal gas filled with spherical amorphous solid particles and explained the behaviour of radius of shock front and total energy with respect to time.

In this chapter, we successfully constructed the analytical solution for a weak shock wave problem for cylindrically symmetric flow in non-ideal magnetogasdynamics and investigated how the magnetic field and non-ideal parameter affects the physical characteristics of the flow (density, velocity, pressure and magnetic pressure) and total energy which exhibiting dependency on space-coordinate and time. Furthermore, the problem is formulated under the assumption that the mass density distribution varies in accordance with the power law of the radial distance from the explosion point. To obtain the precise solution using a straightforward analytical technique proposed by Murata (2006), we tried to fill the gap of analytical solution associated with weak shock problem in a non-ideal medium under the influence of axial and azimuthal magnetic field because there is no literature available that provides the analytical solution of weak shock problem in non-ideal magnetogasdynamics using Murata's technique. The effect of Mach number, magnetic field and non-ideal parameter on the behavior of the total energy carried by the weak shock wave with respect to time has been analyzed via illustrated figures.

10.2 MATHEMATICAL MODEL AND SHOCK-JUMP BOUNDARY CONDITIONS

The fundamental set of compressible hyperbolic Euler's conservation equations (mass balance, momentum balance, energy and magnetic pressure balance) that governs the one-dimensional non-viscous unsteady cylindrically symmetric adiabatic motion of a van der Waals gas under the influence of generalized magnetic field (axial and azimuthal) can be written in a conservative form as [Pai (1963), Zel'dovich and Raizer (1967), Whitham (1974)]

$$\frac{\partial F(U)}{\partial t} + \frac{\partial G(U)}{\partial x} = B(U), x > 0, t > 0,$$ (10.1)

where

$U = (\rho, u, p, h)^{tr}$ and the vectors F, G, and B are given by:

$$F = \left(\rho, \rho u, \frac{(1-b\rho)p}{(\gamma-1)} + \frac{\rho u^2}{2} + h, (h)^{\frac{1}{2}} \right)^{tr},$$

$$G = \left(\rho u, \rho u^2 + p + h, \frac{(\gamma - b\rho)pu}{(\gamma-1)} + \frac{\rho u^3}{2} + 2uh, u(h)^{\frac{1}{2}} \right)^{tr},$$

$$B = \left(-\frac{\rho u}{x}, -\frac{\rho u^2}{x} - \frac{2nh}{x}, \frac{(\gamma-b\rho)}{(\gamma-1)}\frac{pu}{x} - \frac{\rho u^3}{2x} - \frac{2nhu}{x} - \frac{2(1-n)hu}{x}, -\frac{(1-n)u(h)^{\frac{1}{2}}}{x} \right)^{tr},$$

Here, $t > 0$ is the time, $x \in \iota$ is the cylindrical radial coordinate, $\rho(x,t)$ denotes the density, $u(x,t)$ is the velocity in x direction, $p(x,t)$ is the pressure, the magnetic pressure h defined as $h = \mu H^2/2$ with μ as the magnetic permeability and H the transverse magnetic field. The entity $\gamma = c_p/c_v$ is the adiabatic exponent (lies between 1 and 2) and thermodynamic constants c_p and c_v denotes the constant pressure-specific heat, constant volume-specific heat respectively. The constant, $n = 0,1$ exhibits axial and azimuthal magnetic fields, respectively. The system for compressible ideal gas dynamics for cylindrically symmetric flow can be recovered by setting $h \equiv 0$ in the fundamental set of system (10.1).

For motion in non-ideal gas dynamic media, the van der Waals gas equation (equation of state) has been taken in the following form:

$$p = \frac{\rho \Re T}{(1 - b\rho)}, \tag{10.2}$$

which quite accurately describes the state of real gases and the parameters of intermolecular interactions b in them. Here, $\Re$ is the specific gas constant and T is the temperature.

Usually to study analytic solutions of the quasi-one-dimensional basic set of Equation (10.1), we write the given system (10.1) in primitive variable form, which is as follows

$$\frac{\partial \rho}{\partial t} + u\frac{\partial \rho}{\partial x} + \rho\frac{\partial u}{\partial x} + \frac{\rho u}{x} = 0, \tag{10.3}$$

$$\rho\left(\frac{\partial u}{\partial t} + u\frac{\partial u}{\partial x}\right) + \frac{\partial p}{\partial x} + \frac{\partial h}{\partial x} + \frac{2nh}{x} = 0, \tag{10.4}$$

$$\frac{\partial p}{\partial t} + u\frac{\partial p}{\partial x} + \rho a^2\left(\frac{\partial u}{\partial x} + \frac{u}{x}\right) = 0, \tag{10.5}$$

$$\frac{\partial h}{\partial t} + u\frac{\partial h}{\partial x} + 2h\left(\frac{\partial u}{\partial x} + \frac{(1-n)u}{x}\right) = 0, \tag{10.6}$$

where "a^2" represents the square of the equilibrium sound speed defined as $a^2 = \dfrac{\gamma p}{\rho(1 - b\rho)}$. Let us suppose that in the inhomogeneous medium given by

$$u_0 = 0, \; p_0 = \text{constant}, \; \rho_0 = \rho_0(x) \text{ and } h_0 = h_0(x), \tag{10.7}$$

the shock propagation velocity is $V = \dfrac{dR}{dt}$, where $R = R(t)$ be the shock-location at time t. Here and throughout the numeric subscript ' 0 'with respect to flow parameters indicates the values in the undisturbed region. It is also assumed that in order to obtain the exact solution, the weak shock propagation obeys the power law given as

$$\rho_0 = \rho_\xi R^\theta,$$

in which the density profile of the undisturbed region ρ_0 varies with respect to the radius of shock. Here, ρ_ξ denotes a dimensional constant, whereas θ be a constant whose value is to be determined in further analysis.

As we know that the boundary conditions across the shock front are basically a functional dependence of the flow parameters before the shock, known as Rankine-Hugoniot jump conditions. Hence, the R-H jump conditions for describing the flow across the shock front are given as follows:

$$\rho = \frac{(\gamma+1)}{(\gamma-1)}\left\{1 - \frac{2\beta}{(\gamma-1)} - \frac{2}{(\gamma-1)}\frac{1}{M^2}\right\}\rho_0,$$ (10.8)

$$u = \frac{2}{(\gamma+1)}\left\{(1-\beta) - \frac{1}{M^2}\right\}V,$$ (10.9)

$$p = \frac{2}{(\gamma+1)}\left\{(1-\beta) - \frac{(\gamma-1)}{2\gamma}\frac{1}{M^2}\right\}\rho_0 V^2$$
$$- \frac{1}{2}\left\{\frac{(\gamma+1)}{(\gamma-1)}\right\}^2\left\{1 - \frac{4\beta}{(\gamma-1)} - \frac{4}{(\gamma-1)}\frac{1}{M^2}\right\}C_0\rho_0 V^2,$$ (10.10)

$$h = \frac{1}{2}\left\{\frac{(\gamma+1)}{(\gamma-1)}\right\}^2\left\{1 - \frac{4\beta}{(\gamma-1)} - \frac{4}{(\gamma-1)}\frac{1}{M^2}\right\}C_0\rho_0 V^2,$$ (10.11)

where $\beta = (\gamma-1)b\rho_0$ and $C_0 = \frac{2h_0}{\rho_0 V^2}$.

It is well known that the energy produced during the propagation of a weak shock wave in any gas medium is equal to the sum of the kinetic energy and internal energy of the gas and is constant. The balance equation for the total energy ε in real magneto-gasdynamic media supplemented with the simplified van der Waals equation of state is given by

$$\varepsilon = \int_0^R\left[\frac{1}{2}u^2 + \frac{1}{(\gamma-1)}\left\{\frac{(1-b\rho)p}{\rho} - \frac{(1-b\rho_0)p_0}{\rho_0}\right\} + \left(\frac{h}{\rho} - \frac{h_0}{\rho_0}\right)\right]\rho x\,dx.$$ (10.12)

In view of the relation $\int_0^R \frac{\rho}{\rho_0}x\,dx = \frac{R^2}{2}$ obtained from Lagrangian equation of continuity, Equation (10.12) yields the following expression for total energy

$$\varepsilon = \int_0^R\left[\frac{1}{2}\rho u^2 + \frac{(1-b\rho)p}{(\gamma-1)} + h\right]x\,dx - \frac{p_0(1-b\rho_0)}{(\gamma-1)}\frac{R^2}{2} - h_0\frac{R^2}{2}.$$ (10.13)

10.3 EXACT SOLUTION TO WEAK SHOCK WAVE PROBLEM IN NON-IDEAL MAGNETO-GASDYNAMICS FLOW

In this section to accomplish the current investigation, we obtain an analytical solution for compressible Euler's partial differential equations for a quasi-one-dimensional unsteady non-ideal magneto-gas-dynamic flow described by system (10.3)–(10.6), by using a straight forward analytical approach proposed by Murata (2006).

To find an exact solution, subject to the Rankine-Hugoniot's ratios (10.8)–(10.11), we establish an expression immediately behind the shock front for the physical characteristics 'pressure and magnetic pressure' of the flow, in terms of other thermo-dynamical variables 'density and flow velocity' given as

$$p = \frac{\left[4(\gamma-1)^2\left\{(1-\beta)-\frac{(\gamma-1)}{2\gamma}\frac{1}{M^2}\right\}-C_0(\gamma+1)^3\left\{1-\frac{4\beta}{(\gamma-1)}-\frac{4}{(\gamma-1)}\frac{1}{M^2}\right\}\right]}{8(\gamma-1)\left\{1-\frac{2\beta}{(\gamma-1)}-\frac{2}{(\gamma-1)}\frac{1}{M^2}\right\}\left\{(1-\beta)-\frac{1}{M^2}\right\}^2}\rho u^2,$$

$$\tag{10.14}$$

$$h = \frac{C_0(\gamma+1)^3\left\{1-\frac{4\beta}{(\gamma-1)}-\frac{4}{(\gamma-1)}\frac{1}{M^2}\right\}}{8(\gamma-1)\left\{1-\frac{2\beta}{(\gamma-1)}-\frac{2}{(\gamma-1)}\frac{1}{M^2}\right\}\left\{(1-\beta)-\frac{1}{M^2}\right\}^2}\rho u^2. \tag{10.15}$$

Plugging in relations (10.14) and (10.15) the basic set of fundamental balance Equations (10.3)–(10.6) can be reduced in the form

$$\frac{\partial u}{\partial t}+u\frac{\partial u}{\partial x}-\frac{\frac{(\gamma-1)u}{2}\left\{(1-\beta)-\frac{(\gamma-1)}{2\gamma}\frac{1}{M^2}\right\}}{\left\{1-\frac{2\beta}{(\gamma-1)}-\frac{2}{(\gamma-1)}\frac{1}{M^2}\right\}\left\{(1-\beta)-\frac{1}{M^2}\right\}^2}$$

$$\tag{10.16}$$

$$\left[\left(\frac{1}{\rho}\frac{\partial \rho}{\partial t}-\frac{\partial u}{\partial x}+\frac{u}{x}\right)-\frac{nC_0(\gamma+1)^3\left\{1-\frac{4\beta}{(\gamma-1)}-\frac{4}{(\gamma-1)}\frac{1}{M^2}\right\}u}{2x(\gamma-1)^2\left\{(1-\beta)-\frac{(\gamma-1)}{2\gamma}\frac{1}{M^2}\right\}}\right]=0,$$

$$\frac{\partial u}{\partial t} + u \frac{\partial u}{\partial x} + \frac{\left[(\gamma-1)^2 + b(\gamma+1)\left\{ 1 - \frac{2\beta}{(\gamma-1)} - \frac{2}{(\gamma-1)} \frac{1}{M^2} \right\} p_0 \right] u}{2\left[(\gamma-1) - b(\gamma+1)\left\{ 1 - \frac{2\beta}{(\gamma-1)} - \frac{2}{(\gamma-1)} \frac{1}{M^2} \right\} p_0 \right]}$$

(10.17)

$$\left(\frac{\partial u}{\partial x} + \frac{u}{x} \right) = 0,$$

$$\frac{\partial u}{\partial t} + u \frac{\partial u}{\partial x} + \frac{u}{2} \left\{ \frac{\partial u}{\partial x} + \frac{(1-2n)u}{x} \right\} = 0.$$

(10.18)

In view of Equations (10.16) and (10.17), and performing some algebraic manipulations, we get the resulting equation as

$$S(t) = \rho.(u)^{2-K}.(x)^{N-K},$$

(10.19)

where $S(t)$ denotes the single function of time and value of the constants K and N are given as

$$K = \frac{\left\{ 1 - \frac{2\beta}{(\gamma-1)} - \frac{2}{(\gamma-1)} \frac{1}{M^2} \right\} \left[(\gamma-1)^2 + b(\gamma+1)\left\{ 1 - \frac{2\beta}{(\gamma-1)} - \frac{2}{(\gamma-1)} \frac{1}{M^2} \right\} p_0 \right] \left\{ (1-\beta) - \frac{1}{M^2} \right\}^2}{(\gamma-1)\left\{ (1-\beta) - \frac{(\gamma-1)}{2\gamma} \frac{1}{M^2} \right\} \left[(\gamma-1) - b(\gamma+1)\left\{ 1 - \frac{2\beta}{(\gamma-1)} - \frac{2}{(\gamma-1)} \frac{1}{M^2} \right\} p_0 \right]}$$

and

$$N = \frac{n C_0 (\gamma+1)^3 \left\{ 1 - \frac{4\beta}{(\gamma-1)} - \frac{4}{(\gamma-1)} \frac{1}{M^2} \right\}}{2(\gamma-1)^2 \left\{ (1-\beta) - \frac{(\gamma-1)}{2\gamma} \frac{1}{M^2} \right\}}$$

Using Equation (10.19) and after some analytical steps, the mass-balance Equation (10.1) of compressible Euler's system can be transformed in the form

$$\frac{(2-K)}{u} \frac{\partial u}{\partial t} + (1-K) \frac{\partial u}{\partial x} - (K-N+1)\frac{u}{x} - \frac{1}{S(t)} \frac{dS}{dt} = 0.$$

(10.20)

In solving Equations (10.20) and (10.17), we have

$$u = -W \frac{x}{S(t)} \frac{dS}{dt},$$ (10.21)

where the value of the constant W is given as

$$W = \frac{1}{\left[\{(K+1)+(1-N)\} + (2-K) \dfrac{\left[(\gamma-1)^2 + b(\gamma+1)\left\{ \rho_0 - \dfrac{2\beta\rho_0}{(\gamma-1)} - \dfrac{2}{(\gamma-1)} \dfrac{\rho_0}{M^2} \right\} \right]}{\left[(\gamma-1) - b(\gamma+1)\left\{ \rho_0 - \dfrac{2\beta\rho_0}{(\gamma-1)} - \dfrac{2}{(\gamma-1)} \dfrac{\rho_0}{M^2} \right\} \right]} \right]}$$

By substituting the value of velocity parameter from Equation (10.21) in Equation (10.20) and after simplification we have

$$S(t) = S_0 . (t)^{-P},$$ (10.22)

where S_0 is an arbitrary constant and value of P is given as

$$P = \frac{\left[(K+2-N) + \dfrac{(2-K)\left[(\gamma-1)^2 + b(\gamma+1)\left\{ \rho_0 - \dfrac{2\beta\rho_0}{(\gamma-1)} - \dfrac{2}{(\gamma-1)} \dfrac{\rho_0}{M^2} \right\} \right]}{\left[(\gamma-1) - b(\gamma+1)\left\{ \rho_0 - \dfrac{2\beta\rho_0}{(\gamma-1)} - \dfrac{2}{(\gamma-1)} \dfrac{\rho_0}{M^2} \right\} \right]} \right]}{\left[1 + \dfrac{\left[(\gamma-1)^2 + b(\gamma+1)\left\{ \rho_0 - \dfrac{2\beta\rho_0}{(\gamma-1)} - \dfrac{2}{(\gamma-1)} \dfrac{\rho_0}{M^2} \right\} \right]}{\left[(\gamma-1) - b(\gamma+1)\left\{ \rho_0 - \dfrac{2\beta\rho_0}{(\gamma-1)} - \dfrac{2}{(\gamma-1)} \dfrac{\rho_0}{M^2} \right\} \right]} \right]}.$$

By using the shock-jump condition (10.8) and (10.9), we obtain, explicit expressions for the radius of shock front R and constant θ as

$$R = (t)^{\dfrac{(\gamma+1)\left[(\gamma-1) - b(\gamma+1)\left\{ \rho_0 - \frac{2\beta\rho_0}{(\gamma-1)} - \frac{2\rho_0}{(\gamma-1)M^2} \right\} \right]}{\left\{ 2(1-\beta) - \frac{2}{M^2} \right\}\left[\left[(\gamma-1) - b(\gamma+1)\left\{ \rho_0 - \frac{2\beta\rho_0}{(\gamma-1)} - \frac{2\rho_0}{(\gamma-1)M^2} \right\} \right] + \left[(\gamma-1)^2 + b(\gamma+1)\left\{ \rho_0 - \frac{2\beta\rho_0}{(\gamma-1)} - \frac{2\rho_0}{(\gamma-1)M^2} \right\} \right] \right]}}$$

(10.23)

and

$$\theta = \frac{\left[\begin{array}{l}\left[(2-2P-N)-(4-2K-2P)\left(\beta+\dfrac{1}{M^2}\right)\right]\left[(\gamma-1)-b(\gamma+1)\left\{\rho_0-\dfrac{2\beta\rho_0}{(\gamma-1)}-\dfrac{2}{(\gamma-1)}\dfrac{\rho_0}{M^2}\right\}\right] \\ +2\{2-(K+P)\}\left\{(1-\beta)-\dfrac{1}{M^2}\right\}\left[(\gamma-1)^2+b(\gamma+1)\left\{\rho_0-\dfrac{2\beta\rho_0}{(\gamma-1)}-\dfrac{2}{(\gamma-1)}\dfrac{\rho_0}{M^2}\right\}\right]\end{array}\right]}{(\gamma+1)\left[(\gamma-1)-b(\gamma+1)\left\{\rho_0-\dfrac{2\beta\rho_0}{(\gamma-1)}-\dfrac{2}{(\gamma-1)}\dfrac{\rho_0}{M^2}\right\}\right]}$$

(10.24)

Finally, subject to the boundary jump conditions (10.8)–(10.11), the analytical solution of the governing system formulated in the prior section is given as

$$\rho = \frac{S_0.(x)^{2K-N-2}.(t)^{2-(K+P)}}{(PW)^{2-K}},$$

(10.25)

$$u = \frac{\left[\left[(\gamma-1)-b(\gamma+1)\left\{\rho_0-\dfrac{2\beta\rho_0}{(\gamma-1)}-\dfrac{2\rho_0}{(\gamma-1)M^2}\right\}\right]\right]x}{\left[\left[(\gamma-1)-b(\gamma+1)\left\{\rho_0-\dfrac{2\beta\rho_0}{(\gamma-1)}-\dfrac{2\rho_0}{(\gamma-1)M^2}\right\}\right]+\left[(\gamma-1)^2+b(\gamma+1)\left\{\rho_0-\dfrac{2\beta\rho_0}{(\gamma-1)}-\dfrac{2\rho_0}{(\gamma-1)M^2}\right\}\right]\right]t}$$

(10.26)

$$p = \frac{S_0(x)^{2K-N}\left[4(\gamma-1)^2\left\{(1-\beta)-\dfrac{(\gamma-1)}{2\gamma M^2}\right\}-C_0(\gamma+1)^3\left\{1-\dfrac{4\beta}{(\gamma-1)}-\dfrac{4}{(\gamma-1)M^2}\right\}\right]}{8(\gamma-1)(t)^{(K+P)}\left\{1-\dfrac{2\beta}{(\gamma-1)}-\dfrac{2}{(\gamma-1)M^2}\right\}\left\{(1-\beta)-\dfrac{1}{M^2}\right\}^2(PW)^{-K}},$$

(10.27)

$$b = \frac{S_0(x)^{2K-N}C_0(\gamma+1)^3\left\{1-\dfrac{4\beta}{(\gamma-1)}-\dfrac{4}{(\gamma-1)}\dfrac{1}{M^2}\right\}}{8(\gamma-1)(t)^{(K+P)}\left\{1-\dfrac{2\beta}{(\gamma-1)}-\dfrac{2}{(\gamma-1)}\dfrac{1}{M^2}\right\}\left\{(1-\beta)-\dfrac{1}{M^2}\right\}^2(PW)^{-K}}.$$

(10.28)

By applying the values of the physical flow parameters, the explicit cumbersome solution for total energy ε is given as

$$\varepsilon = \varepsilon_0.(t)^{\frac{PW(\gamma+1)\{2(K+1)-N\}-2(K+P)\left\{(1-\beta)-\frac{1}{M^2}\right\}}{2\left\{(1-\beta)-\frac{1}{M^2}\right\}}},$$

(10.29)

where

$$\varepsilon_0 = S_0$$

$$
\frac{\left[\dfrac{C_0}{(\gamma+1)^{-3}} \left\{ \dfrac{\{(\gamma-1)-4\beta\}M^2-4}{(\gamma-1)M^2} \right\} \left[(\gamma-2) + \dfrac{(\gamma+1)b}{(\gamma-1)} \left\{ \rho_0 - \dfrac{2\beta\rho_0}{(\gamma-1)} - \dfrac{2\rho_0}{(\gamma-1)M^2} \right\} \right] - \right.}{}
$$

$$
\frac{2(\gamma+1)}{(\gamma-1)^{-3}(2K-N+2)^{-1}} \left\{ \dfrac{(1-b\rho_0)^2}{2\gamma(\gamma-1)M^2} + \dfrac{C_0}{4} \right\} + \dfrac{4}{(\gamma-1)^{-2}} \left[\left\{ \dfrac{[(\gamma-1)-2\beta]M^2-2}{(\gamma-1)M^2} \right\} \right.
$$

$$
\left. \left. \left\{ \left(1-\beta\right) - \dfrac{1}{M^2} \right\}^2 - \dfrac{(\gamma+1)b}{(\gamma-1)} \left\{ (1-\beta)\rho_0 - \dfrac{(\gamma-1)\rho_0}{2\gamma M^2} \right\} \right\} + \left\{ (1-\beta) - \dfrac{(\gamma-1)}{2\gamma M^2} \right\} \right]
$$

$$
\overline{8(\gamma-1)^2(2K-N+2)\left\{ \dfrac{[(\gamma-1)-2\beta]M^2-2}{(\gamma-1)M^2} \right\} \left\{ (1-\beta) - \dfrac{1}{M^2} \right\}^2 (PW)^{-K}}
$$

10.4 RESULTS AND DISCUSSION

The formulation of precise and analytical solutions is of great importance in applied mathematics and mathematical physics as these solutions are very helpful in understanding and classifying the physical phenomena involved. Equations (10.25)–(10.28) represent the analytical solution of physical flow characteristics (such as density, velocity, pressure and magnetic pressure) and total energy, to the cylindrical symmetric weak shock wave problem in a non-ideal supersonic magnetogasdynamic flow. It is worth pointing out here that the obtained analytical solution for the thermodynamic variables density, velocity and pressure is greatly affected by the magnetic field (axial and azimuthal) and van der Waals excluded volume. However, the solutions obtained for velocity profile and shock front radius remain unchanged under the influence of the magnetic field while the effect of the parameter of non-idealness significantly appears. For sake of convenience, the effect of magnetic field (h) and van der Waals excluded volume (b) has been assessed in the form of Shock-Cowling number (C_0) and parameter of non-idealness (β) respectively. For $C_0 = 0$ and $\beta = 0$, the explicit solution corresponds to ideal gas-dynamical flow. It is worth mentioning here that, the reported results elucidate a closer agreement with the results carried out by many authors in various gas-dynamic regimes using different analytical and numerical approaches.

The impact of Mach number, shock cowling number and van der Waals excluded volume on the cumbersome solution of total energy for the cylindrically symmetric flow of weak shock have been shown in Figures 10.1–10.5 respectively.

Figures 8.1 and 8.2 show the variation of Mach number M on the behavior of total energy carried by a cylindrical weak shock wave in the non-ideal medium with respect to time t for axial $(n = 0)$ and azimuthal $(n = 1)$ magnetic field respectively. For computational work the value of the constants involved has been taken as: $\rho_0 = 1$, $b = 0.0009$, $\beta = 0.0125$ and $C_0 = 0.01$.

From Figures 10.1 and 10.2, it is evident that an increase in the value of Mach number of the energy of the cylindrical weak shock waves under the influence of the magnetic field causes it to increase much faster with respect to time. Moreover, it may remarkable here that for the case of axial magnetic field $(n = 0)$, the values of the total energy with respect to time for varying Mach number enhances much faster as compared to the azimuthal magnetic field $(n = 1)$, which confirms the corresponding theoretical results.

Figure 10.3 shows the variation of Shock-Cowling number C_0 on the behavior of total energy carried by a cylindrical weak shock wave in the non-ideal medium with respect to time t for azimuthal $(n = 1)$ magnetic field. For computational work the value of the constants involved has been taken as: $\rho_0 = 1$, $b = 0.0009$, $\beta = 0.0125$ and $M = 1.30$. From Figure 10.3, it is evident that the energy of the cylindrical weak shock waves increases much slower for an increasing value of Shock-Cowling number with respect to time. Moreover, it is remarkable here that for the case of axial magnetic field $(n = 0)$, for increasing values of Shock-Cowling number the values of the total energy with respect to time are much closer, which makes the conclusion that as compared to the axial magnetic field the effect of the azimuthal magnetic field has considerable impact on the behavior of energy with respect to time.

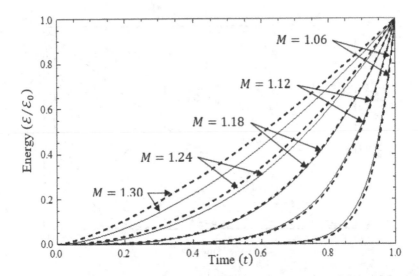

Figure 10.1 Variation of energy $(\varepsilon/\varepsilon_0)$ with time at various values of Mach number (M) for axial magnetic field (Dash line and solid lines corresponds for $\gamma = 1.4$ and $\gamma = 1.67$.)

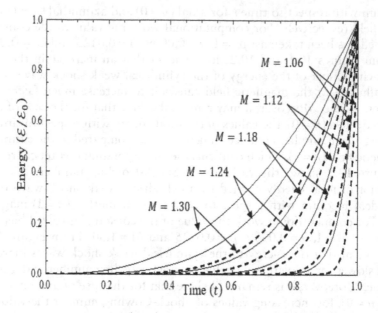

Figure 10.2 Variation of energy $(\varepsilon/\varepsilon_0)$ with time at various values of Mach number (M) for azimuthal magnetic field (Dash line and solid lines corresponds γ = 1.4 and γ = 1.67.)

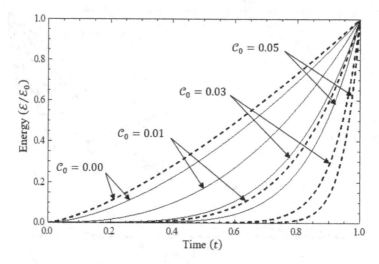

Figure 10.3 Variation of energy $(\varepsilon/\varepsilon_0)$ with time at various values of Shock-Cowling number (C_0) for azimuthal magnetic field (Dash and solid lines corresponds γ = 1.4 and γ = 1.67.)

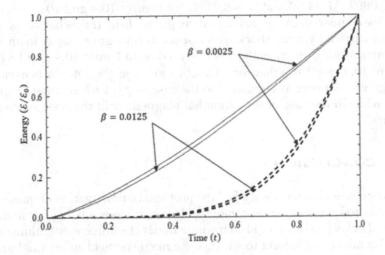

Figure 10.4 Variation of energy $(\varepsilon/\varepsilon_0)$ with time at various values of the non-ideal parameter (β) for adiabatic exponent $\gamma = 1.4$. (Dash and solid lines corresponds $n = 1$ and $n = 0$.)

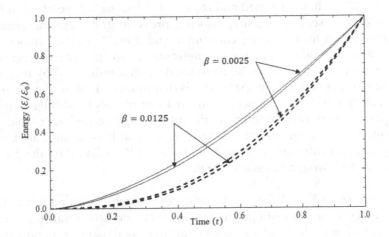

Figure 10.5 Variation of energy $(\varepsilon/\varepsilon_0)$ with time at various values of the non-ideal parameter (β) for adiabatic exponent $\gamma = 1.67$. (Dash and solid lines corresponds $n = 1$ and $n = 0$.)

Figures 10.4 and 10.5 shows the variation of parameter of non-idealness β on the behavior of total energy carried by cylindrical weak shock wave in the non-ideal medium with respect to time t for axial $\{(n = 0) ; \text{Solid line}\}$ and azimuthal $\{(n = 1) ; \text{Dash line}\}$ magnetic field respectively. For computational work the value of the constants involved has been taken as: $\rho_0 = 1$,

$b = 0.0009$, $M = 1.30$ and $C_0 = 0.01$. From Figures 10.4 and 10.5, it is clear that as compared to the azimuthal magnetic field the behavior of total energy carried by weak shock with respect to time grows faster in an axial magnetic field. Consequently, it is observed from Figure 10.4 $(\gamma = 1.4)$ and Figure 10.5 $(\gamma = 1.67)$ that for adiabatic exponent $\gamma = 1.4$, the behavior of energy grows faster as compared to the case of $\gamma = 1.67$ an axial magnetic field while in the case of the azimuthal magnetic field the reverse behavior occurs.

10.5 CONCLUSION

In the present chapter, we studied the problem of propagation of quasi-one-dimensional unsteady state compressible non-viscous adiabatic cylindrical weak shock wave in a non-ideal medium under the influence of infinite electrical conductivity subject to a transverse magnetic field either axial or azimuthal. The considered plasma is assumed to be a simplified van der Waals gas which is more realistic in nature and provides a better insight into propagation dynamics of weak disturbances as compared to the ideal gas model. For this purpose, a simple and efficient analytical technique proposed by Murata (2006) has been used and the exact solutions for the physical flow characteristics such as density, velocity, pressure and magnetic pressure are obtained in form of space-coordinate and time. Also, the cumbersome expression for total energy (sum of kinetic energy and potential energy) carried by weak shock flow in the considered plasma influenced by magnetic field having dependency on space-time is determined. To describe the effect of Mach number, the magnetic field (in form of shock cowling number) and parameter of non-idealness on the total energy carried are graphically presented. All the computations for this purpose have been done by the computational software package MATHEMATICA. Based on the present study, the following conclusions are made:

1. Figures 10.1–10.5 depicts the behavior of energy with respect to time under the influence of magnetic field (axial and azimuthal) and parameter of non-idealness for various varying gas-dynamical parameters.
2. An increase in the value of Mach number the energy of the cylindrical weak shock waves under the influence of the magnetic field causes to increase much faster with respect to time. Also, it is observed that as compared to the azimuthal magnetic field the value of the total energy with respect to time for varying Mach number for axial magnetic field enhances faster.
3. For an increasing value of Shock-Cowling number with fixed values of ρ_0, b, β and M subject to time the energy increases comparatively slower. Subsequently, the azimuthal magnetic field greatly affected the total energy as compared to the axial magnetic field.

4. As compared to the azimuthal magnetic field, the behavior of total energy carried by weak shock with respect to time grows faster in the axial magnetic field. Consequently, the behavior of energy grows faster as compared to the case of $\gamma = 1.67$ in axial magnetic field while in the case of the azimuthal magnetic field the reverse behavior occurs.

REFERENCES

Ali, G., & Hunter, J. K. (1998). Wave interactions in magnetohydrodynamics. *Wave Motion*, 27, 257–277.

Avramenko, A. A., Shevchuk, I. V., & Dmitrenko, N. P. (2022). Shock wave in van der walls gas. *Journal of Non-Equilibrium Thermodynamics*, 47, 255–267.

Bazer, J., & Ericson, W. B. (1959). Hydromagnetic shocks. *Astrophysical Journal*, 129, 758–785.

Bazer, J., & Fleischman, O. (1959). Propagation of weak hydromagnetic discontinuities. *Physics of Fluids*, 2, 366–378.

Bertram, L. A. (1973). Unified geometric and analytical treatment of magnetogasdynamics shocks. Part 1. General solutions and theorems. *Journal of Fluid Mechanics*, 58, 313–336.

Chaudhary, J. P., & Singh, L. P. (2018). Exact solution of the weak shock wave in non-ideal gas. *International Journal of Applied & Computational Mathematics*, 4, 136(1)–136(9).

Chauhan, A., Arora, R., & Tomar, A. (2018). Convergence of strong shock waves in non-ideal magnetogasdynamics. *Physics of Fluids*, 30, 116105(1)–116105(10).

Ferraro, V. C. A., & Plumpton, C. (1966). *An introduction to magneto-fluid mechanics*, Oxford, Clarendon Press.

Friedel, H. (1967). Discontinuities in non-ideal magnetogasdynamics. *Journal of Mathematical Physics*, 8, 2234–2236.

Genot, V. (2009). Analytical solutions for anisotropic MHD shocks. *Astrophysics & Space Sciences Transactions*, 5, 31–34.

Gordan, A. R., & Helliwell, J. B. (1967). Ionizing detonation waves in transverse electromagnetic fields. *Journal of Plasma Physics*, 1, 219–228.

Greenspan, H. P. (1962). Similarity solution for a cylindrical shock-magnetic field interaction. *Physics of Fluids*, 5, 255–259.

Helliwell, J. B. (1963). Magnetogasdynamic deflagration and detonation waves with ionization. *Journal of Fluid Mechanics*, 16, 243–261.

Husain, A., Haider, S. A., & Singh, V. K. (2021). Efficient exact solution of blast waves in magneto-gas-dynamic flow at stellar surfaces. *Advances & Applications in Mathematical Sciences*, 20, 1599–1608.

Husain, A., Haider, S. A., & Singh, V. K. (2022). Solution of one-dimensional weak shock wave problem in a non-ideal dusty medium: An analytical approach. *Materials Today: Proceedings*, 62, 1613–1619.

Husain, A., Singh, V. K., & Haider, S. A. (2020). An efficient analytical solution of blast wave problem in real gas flow under the influence of dust-laden particles. *International Journal of Engineering & Advanced Technology (IJEAT)*, 9, 2490–2494.

Kanwal, R. P. (1960). On magnetohydrodynamic shock waves. *Journal of Mathematics & Mechanics*, 9, 681–695.

Lock, R. M., & Mestel, A. J. (2008). Annular self-similar solutions in ideal magnetogasdynamics. *Journal of Plasma Physics*, 74, 531–554.

Murata, S. (2006). New exact solution of the blast wave problem in gas dynamics. *Chaos, Solitons & Fractals*, 28, 327–330.

Murray, J. D. (1961). Strong cylindrical shock waves in magnetogasdynamics. *Mathematika*, 8, 99–120.

Nath, G. (2021). Analytical solution for unsteady flow behind ionizing shock wave in a rotational axisymmetric non-ideal gas with azimuthal or axial magnetic field. *Zeitschrift für Naturforschung A*, 76, 265–283.

Nath, G., & Devi, A. (2021). A self-similar solution for unsteady adiabatic and isothermal flows behind the shock wave in a non-ideal gas using Lie group analysis method with azimuthal or axial magnetic field in rotating medium. *European Physical Journal Plus*, 136, 477(1)–477(26).

Nath, O. (1989). A study of self-similar cylindrical MHD shock waves in monochromatic radiation. *Astrophysics & Space Science*, 155, 163–167.

Nath, T., Gupta, R. K., & Singh, L. P. (2017). Evolution of weak shock waves in non-ideal magnetogasdynamics. *Acta Astronautica*, 133, 397–402.

Oliveri, F., & Speciale, M. P. (2005). Exact solutions to the ideal magnetogasdynamics equations through Lie group analysis and substitution principles. *Journal of Physics A*, 38, 8803–8820.

Pai, S. I. (1963). *Magnetogasdynamics and plasma dynamics*, Vienna, Springer-Verlag, Prentice-Hall.

Radha, C., & Sharma, V. D. (1995). High and low frequency small amplitude disturbances in a perfectly conducting and radiating gas. *International Journal of Engineering Science*, 33, 2001–2010.

Ram, S. D., Singh, R., & Singh, L. P. (2013). An exact analytical solution of the strong shock wave problem in nonideal magnetogasdynamics. *Journal of Fluids*, 2013, 810206(1)–810206(4).

Ray, G. D. (1971). Propagation of cylindrical blast waves in magneto-gas dynamics. *Journal of Applied Mathematics & Physics (ZAMP)*, 22, 915–920.

Rosenau, P., & Frankenthal, S. (1976). Equatorial propagation of axisymmetric magnetohydrodynamic shocks. *Physics of Fluids*, 19, 1889–1899.

Shah, S., & Singh, R. (2019). Propagation of non-planar weak and strong shocks in a non-ideal relaxing gas. *Ricerche di Mathematica*, 70, 371–393.

Sharma, V. D. (1981). Propagation of discontinuities in magnetohydrodynamics. *Physics of Fluids*, 24, 1386–1387.

Sharma, V. D., Singh L. P., & Ram, R. (1987). The progressive wave approach analysing the decay of a sawtooth profile in magnetogasdynamics. *Physics of Fluids*, 30, 1572–1574.

Singh, L. P., Husain, A., & Singh, M. (2010). An analytical study of strong non-planar shock waves in magnetogasdynamics. *Advances in Theoretical& Applied Mechanics*, 6, 291–297.

Singh, L. P., Husain, A., & Singh, M. (2011a). A self-similar solution of exponential shock waves in non-ideal magnetogasdynamics. *Meccanica*, 46, 437–445.

Singh, L. P., Ram, S. D., & Singh, D. B. (2011b). Exact solution of planar and no planar weak shock wave problem in gas dynamics. *Chaos, Solitons & Fractals*, 44, 964–967.

Singh, D., & Arora, R. (2021a). Propagation of shock waves in a non-ideal gas under the action of magnetic field. *Mathematical Methods in Applied Sciences*, 44, 1514–1528.

Singh, M., & Arora, A. (2021b). Propagation of one-dimensional planar and non-planar shock waves in nonideal radiating gas. *Physics of Fluids*, 33, 046106(1)–046106(17).

Singh, R., Singh, L. P., & Ram, S. D. (2014). Acceleration waves in non-ideal magnetogasdynamics. *Ain Shams Engineering Journal*, 5, 309–313.

Taylor, M. G. G. T., & Cargill, P. J. (2001).A general theory of self-similar expansion waves in magnetohydrodynamic flows. *Journal of Plasma Physics*, 66, 239–257.

Whitham, G. B. (1974). *Linear and nonlinear waves*, New York, John Wiley & Sons.

Zel'dovich, Y. B., & Raizer, Y. P. (1967). *Physics of shock waves and high temperature hydrodynamic phenomena, vol. II*, New York, Academic Press.

Chapter 11

Investigation of the boundary layer flow using glass box testing scheme

Alok Dhaundiyal

Institute for Energy Security and Environmental Safety,
Centre for Energy Research, Budapest, Hungary

Laszlo Toth

Institute of Technology, Hungarian University of
Agriculture and Life Sciences, Godollo, Hungary

CONTENTS

11.1 INTRODUCTION

In recent years, the cognizance related to the application of renewable sources among the masses has drastically increased. The reason for this awareness is the increasing dependency on gas and oil, and the collateral damage to the world's economy due to bull and bear markets (energy stock market). The food shortage and the depreciation of the local currency also have not remained untouched. All these uncertainties about the conventional source of energy have led to the global energy crisis. To avoid the domino effect of an energy shortage, clean energy should be brought into use in different public and private sectors. One of such clean sources is the sun and the utilization of its thermal potential to dry agricultural products. The other aspects related to solar energy are discussed in subsequent paragraphs.

DOI: 10.1201/9781003367420-11

The quandary of how to dry perishable agricultural products can be sorted out if proper application of solar energy is envisaged by the researchers. In the bygone era, it was usually carried out by laying the product on the open ground. But the time and rate of drying were quite uncertain. Later, a direct gain approach was adopted to facilitate the drying process in a swift and controlled manner at a pilot scale. Some additional changes in this methodology were seen with time. The mode of providing thermal energy was slightly changed and the natural circulation of air was replaced by a forced circulation with the help of a blower. But it was noticed that the change in the flow pattern caused perturbation in the temperature distribution of the dried product. Thus, a new scheme was chalked out and direct drying was supplanted, and another type of indirect active device was introduced.

Recent progress related to geometry and material selection was thoroughly investigated in the literature. A corrugated aluminium alloy plate was used for the assessment of the dual-mode system. It was seen that the rise in the mass flow rate of air by 33% would lead to an increase in the thermal efficiency of the unit by 27% [1]. But the simultaneous effect of pressure and temperature on the performance of the solar air heater was omitted. Ural (2019) carried out a thermodynamic analysis of a textile-based solar air collector (TB-SAC). According to his study, the available energy of the system was enhanced by 32% when TB-SAC was compared with the flat plate solar air collector. Albeit the pressure drop was remarkably elevated. The reason for this rapid change in pressure was loosely covered in this study. It was also observed that the highest value of instantaneous collector efficiency (62%) was computed at a mass flow rate of 0.62 kg/s. The intrinsic factor that might influence the collector efficiency was not discussed as well [2]. Wang and Wang (2021) proposed a semi-circular baffle solar air collector. The geometric aspect of the semi-circular baffle was discussed, and it was reported that the increasing diameter of the semicircle augmented the collector efficiency by 3.68%–5.12%. However, the outlet temperature of the air also surged by 0.86% as the diameter of the semicircle increased from 20 to 40 mm. The analysis was based on the effect of the ratio of baffle spacing (D) to the diameter of the semicircle (d). The work also discussed the pressure loss across the collector, but the effect of velocity regime and friction drag was not considered in their work [3]. In the same way, the Moses and Carson correlation was used to determine the effect of natural draught on the drying model. Rather than focusing on a single terminology, pressure drop, a detailed pressure profile was drawn for a flat plate collector. The effect on sudden enlargement and contraction was also evaluated, which is often overlooked by the researchers, but this detailed work omitted the hydrodynamic behaviour of air near the collector surface [4]. Dhaundiyal and Atsu (2021) used a thermal boundary model to explain the effect of surface temperature on the thermal boundary layer and velocity gradient, but it is merely a single aspect of the thermal sciences.

They did not discuss the simultaneous effect on fluid dynamics of flowing air on the smooth collector surface [5]. In another work, the Pohlhausen method was applied to estimate the effect of wind flow over the PV surface so that the best orientation of PV modules could be estimated. The scheme was fascinating, and it provided a promising solution for the PV modules constructed by different commercial companies. The modeling approach adopted was confined to a quadratic approximation of velocity ratio (u/U) [6]. In another study, the energy absorption efficiency of two flat plate collectors and a cylindrical solar collector was examined. It was reported that the increasing concentration of nanofluid augmented the energy absorption efficiency of flat plate collectors with a smaller surface area. The objective is largely focussed on the quantitative approach to improving the performance of the solar collectors using the nanofluid particles, but the thermal interaction of nanofluid (service fluid) with the water was not discussed [7]. Assadeg et al. (2021) performed the exergy and energy analysis of four different arrangements of double pass solar collector (with fins and phase change material (PCM), with fins only, with PCM only and without fins and PCM). It was concluded in their study that a solar collector comprising fins and PCM provided the optimum results in the context of energy performance. The highest efficiency acquired by the collector was estimated to be 66.5%. But their work was devoid of the qualitative aspect of the proposed design. The aim was merely to draw a comparison among different materials without involving state variables (pressure, temperature, internal energy, and enthalpy) [8]. However, this work was covered in another work, where a simple flat plate collector was examined at the microscopic level. The exergy and its effectiveness were determined for each component and the conclusion was based on the facts derived from the analysis report [9]. Kizildag et al. (2022) used a flat plate collector with transparent insulating material (TIM) to protect the prototype model from overheating and commercialization of the new product. Like others, they also focused on the performance of the unit, but their work was much more pivoted on the complications faced while operating the solar collector. The cost of production was noticed to increase by 30%–40% as compared to the conventional one. The energy collector for a system with TIM was estimated to be 2.5–1.4 times higher than the standard collector in winter [10]. The evacuated tube solar collector (ETC) was investigated in another study, and it was noticed inadequate vacuum inside ETC influenced the boiling and condensation regimes, which in turn affected the heat transferring ability of the working fluid. They developed a one-dimensional steady-state model to simulate the thermal performance of ETC. It was suggested that the two-phase heat transfer model would be able to impart accurate simulation and prediction in terms of outlet temperature and efficiency of the collector. A significant deviation between the numerical solution and experimental data was seen when the percentage of vacuum level was less than 0.001 Pa [11]. The endeavour to reduce the simulation time and the effect of ambient temperature to predict temperature distribution inside the solar collector was carried out using the Lattice Boltzmann model (LBM). The simulation was performed for the whole working cycle, and it was found that the predicted average temperature of the fin provided an error of 5.16% while comparing numerical solutions to experimental data. Their work used LBM, but they did not cover other properties of the system that might influence the numerical stability [12]. The four different designs of Solar air collectors (SACs) (Flat plate, Das model, revised Das model and Z-type) were examined, and the designed 3D model of SACS was examined with ANSYS FLUENT 18.1, and the estimated error was noticed to be less than 1%. Among different collectors, the Z-type collector was noticed to have the highest collector efficiency. But the flow regime was assumed to be constant, however, it is the function of dynamic viscosity that changes with the temperature of fluid [13]. Hu et al. (2020) used swirling motion inside the solar air collector to determine its impact on thermal performance. According to their numerical solution, it was established that the active swirling flow could provide relatively promising results as compared to passive swirling flow. The maximum thermal efficiency growth rate (TEGR) of active swirling flow was found to be 48.65% higher than that obtained for passive swirling motion in the proposed model, which can be enhanced by 13.24% if the experiments are conducted at lower mass flow rates. The gap between the model and experimentally determined data could be dilated for a higher flow regime (Turbulent flow) [14]. Özcan et al. (2021) used a parabolic trough solar collector for water heating application. The work focused on the material that should be used to construct an absorber tube. It was found that the maximum efficiency values derived for copper trough were 2.47% higher than that obtained from aluminium tubes. A similar solution was also tried to be established through the CFD model. The temperature at the outlet of the absorber plate exhibited a difference of 6.5% from the experimental results [15]. Pandey et al. (2021) used an arc plug inside the absorber plate for testing the thermal performance of the solar parabolic trough collector (PTC). To simulate the thermodynamic and hydraulic properties, even different cases were investigated. They used SOL TRACE-7.9 to determine the heat flux boundary condition. For a ratio of radii of arc plug to receiver tube ($R=1$ and 0.879), PTC showed the highest thermal efficiency, whereas PTC had the highest heat transfer performance at $R=1.606$. The highest thermal enhancement index (TEI) was found to be 1.1039 and 1.3016 for factors $R=1$ and $R=1.606$, respectively. They concluded that it is not essential that a rise in TEI would also lead to higher thermal efficiency [16]. The effect of natural draught on the performance of the flat plate collector was also studied and it was noticed that the natural draught would undoubtedly improve the convective heat transfer, but at the same time, the heat gain would be severely impacted as well. The static pressure change was negligibly increased when a circular chimney was provided with the drying chamber [17]. In another work, the Chebyshev method of collocation was used to examine the influence of physical parameters (slip condition, temperature jump, Brownian motion, thermophoresis, suction (or injection) parameters and Lewis and Prandtl numbers) on the temperature and concentration profiles [18]. Said et al. (2015) studied the effect of thermo-physical properties of short single-wall carbon nanotubes (SWCNTs) on the thermal efficiency of a flat plate solar water heating system. Sodium dodecyl sulphate was used as a dispersant to prepare a stable nanofluid. It was noticed that thermal conductivity varied linearly with the concentration of particles and temperature. However, the viscosity of nanofluids and water was dropped with the increase in temperature and particle loading. The maximum energy and exergy efficiencies of the flat plate collector were drastically increased up to 95.12% and 26.15% after improving the thermo-physical properties of nanofluid, which is 42.07% and 8.77% with water, respectively [19].

In this work, the emphasis is laid on examining the boundary layer formation near the smooth absorber plate surface. The programming script and call functions of the white-box model were developed using the MATLAB interface and the results were compared with the experimental results derived through the measuring instruments.

11.2 MATERIAL AND METHODS

11.2.1 Experimental set-up

The location of the testing unit is the Solar laboratory of the Hungarian University of Agriculture and Life Sciences, Godollo, Hungary. The latitude and longitude of the site are 47°35′24″ N and 19°21′36″ E. The flat plate solar collector was connected to the drying chamber through the circular duct. The provision of incorporating the chimney at the exhaust end of the drying chamber was provided to create a draught. The surface of the absorber plate was selectively coloured with matt black paint. The schematic diagram of the unit is illustrated in Figure 11.1. The geometrical detail of the solar collector is provided in Table 11.1. The input parameters data used for modeling purpose is tabulated in Table 11.2. The experimental data used for comparison was retrieved between 25th September 2019 and 30th September 2019. The t-type thermocouple was considered for experimental work. The duration of the experiment was from 10:00 AM to 3:00 PM each day. The air velocity at the inlet of the collector was measured by a digital anemometer (Eurochron EC-MR, Austria) with an accuracy of ±0.3%. The orientation of the solar collector is true south. The collector was mounted on the titled rack at an angle of 45°. The data acquisition was performed using the ADAMS4018 (Advantech, Taipei, Taiwan). The height of chimney was 2 m.

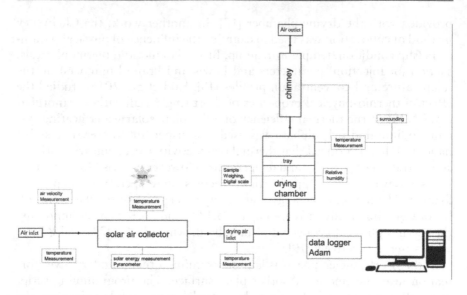

Figure 11.1 Schematic diagram of solar drying unit [4].

Table 11.1 Geometrical Detail of the Flat Plate Solar Collector [17]

Material	Dimension	Thickness
Glass cover	1,160×460 mm²	4 mm
Plywood box	1,200×500×150 mm³	20 mm (side), 2 mm (bottom), 2 mm (reveals approx.)
Copperplate	1,160×460 mm²	1.2 mm
Polystyrene	1,160×460 mm²	80 mm
Air space	1,160×460 mm²	60 mm
PVC duct	200×45 mm²	3.17 mm

Table 11.2 The Information about the Miscellaneous Parameters Used for Modeling Purposes

Parameters	Draught	Conventional
C_p [20]	0.39 kJ·kg/K	
^aC_f [20]	1.01 kJ·kg/K	1.01 kJ·kg/K
^av [20]	1.77×10⁻⁵ m²/s	1.78×10⁻⁵ m²/s
h_{fp} [17]	2.49 W/m²/K	0.71 W/m²/K
h_{fc} [17]	2.49 W/m²/K	0.71 W/m²/K
Q [17]	225.88 W	
ρ [21]	8,960 kg/m³	
$T_p (0)$ [17]	373.29 K	489.66 K
$^a\rho_f$ [20]	1.10 kg/m³	1.10 kg/m³

^a At atmospheric pressure.

11.2.2 Numerical methodology

The Blasius equation- first-order boundary layer and time-dependent convective equations were examined with the help of the numerical method based on the Runge-Kutta-Fehlberg method (RKF45), whereas the non-steady and one-dimensional heat conduction for the absorber plate (Copperplate) was solved using the spatial discretization of the parabolic equation in one space variable. For mathematical simulation, MATLAB 2015a (Math Works, California, the USA) was used to write down the algorithm for the laminar flow across a flat plate collector. The flow pattern is shown in Figure 11.2.

According to Prandtl, the pressure gradient (along Y-axis) and shear rate (along X-axis) are negligibly small and they can be dropped from momentum equations. Thus, we get Equation (11.1) along X-direction as follows

$$u\frac{\partial u}{\partial x}+v\frac{\partial u}{\partial y}=v\frac{\partial^2 u}{\partial y^2}+\frac{1}{\rho}\frac{dp}{dx} \tag{11.1}$$

Similarly, the momentum equation for Y-direction can be given by Equation (11.2)

$$u\frac{\partial v}{\partial x}+v\frac{\partial v}{\partial y}=v\frac{\partial^2 v}{\partial y^2} \tag{11.2}$$

For constant fluid properties and zero pressure gradient $\frac{dp}{dx}=0$, Equation (11.1) can further boil down to the Blasius equation (Equation 11.3). It is to

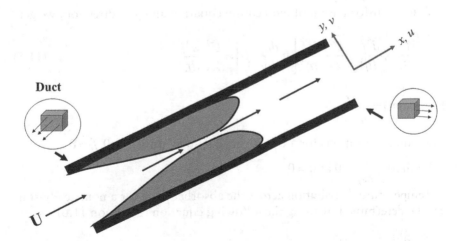

Figure 11.2 The representation of the flow direction inside the flat plate solar collector.

be noted that the Blasius solution is not unique [22], therefore a particular set of eigenvalues are considered.

$$u\frac{\partial u}{\partial x} + v\frac{\partial u}{\partial y} = v\frac{\partial^2 u}{\partial y^2} \tag{11.3}$$

Since equations and boundary conditions remain invariant under the transformation, therefore an appropriate scaling of explanatory and response variables might not influence the numerical solution of Equation (11.3). As the velocity profile along the collector surface is geometrically the same, so it might only differ by a stretching factor along the Y-direction. In other words, the dimensionless ratio of velocities u/U can be defined as the function of dimensionless distance from the wall $\chi\left(\dfrac{y}{\delta}\right)$

Near the boundary layer, the momentum equation along the X-direction provides the following relationship between local Reynold's number (Re_x) and boundary layer thickness, δ

$$\frac{\delta}{x} = \frac{1}{\sqrt{Re_x}} \tag{11.4}$$

In this way,

$$\frac{u}{U} = \chi\left(\frac{y\sqrt{Re_x}}{x}\right) = \chi(\eta)$$

where $\eta = y\left(\sqrt{\dfrac{U}{vx}}\right)$ represents the stretching factor

After transformation of momentum equation along X-direction, we get

$$-\frac{U^2}{2x}\eta\frac{d^3f}{d\eta^3} + \frac{U^2}{2x}\frac{d^2f}{d\eta^2}\left[\eta\frac{df}{d\eta} - f\right] = \frac{U^2}{2x}\frac{d^3f}{d\eta^3} \tag{11.5}$$

Here, $f(\eta) = \dfrac{\psi}{\sqrt{Uvx}}$

The initial conditions for a transformed system are at $\eta = 0, f = 0$.

Similarly, $\dfrac{df}{d\eta} = 0$ at $\eta = 0$

Temperature distribution across the absorber plate for a non-steady-state can be determined by using the following equation (Equation 11.6)

$$\kappa\frac{\partial^2 T}{\partial x^2} = \rho C\frac{\partial T}{\partial t} + Q + h_e(T_{ext} - T) \tag{11.6}$$

The boundary conditions for the copper plate can be derived from the following equation (Equation 11.7)

$$q(x,T,t)+p(x,t)R\left(x,T,t,\frac{dT}{dx}\right)=0 \tag{11.7}$$

In the same way, the temperature distribution across the carrier fluid can be obtained from the energy analysis of the flat plate collector [23] (Equation 11.8)

$$\dot{m}C_p\frac{dT_f}{dx}=h_{fp}L\left(T_p-T_f\right)+h_{fc}L\left(T_c-T_f\right) \tag{11.8}$$

The initial condition assumed for Equation (11.8) is $T_f=T_p$ at $y=0$.

11.3 RESULTS AND DISCUSSION

The momentum and heat transfer equation are solved using the numerical method. The flat plate collector with and without natural draught is examined mathematically, and the discussion is based on the facts that make the same system deviate from the right course due to the addition of an auxiliary component. The two-dimensional flow with unit width was considered while evaluating the numerical solution.

11.3.1 Hydrodynamic analysis of the air

The dimensionless velocity (u/U) relationship with the stretching factor η is shown in Figure 11.3. It can be observed from the numerical solution that the dimensionless velocity (u) inside the boundary layer is relatively high while running the drying plant without any auxiliary unit. From the numerical solution, it was noticed that the collector equipped with the chimney influenced the boundary layer formation near the surface of the absorber plate. At the constant value of η, the boundary layer velocity, u, dropped by 9.65% for the same collector when it was operated with the natural draught system. The standard deviation in the numerical solution for conventional (without chimney) was increased by 7.22% as compared to the solution retrieved for the natural draught system. It shows that the relative change in the dimensionless velocity (u/U) was not drastically drifted around its mean value for a system with the chimney and the flow would be more uniform than that estimated in the case of the conventional system. According to the statistical analysis, both the numerical solutions are negatively skewed and the skewness in the dimensionless velocity distribution was seen to be 53.76% lower than that obtained for the natural

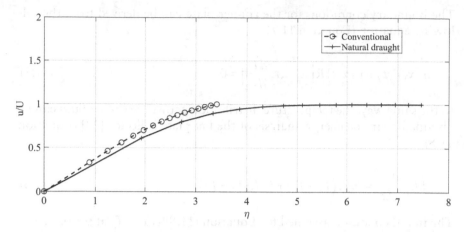

Figure 11.3 The change in the ratio of local to mainstream velocity to the stretching factor (η).

draught unit. It could be concluded that the relative change in the mean values of dimensionless velocity to its median for the conventional system would be relatively low. While comparing the kurtosis (Kr) for both the systems, it dropped by 59.57% for the conventional unit as compared to the corresponding value derived for the natural draught system. The distribution behaviour of the numerical solution is super-Gaussian in nature. As compared to the conventional system, the relative variability in the velocity distribution near the absorber plate surface would be much higher for a collector running with the natural draught.

So, after analysing the velocity distribution for both the systems, the inference can be drawn that the predictability to achieve the mainstream velocity would be higher for the system operated with the natural draught. The large changes in the velocity distribution would be more conspicuous with the chimney than without it. However, some other parameters may also require to validate whether the numerical solution of Blasius ODE would converge faster for a natural draught or not.

The stream function (ψ) derived for given systems is shown in Figure 11.4. While estimating the stream functions for both natural draught and conventional ones, the distribution pattern was noticed to be more negatively skewed when the collector was coupled with a drying chamber equipped with the chimney than in the case of the conventional unit. The skewness percentage plunged by 73.76% when it was equated with the skewness of the numerical data derived for the natural draught. Likewise, kurtosis of ψ derived for the conventional collector was also dropped by 10.12%. The distribution pattern of the stream function in both cases is mesokurtic. Unlike the distribution behaviour derived for the dimensionless velocity (u/U), the probability of the existence of extreme numerical values of ψ is

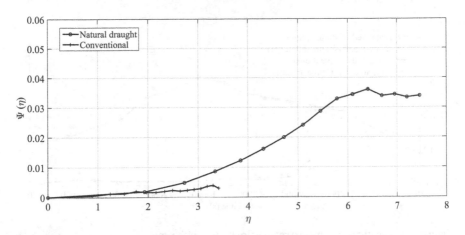

Figure 11.4 The numerical solution for the stream function, Ψ to stretching factor, η.

much more likely the same as could be derived through normal distribution. The standard deviation in the numerical solution of ψ was reduced by 65.01% for a system without a chimney. In the case of natural draught, the maximum value ψ was 5.74 (at $\eta=6.94$), whereas it was noticed to be 1.97. During the initial iteration of RKF45, it could be seen that the deviation in the ψ values was marginally low until $\eta=1.96$. Adrift in the numerical solution was seen after $\eta=2.74$. As the stretching factor and time increase, the significant rise in the numerical solution would be seen in a flat plate solar collector operating under the influence of natural draught. A sudden plunge in the stream function (ψ) was encountered at $\eta=6.67$ while running the unit with a circular chimney. The variabilities in numerical solution would not much that can influence the velocity gradient along X and Y directions.

The comparison between the measured and numerical obtained velocity with time for both natural draught and conventional ones is demonstrated in Figure 11.5. The standard deviation in the numerical solution for natural draught was relatively increased by 26% when it was compared with the corresponding value measured experimentally with the help of an anemometer. Conversely, as compared to the experimentally obtained velocity for the conventional system, it dropped by 58.83% for the predicted solution of the same. The reason for this relative fluctuation is the initial condition considered for the numerical solution. It was assumed to be zero, but there is an existence of a no-slip condition on the solid boundary of the absorber plate operating without a circular chimney. However, in the case of a natural draught system, the prediction of experimentally derived velocity through the numerical solution fared well. The natural velocity distribution experimentally is also different for both cases. The predicted solution and experimentally measured data are negatively skewed, whereas it is the opposite for the conventional unit. The positively skewed distribution was

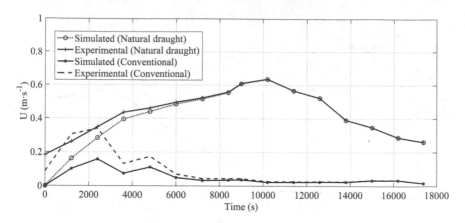

Figure 11.5 The change in mainstream velocity (U) with time.

estimated for simulated as well as experimental data. Corresponding to the experimentally recorded velocity, the skewness in the distribution function of velocity was dropped by 68.86% for the prediction solution of natural draught. In the case of the conventional unit, the skewness in the numerical solution, as compared to the measured value, was elevated by 17%. The excessive kurtosis was noticed in the conventional data sets, whereas it was seen to be mesokurtic for the velocity distribution function of the natural draught. It implies the variability in the numerical solution of velocity would be fairly low for the natural draught unit, whereas it is not the case for the conventional drying unit (without a chimney). It was seen that the presence of the natural draught also influences the distribution characteristic of measured velocity. The reason for the deviation can be the recirculation of the flow near the edge of the collector plate for the conventional unit. As compared to the experimentally derived values, the kurtosis for the velocity distribution function of numerical solution (natural draught) was seen to be increased by 53.64%, whereas it was dropped by 5.93% for the conventional unit. The maximum airflow velocity derived from the numerical solution is 0.64 m/s, which is insignificantly lower (0.15%) than the corresponding value measured from the anemometer. Correspondingly, the same difference gets widened by 53.74% when the numerical solution for the conventional unit was compared with the recorded values measured by the anemometer. The various factors such as the skin friction coefficient (Cf_x), momentum thickness (θ), energy thickness (δ^{**}) and the energy loss/gain, might influence the experimental/numerical result. This drift in the conventional system is mainly seen at the leading edge of the plate.

The change in gradient of velocity components (u and v) along X and Y with time is shown in Figure 11.6. The minimum gradient of $u_x\left(u_x = -\dfrac{\partial u}{\partial x}\right)$ along the length of the absorber plate was estimated to be $0.69\,\mathrm{s}^{-1}$, which is

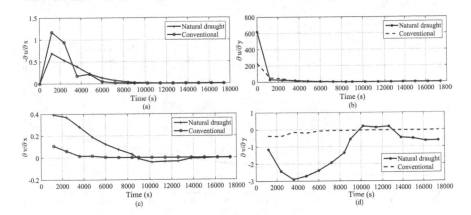

Figure 11.6 The variation in the velocity gradient with time (a, b: gradient of *u* along with X and Y directions, respectively, c, d: gradient of v along with X and Y directions, respectively).

41.08% higher than the obtained value of gradient for the conventional system at the same time. This change was noticed to subdue as time proceeds. However, at $t=3,600$ s, a drastic drop of 55.5% in u_x was observed for the conventional system. Around $t=4,800$ s, both the systems would have the same velocity gradient, u_x. The shear rate or the change in the velocity $\left(v_y = -\dfrac{\partial v}{\partial y}\right)$ with transverse distance y from the solid boundary is highest for the natural draught unit. It implies that the shear stress (τ) offered to the airflow would be 0.011 N/m² at the solid boundary. Correspondingly, it was estimated to be 0.0041 N/m² while running the unit without a circular chimney, which is 62.38% lower than the computed value of τ for a unit operating with a circular chimney. However, the drastic drop in the shear rate was noticed with time, and an air gap was provided between the cover and absorber plates. With the increase in y, as compared to the corresponding value of u_y for the conventional unit, it was reduced by 69.75% at the same time for the natural draught unit. The change in the gradient of the vertical component of velocity, v along the length of the collector was remarkably high for the system operated with the help of the circular chimney. On the other hand, $v_x\left(v_x = \dfrac{\partial v}{\partial x}\right)$ varied uniformly with the length of the collector and time for the conventional system. At the same time, the maximum value of v_x computed for the conventional system was found to be 74.35% lower than the corresponding value computed numerically for the natural draught unit. At 9,000 s, the numerical value of v_x for both the cases would be the same. For a time interval of 9,000–13,800 s, the velocity gradient, v_x, for the natural draught would change its direction. The point of inflexion was calculated at 9,000 and 13,800 s. The derived values of

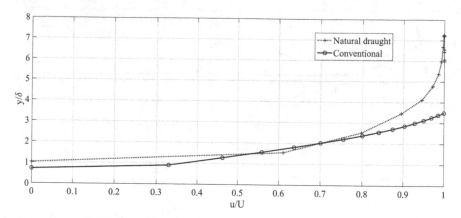

Figure 11.7 The change in the laminar flow behaviour with the numerical solution of the Blasius equation.

v_x at inflexion points, 9,000 and 13,800 s are 0.0021 s⁻¹ and 9.38×10⁻⁴/s, respectively. Similarly, the inflexion points obtained for v_y $\left(v_y = -\dfrac{\partial v}{\partial y} \right)$ in between the time intervals of 9,000–10,200 s and 12,600–13,800 s for the flat plate collector operated with the help of the natural draught. The lowest value of v_y measured numerically for the natural draught system is −2.90 s⁻¹, whereas the maximum value is 0.204 s⁻¹. The lowest value of v_y for the conventional system is −0.36 s⁻¹, which is 87.58% higher than the corresponding value estimated after incorporating a circular chimney (natural draught). A change in the direction and magnitude of the gradient v_x and v_y for natural draught is relatively high as compared to the corresponding gradient of v for the conventional. Conversely, it is just the opposite of the characteristics of u_x and u_y evaluated for the natural draught.

The change in the dimensionless transverse distance from the solid boundary of the absorber plate with the dimensionless velocity is shown in Figure 11.7. The slope of the curve defines the flow characteristic of air in either case. The flow remains laminar near the surface of the absorber plate. With the increase in the transverse distance, y, a steep rise in the velocity was noticed when the air was induced with the help of a natural draught.

The dimensionless velocity was seen to vary uniformly inside the boundary layer. A sudden drift in the dimensionless velocity was noticed at $\dfrac{y}{\delta} = 1.5$. The slope was elevated by 54.34% using natural draught. The flow behaviour did not drastic within the precinct of boundary layer thickness, $y = \delta$. For $y \geq \delta$, a small perturbation was noticed in the flow behaviour of air without the presence of any external draught system. Conversely, a significant drift in the dimension velocity (u/U) was noticed for a system operated with the natural draught unit at $y \geq 2.528\delta$ and the tendency of flow to become transient.

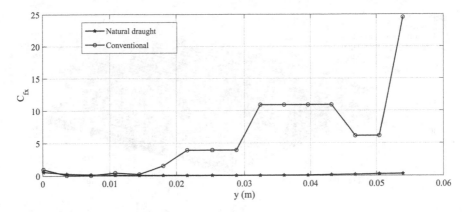

Figure 11.8 The variation in the skin friction coefficient near the collector surface for a different arrangement.

The variation in the skin-friction coefficient (Cf_x) with the transverse distance (y) from the solid boundary is illustrated in Figure 11.8. Operating a flat plate collector with a conventional drying chamber will offer a significant drag force to the air flowing with the increase in the air passage between the cover and absorber plates. The skin-friction coefficient (Cf_x) was found to be nominal if the air gap thickness varies in the open interval of 0–0.014 m. The air gap between the parallel plates of more than 0.014 m would not be effective if the drying chamber is operated without the circular chimney. The Cf_x can reach up to 24.54 for the given conventional unit, whereas this value cannot be more than 0.291 if the drying chamber connected with the flat plate collected is equipped with a circular chimney. The drag force for the natural draught system would be fairly constant, whereas it would be constant in a piecewise manner for the conventional unit. In the air gap interval of 0.02–0.03 m, the value of Cf_x would be 3.93. Similarly, it increased by 177.60% for air gap thickness of 0.03–0.04 m. A sudden fall of 43.76% in the value of Cf_x was noticed for a short interval of 0.046–0.050 m. The average drag force for the draught system is 0.0086 N, which is 22.47% lower than the average drag force exerted by the plate on the airflow in the conventional unit.

11.3.2 Heat analysis of the flat plate collector

The temperature distribution was derived by spatial discretisation of the parabolic heat equation. The comparison between the numerical and experimentally obtained solutions is illustrated in Figure 11.9. The standard deviation in the numerical solution of absorber plate temperature is 1.03×10^{-5} if the flat plate solar collector is operated with the natural draught system. The mean temperature calculated through the numerical method is 0.052% lower than the experimentally measured temperature for the flat

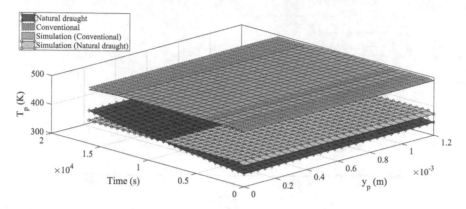

Figure 11.9 Comparison of numerical solution of copper plate temperature with the experimental values.

plate collector provided with the natural draught (Table 11.3). Conversely, it overestimated the value of T_p for the conventional system. It is marginally augmented by 0.01% when it was compared with the instrumental data set for the conventional system. For the conventional unit, the distribution function of absorber plate temperature T_p is positively skewed, whereas the experimental data set has a skewness of –1.0924. On contrary, in the case of the natural draught unit, the temperature distributions obtained numerically and experimentally are negatively skewed. However, the distribution function of T_p (obtained numerically as well as experimentally) exhibits high kurtosis for both systems. That implies temperature distribution across the absorber plate is strongly influenced by the temperature gradient formed at the front and back sides of the absorber plate.

The change in the carrier fluid temperature (air) (T_f (K)) with time and length of the collector plate is shown in Figure 11.10. The mean temperature of air derived for both natural draught and conventional units is within a predicted domain. The mean value of T_f dwindled by 0.76% when it was compared with the experimental mean value of T_f for the natural draught unit. Similarly, the deviation in the mean values of T_f is 1.65% if the flat plate collector is operated without a chimney. The kurtosis of the numerical data set was seen to be diminished by 23.62% while examining the distribution function of the carrier fluid temperature.

Conversely, it is not true for a conventional unit. Corresponding to the kurtosis (Kr) estimated for the experimental dataset, an increase of 30.9% was observed in the kurtosis of the distribution function of air temperature derived from the numerical solution for the conventional unit. Both numerically and experimentally derived data sets are negatively skewed in the case of a natural draught system. However, it differs in the case of the conventional unit. The predicted air temperature has a positive skewness,

Table 11.3 Comparison between the Numerical Solution and Experimental Results (Average Values)

Parameters		Simulated	Experimental
U	Natural draught	0.40 m/s	0.43 m/s
	Conventional	0.05 m/s	0.08 m/s
u	Natural draught	0.40	–
	Conventional	0.05	
v	Natural draught	0.32×10^{-2} m/s	–
	Conventional	7.46×10^{-4} m/s	
$-u_x$	Natural draught	$0.13\,s^{-1}$	–
	Conventional	$0.16\,s^{-1}$	
u_y	Natural draught	$44.59\,s^{-1}$	–
	Conventional	$21.69\,s^{-1}$	
v_x	Natural draught	$0.084\,s^{-1}$	–
	Conventional	$0.014\,s^{-1}$	
$-v_y$	Natural draught	$1.074\,s^{-1}$	–
	Conventional	$0.10\,s^{-1}$	
δ	Natural draught	4.60×10^{-3} m	–
	Conventional	15.60×10^{-3} m	
δ^*	Natural draught	4.59×10^{-3} m	–
	Conventional	25.78×10^{-3} m	
θ	Natural draught	37.40×10^{-7} m	–
	Conventional	-16.82×10^{-3} m	
δ^{**}	Natural draught	37.50×10^{-7} m	–
	Conventional	-5.84×10^{-3} m	
E_b	Natural draught	2.12×10^{-7} J/m	-153.89×10^{-7} J/m
C_{fx}	Conventional	0.16	5.97
T_p	Natural draught	379.93 K	380.13 K
	Conventional	493.80 K	493.75 K
T_f	Natural draught	339.23 K	341.86 K
	Conventional	394.95 K	401.58 K

whereas its experimentally determined counterpart has asymmetry in the opposite direction. The presence of an outlier in the numerical solution would appear during the last stages of the iteration steps. The variability would be high at the outlet of the collector. In contrast, the instrumental measurement might show more variability at the inlet duct of the collector. The average values of experimentally and numerically derived parameters are tabulated in Table 11.3. Corresponding to the boundary layer thickness (δ) derived for the conventional drying unit, the thickness of the boundary layer (δ) plummeted by 70.51% while operating the drying unit with a chimney. Similarly, in the case of the conventional unit, the displacement

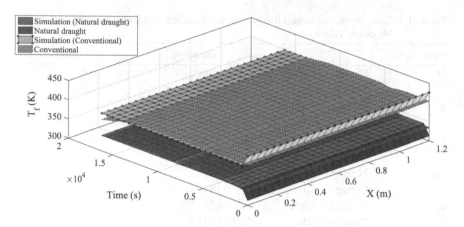

Figure 11.10 Comparison of ODE45 solution of the carrier fluid temperature with experimentally derived values.

thickness (δ^*) of airflow was phenomenally increased by 461.65% when it was weighed up against the corresponding average value of δ^* for the natural draught system.

The momentum thickness (θ) and energy thickness (δ^{**}) of airflow (natural draught) were noticed to be marginally low, whereas the negative value of the same parameters was computed when the solar collector was operated without the circular chimney. That implies the flow near the absorber plate surface has a tendency to be separated and that eventually leads to an increase in the pressure gradient along the length (x) of the absorber plate. The possibility of recirculation of flow can be relatively high in the case of the conventional system. Mathematically, it implies the flow near the surface is relatively more energized than the free stream, which is not the case. It should also be noted down that the presence of pressure gradient is neglected while solving the Blasius equation. So, the Prandtl equation could provide a better perspective of the air flowing across the flat plate collector with no chimney. The momentum transfers to the lower fluid layer usually happen when the velocity gradient near the surface is very low and yes, it is numerically estimated that the velocity gradient is negative near the leading edge of the absorber plate (Figure 11.6). With the given fact, it could be stated that the flat plate collector operating without a chimney might lead to boundary separation even in the absence of the bluff body at the leading edge of the absorber plate. The energy loss per unit width (E_b) would also be negative that implies the energy of the system will increase near the surface of the absorber plate for the conventional system. However, it is not the case with the natural draught system and the energy loss per unit width would amount to 2.12×10^{-7} J/m. The results were compared with other methodologies and it was found to agree with the estimated values [5,6,9,17].

11.4 CONCLUSION

The airflow behaviour across the parallel duct was numerically examined using the RKF45 model and spatial discretisation of the parabolic heat equation. The two experimental parameters, temperature, and velocity, were compared with the numerical solutions derived from the Blasius equation for laminar flow. The following salient points were obtained after comparing the numerical solution with the instrumental data.

1. The predictability of ODE45 was noticed to be relatively good with the data provided for the natural draught. At the constant value of the stretching factor, the boundary layer velocity, u, was lower than 9.65% for natural than the corresponding value computed for the conventional drying plant. The change in the mean value of dimensionless velocity to its median is rather low for the conventional system. The distribution function of the dimensionless velocity would be super-Gaussian for both cases. The relative variability in the velocity distribution would be higher for the collector equipped with natural draught.

2. The stream function can attain extreme numerical values. The distribution function of the stream function is mesokurtic. The marginal difference in the stream function values was noticed for a small number of iterative steps. The variability in the numerical solution of stream function is relatively low as compared to the numerical solution obtained for the dimensionless velocity.

3. The insignificant deviation was recorded between the numerical solution and instrumental data set for measuring the velocity distribution in case of natural draught, whereas the drift between the numerical and experimental values was increased up to 53.74% for the conventional unit. The boundary layer formation and various factors such as momentum thickness and energy thickness might affect the numerical result. The flow behaviour at the upper layers of carrier fluid may tend to be transient flow. However, both systems will exhibit laminar characteristics within boundary layer thickness. The velocity gradient will be negative at the leading edge of the absorber plate. But with the increase in distance and time, the velocity gradient will change its magnitude and direction.

4. The increase in the air gap thickness will also augment the skin friction drag for the conventional unit. The friction drag was seen to be proportionally increased by 177.60% if the air gap thickness is increased up to 0.04 m. However, it also dropped by 43.76% for the short interval of 0.046–0.050 m. The average drag force estimated in the case of the draught system is 0.0086 N which is 22.47% lesser than the calculated drag force for the conventional model.

5. Numerically, the average temperature of the absorber plate was estimated to be 379.93 K while operating the flat plate collector with the
   circular chimney, whereas it was found to be 493.80 K without incorporating the draught system. Likewise, the predicted carrier fluid temperature was, respectively, 339.23 and 394.95 K for natural draught
   and the conventional systems. The integral of the momentum defect
   was estimated to be negative for the conventional system. Unlike the
   natural draught unit, the airflow near the absorber surface will gain
   energy due to the boundary layer formation.

NOTATION

u	Boundary layer velocity	m/s
U	Mainstream velocity	m/s
η	Stretching factor	–
Ψ	Stream function	–
χ	Function related to transverse distance from the plate	–
f	An integral function of the stretching factor	–
x	Length of the absorber plate	–
y	Air space thickness	–
T_p	The temperature of the absorber plate	K
T_f	The temperature of a carrier fluid	K
v	Kinematic viscosity	m²/s
ρ	The density of the absorber plate	kg/m³
δ	Boundary layer thickness	m
δ^*	Displacement thickness	m
θ	Momentum thickness	m
δ^{**}	Energy thickness	m
u_x, u_y	The gradient of horizontal velocity component along x and y directions, respectively	s⁻¹
V_x and V_y	The gradient of vertical velocity component along x and y directions, respectively	s⁻¹
C_f	Specific heat capacity of a carrier fluid	kJ/kg/K
ρ_f	The density of carrier fluid	kg/m³
Q	The heat rate flow	W
C_p	Specific heat capacity of the absorber plate	kJ/kg/K
h_{fp}	Convective heat transfer coefficient between fluid and plate	W·m²/K
h_{fc}	Convective heat transfer coefficient between fluid and cover plate	W·m²/K
T_c	The temperature of the cover plate	K
t	Duration of experimental work	s
y_p	The thickness of the copper plate	m
C_{fx}	Skin friction coefficient	–
E_b	Energy loss per unit width due to the boundary layer formation	J/m

REFERENCES

[1] Dutta, P., Dutta, P. P., Kalita, P., Goswami, P., & Choudhury, P. K. (2021). Energy analysis of a mixed-mode corrugated aluminium alloy (AlMn1Cu) plate solar air heater. *Materials Today: Proceedings, 47*, 3352–3357. http://doi.org/10.1016/j.matpr.2021.07.156

[2] Ural, T. (2019). Experimental performance assessment of a new flat-plate solar air collector having textile fabric as absorber using energy and exergy analyses. *Energy, 188.* http://doi.org/10.1016/j.energy.2019.116116

[3] Wang, L., & Wang, Y. (2021). Research on the collect heat performance of new type collector. *Energy Sources, Part A: Recovery, Utilization and Environmental Effects.* http://doi.org/10.1080/15567036.2021.1954729

[4] Dhaundiyal, A., Gebremichael, G. H., & Atsu, D. (2021). Comprehensive analysis of a solar dryer with a natural draught. *Energy Sources, Part A: Recovery, Utilization and Environmental Effects.* http://doi.org/10.1080/15567036.2021.1951899

[5] Dhaundiyal, A., & Atsu, D. (2022). The effect of thermo-fluid properties of air on the solar collector system. *Alexandria Engineering Journal, 61*(4), 2825–2839. http://doi.org/10.1016/j.aej.2021.08.015

[6] Dhaundiyal, A., & Atsu, D. (2020). The effect of wind on the temperature distribution of photovoltaic modules. *Solar Energy, 201*, 259–267. http://doi.org/10.1016/j.solener.2020.03.012

[7] Ahmadlouydarab, M., Anari, T. D., & Akbarzadeh, A. (2022). Experimental study on cylindrical and flat plate solar collectors' thermal efficiency comparison. *Renewable Energy, 190*, 848–864. http://doi.org/10.1016/j.renene.2022.04.003

[8] Assadeg, J., Al-Waeli, A. H. A., Fudholi, A., & Sopian, K. (2021). Energetic and exergetic analysis of a new double pass solar air collector with fins and phase change material. *Solar Energy, 226*, 260–271. http://doi.org/10.1016/j.solener.2021.08.056

[9] Dhaundiyal, A., & Gebremicheal, G. H. (2022). The effect of psychrometry on the performance of a solar collector. *Environmental Science and Pollution Research, 29*(9), 13445–13458. http://doi.org/10.1007/s11356-021-16353-5

[10] Kizildag, D., Castro, J., Kessentini, H., Schillaci, E., & Rigola, J. (2022). First test field performance of highly efficient flat plate solar collectors with transparent insulation and low-cost overheating protection. *Solar Energy, 236*, 239–248. http://doi.org/10.1016/j.solener.2022.02.007

[11] Saikia, S. S., Nath, S., & Bhanja, D. (2019). Effect of vacuum deterioration on thermal performance of coaxial evacuated tube solar collector considering single- and two-phase flow modelling: A numerical study. *Solar Energy, 177*, 127–143. http://doi.org/10.1016/j.solener.2018.10.089

[12] Nokhosteen, A., & Sobhansarbandi, S. (2021). Numerical modeling and experimental cross-validation of a solar thermal collector through an innovative hybrid CFD model. *Renewable Energy, 172*, 918–928. http://doi.org/10.1016/j.renene.2021.03.070

[13] Alic, E., Das, M., & Akpinar, E. K. (2021). Design, manufacturing, numerical analysis and environmental effects of single pass forced convection solar air collector. *Journal of Cleaner Production, 311.* http://doi.org/10.1016/j.jclepro.2021.127518

[14] Hu, J., Guo, M., Guo, J., Zhang, G., & Zhang, Y. (2020). Numerical and experimental investigation of solar air collector with internal swirling flow. *Renewable Energy, 162*, 2259–2271. http://doi.org/10.1016/j.renene.2020.10.048

[15] Özcan, A., Devecioğlu, A. G., & Oruç, V. (2021). Experimental and numerical analysis of a parabolic trough solar collector for water heating application. *Energy Sources, Part A: Recovery, Utilization and Environmental Effects.* http://doi.org/10.1080/15567036.2021.1924317

[16] Pandey, M., Padhi, B. N., & Mishra, I. (2021). Numerical simulation of solar parabolic trough collector with arc-plug insertion. *Energy Sources, Part A: Recovery, Utilization and Environmental Effects, 43*(21), 2635–2655. http://doi.org/10.1080/15567036.2020.1822467

[17] Dhaundiyal, A., & Gebremicheal, G. H. (2022). The stack effect on the thermal-fluid behaviour of a solar collector. *Energies, 15*(3). http://doi.org/10.3390/en15031188

[18] He, J. H., & Abd Elazem, N. Y. (2021). Insights into partial slips and temperature jumps of a nanofluid flow over a stretched or shrinking surface. *Energies, 14*(20). http://doi.org/10.3390/en14206691

[19] Said, Z., Saidur, R., Sabiha, M. A., Rahim, N. A., & Anisur, M. R. (2015). Thermophysical properties of single wall carbon nanotubes and its effect on exergy efficiency of a flat plate solar collector. *Solar Energy, 115*, 757–769. http://doi.org/10.1016/j.solener.2015.02.037

[20] Arora, C. P (1981). *Refrigeration and air conditioning: (in SI units).* TataMcGraw-Hill Pub. Co, New Delhi.

[21] ASM International (2002) 'Thermal properties of metals', ASM Ready Reference, pp. 5–7.

[22] Van Dyke, M., & Rosenblat, S. (1976). Perturbation method in fluid mechanics. *Journal of Applied Mechanics, 43*(1), 189–190. https://doi.org/10.1115/1.3423785

[23] Whillier, A. (1964). Performance of black-painted solar air heaters of conventional design. *Solar Energy, 8*(1), 31–37. https://doi.org/10.1016/0038-092X(64)90008-8

Chapter 12

Mixing of methanol and water in a micro wavy channel

Chandra Bhushan Vishwakarma, Ananya Dwivedi, and Devendra Yadav

Galgotias College of Engineering and Technology, Greater Noida, Uttar Pradesh, India

CONTENTS

12.1 INTRODUCTION

The world is looking for alternative environment-friendly energy sources that can reduce our reliance on fossil fuels. Fossil fuel combustion results in the release of greenhouse gases that contribute to both global warming and ozone layer depletion (Liu, 2010). As shown in Figure 12.1, fossil fuels account for more than 80% of energy consumption, and it's expected to be exhausted within 100 years if consumed at the same pace. Many countries are currently focusing on renewable energy sources to meet the rising energy demand. However, fossil fuels still continue to be a primary source of energy production, and renewable energy still makes up a small portion of total production (Lund, 2022). Since it is expected that fossil fuel reserves will run out in the coming years, our desire to develop alternate energy sources has grown significantly. One of the alternatives to fossil fuels is renewable energy resources, since they are influenced by climatic and environmental factors, they cannot be relied upon for an extended period (Wang et al., 2021)

DOI: 10.1201/9781003367420-12

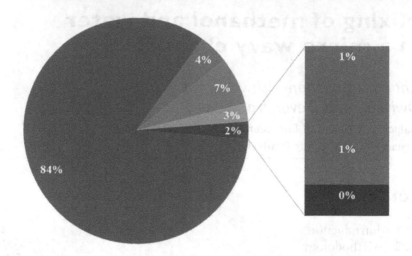

■ Fossile Fuel ■ Nuclear ■ Hydropower ■ Wind ■ Solar ■ Biofuels ■ Other Renewables

Figure 12.1 World Energy Consumption 2020 (TWh). (Energy mix - Our World in Data n.d.)

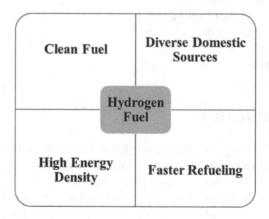

Figure 12.2 Hydrogen feature as a fuel for transportation.

As illustrated in Figure 12.2, there are several special advantages of hydrogen fuel that make it preferable to alternative renewable energy sources. Hydrogen fuel is considerably clear, refills quickly, and has a high energy density, making it an ideal fuel for a vehicle.

Hydrogen is the most abundant and lightest element in the universe. Research and technology development in hydrogen-based fuel has great potential for addressing three major energy challenges: (1) to eliminate pollutants and decouple the link between greenhouse gas emissions and energy utilization in the end-use system, (2) to supply more clean fuel to meet the rising demand for gaseous and liquid fuels, as well as electricity. (3) Increased energy efficiency is required for the production of fuel and electricity (Satterfield Cambridge MA (United States). Dept. of Chemical Engineering, n.d.). (4) the cost of hydrogen production needs to be optimized. The world's current energy system is largely based on the use of fossil fuels such as natural gas, coal, and petroleum in stationary and mobile applications. To achieve a net-zero world by 2050, 306 million tons of green hydrogen would be required per year (Chen et al., 2011). The term hydrogen economy is introduced, which describes the ability of hydrogen to store energy and maintain a strong relationship between the sustainable energy economy and sustainable energy technology (Muradov and Veziroğlu, 2005). The cost of hydrogen production needs to be reduced within the next few decades, and we also need to develop the infrastructure to supply hydrogen to users of light-duty vehicles. These are the key issues that are now impeding the development of a hydrogen economy. We must create a fuel cell and a long-lasting, economical, and ecologically safe hydrogen storage system (Edwards et al., 2007). Hydrogen is not readily available on the planet. It needs to be processed from various raw materials before it can be utilized in the industry. Fossil fuels, biological algae, microorganisms, and even the breakdown of water molecules can be used as raw material in the generation of hydrogen. Because hydrogen may be transferred and stored in liquid, gaseous, or in the form of hydrides, it is favored over other types of energy (Schlapbach and Züttel, n.d.). Hydrogen can be produced in a variety of ways, including steam reforming, electrolysis, dark fermentation, photo fermentation, coal gasification, biomass gasification, partial oxidation, electrolysis, pyrolysis, and plasma reforming. Steam reforming produces over 80% of the world's hydrogen at a low cost. Although methane is the most commonly used fuel for hydrogen production, recent research has shown that methanol and ethanol can also be used (Zhen et al., 2020). One of the most significant issues in hydrogen production is the lack of an efficient hydrogen production method. Existing reforming systems have several drawbacks, including a high-production cost, lower efficiency, emissions of greenhouse gases, carbon deposition, and complexity (Rohr, 2002).

Microreactor technology has drawn researcher's attention recently because of its high performance and applicability for onboard applications. The key benefits of microreactor technology are shown in Figure 12.3.

In a microreactor, hydrogen can be produced through steam reforming in a thermally driven catalytic reaction. Microreactors have channels typically in the order of micrometers. The ratio of surface area to volume is much higher which enables microreactors safe handle reactions and reaction parameters.

The microreactor facilitates the use of a wide range of fuels for reforming. Methanol is the most popular fuel for steam reforming in a microreactor to produce hydrogen. The comparison of methane and methanol fuel for hydrogen production is shown in Table 12.1 below.

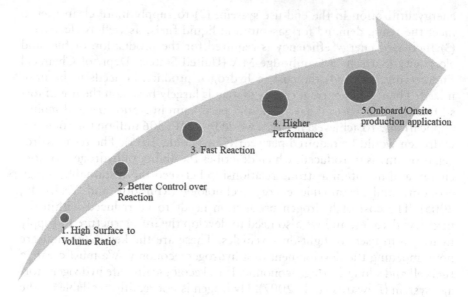

Figure 12.3 Advantages of microreactor.

Table 12.1 Methane vs Methanol for Hydrogen-Fueled Ship Application (Lee et al., 2020)

Factors	Methane	Methanol
$T_{reforming}$ (°C)	700–900	150–350
S/C	3–4	1–2
P_{ref} (Bar)	1–25	1–3.5
Efficiency	Low	High
Fuel Cost	Low	High
Space required	Low	High

Methanol is a unique and advantageous fuel. Methanol has the same hydrogen-to-carbon ratio (4:1) as methane. Unlike methane, methanol has no carbon-carbon bonds that must be broken, so it can be converted to hydrogen at a lower temperature of around 150°C–350°C, whereas another fuel requires >500°C (Chen et al., 2009). It has a low boiling point of up to 65°C, allowing for easy valorization in roughly the same temperature range as water (Ravi et al., 2017). Unlike liquefied petroleum gas (LPG) or methane, it is in the liquid phase at normal environmental temperature and atmospheric pressure. Because methanol is miscible with water, it does not spread over large areas of open water the way oil and gasoline spills do (Palo et al., 2007) Methanol's miscible properties also provide a key advantage in terms of fuel handling in the system, for example, water recycling is not required, and the system can be simplified by providing a single inlet and a single vaporizer. When it comes to fuel reforming, the fact that methanol is inherently a synthetic fuel that does not suffer from sulfur contamination means that the system does not require a front-desulfurization operation or sulfur-tolerant catalysts to operate on methanol. Methanol can be produced from a variety of sources, including natural gas, biomass via a syngas-to-methanol conversion process, and coal. Methanol is primarily formed when steam methane reforming produces a mixture of CO_2, CO, and H_2 at 200°C–300°C over Cu/ZnO catalysts. Syngas is then converted to methanol, as shown by the equations (Hsueh et al., 2009).

$$2CH_4 + 3H_2O \rightarrow CO + CO_2 + 7H_2 \tag{12.1}$$

$$CO_2 + 3H_2 \rightarrow CH_3OH + H_2O \tag{12.2}$$

Many researchers have attempted to examine the performance of micro-reactors in producing hydrogen (Hafeez et al., 2020; Park et al., 2005; Qian et al., 2017). Chen et al. (2011) constructed and numerically analyzed the three-dimensional model for methanol steam reforming a rectangular shape tree-shaped network microreactor. The study claimed that the design gives a higher methanol conversion rate. Hsueh et al. (2009) and Chen et al. (2009) numerically investigated a plate steam methanol microreactor with parallel microchannels. The investigation shows a better H_2 production rate and the methanol conversion rate for the smaller aspect ratio. Kawamura et al. (2006) had a micro reformer with a serpentine microchannel for supplying small Proton-exchange membrane fuel cells (PEMFC) with the help of the Microelectromechanical systems (MEMS) system. Microchannels have many applications such as in cooling devices, MEMS devices, and chemical reactors (Bayareh et al., 2020; Sia and Whitesides, 2003). Aside from the microchannel's quick reaction advantage, the mixing of fluid streams is also critical, as it can affect the overall performance of the microreactor as well as its overall design and size. The design of microreactors has not yet considered the mixing of fluid streams in available studies (Meijer et al., 2009). The role of the microchannel in the microreactor design is the very important part that mixes the incoming fluid streams and possesses the chaotic advection inflow (CFD Module User's Guide, 1998; Sowndarya et al., 2021) by improving the exposure time between the fluid streams. Sowndarya et al. (2021) utilized passive micromixers with a Y-shaped inlet port for the mixing of two-phase fluids. The results revealed that the mixing index (MI) increases with the thickness of the vanes. The microchannel possesses a smaller quantity of mass flow rate and also the velocity at the inlet is very low that the laminar flow condition is fulfilled. The mixing of fluids in a microchannel with rectangular grooves was studied by Alam and Kim (2012). Variations in the Reynolds number were used to analyze a broad variety of geometrical characteristics, including width and depth effects. The results showed that the grooved microchannel enhanced mixing performance and that the mixing index was not sensitive to the depth of grooves, but was sensitive to the width of grooves for some synod ranges. Many researchers showed their interest in the field of methanol steam reforming. Lu et al. (2022) studied designs of microchannel structures for hydrogen supply for methanol steam reforming and found that sinusoidal microchannel with a dimple (SMD) had shown the best property according to HTEF, Nu, f and experimental results. Wang et al. (2020) numerically modeled the microchannel reactors for hydrogen production based on fractal geometry with gradient porous surface and found that the overall performance of microreactors having positive gradient pore-sized surface is relatively better, pressure drop was decreased up to 8% and heat transfer performance was increased by 18%. Herdem et al. (Herdem et al., 2019) designed and arranged the catalyst layer in order to improve the performance of a microchannel in methanol steam reforming and the study found that with less catalyst the segmented coating had increased methanol conversion by nearly 25%.

So far there is no study available on analyzing the mixing of water and methanol for microreactor applications. This work aims at the microchannel application for analyzing the mixture of methanol and water in the wavy channel that could help in obtaining the suitable length while designing the microreactor. For analyzing the homogeneity of the mixer, the concentration at all of the corners and mixing index along the arc length are evaluated. In the analysis, parametric sweeps are undertaken to see how the results change as parameters like reforming temperature and mass flow rate change. The concentration and mixing index were used as output parameters in sweep analysis to measure the performance of species mixing.

12.2 METHODOLOGY

In the microreactor, incoming fluid streams are mixed and vaporized before being transported to the reforming section. The study starts with the design of a microchannel that will facilitate the mixing of incoming fluid streams. Autodesk Fusion 360 was used to create a 3D-simulation model for a microchannel. The geometric dimensionality of the computational domain is directly affected by operating conditions such as temperature and liquid flow rate. The geometry was then imported into COMSOL for simulation, where three physics were added: laminar flow under steady flow in fluid flow, dilute species transport under chemical transport, and heat transfer in the fluid under heat transfer. The steady-state was chosen, as the homogeneity of methanol and water was to be investigated.

12.2.1 Governing equations

To investigate the homogeneity of the mixed fluid in a wavy microchannel, Laminar flow conditions, mixing of diluted species, and heat transfer modules were utilized. Incoming fluid streams are allowed to enter at 10^{-6} kg/s and a temperature of 293°C. The bottom surface of the channel is kept maintained at a constant temperature. The governing equations for solving the problems are shown below:

Laminar flow (CFD Module User's Guide, 1998)

Navier-stokes equation governs the single-phase fluid flow interface which can be expressed in the following equations

$$\rho(u \cdot \nabla)u = \nabla \cdot \left[-pI + K\right] + F \tag{12.3}$$

$$\rho\nabla \cdot u = 0 \tag{12.4}$$

$$K = \mu\left(\nabla u + (\nabla u)^T\right) \tag{12.5}$$

where ρ=density (kg/m³), u=velocity vector (m/s), p=pressure (Pa), T=absolute temperature (K), F=volume force vector (N/m³), C_i=specific heat capacity at constant pressure (J/(kg K)), q=heat flux vector (W/m²).

The three equations shown above are vector equations. The first two equations express mass conservation through momentum conservation and continuity equation, respectively. The third equation deals with energy conservation in terms of temperature.

Transport of diluted species (Chemical Reaction Engineering Module User's Guide, 1998)

This interface offers a predefined modeling environment that makes it possible to investigate the evolution of chemical species using it. Both incoming species are diluted. For one or more species, solving the mass conservation equation by diffusion and convection is the default node attribute for species transport. The diffusion and conversion mechanisms are included in Equations (12.6) and (12.7).

$$\nabla \cdot J_i + u.\nabla c_i = R_i \tag{12.6}$$

$$J_i = -D_i\nabla C_i \tag{12.7}$$

where C_i= concentration of the species (mol/m³), D_i=diffusion coefficient (m²/s), R_i=expression for reaction rate for both the species (mol/(m³s)), u=mass averaged velocity vector (m/s)

Heat Transfer in fluids (Heat Transfer Module User's Guide, 1998)

$$\rho c_p u.\nabla T + \nabla \cdot q = Q + Q_p + Q_{vd} \tag{12.8}$$

where ρ=density of layer (kg/m^3), C_p=specific heat capacity at constant pressure (J/(kg.)), T=temperature (K), u=velocity vector (m/s), q=heat flux by conduction (W/m^2), Q=heat source excluding viscous dissipation (W/m^3)

For heat balance the equation is;

$$\frac{dE_\Omega}{dt} = P_{Str} + Q_{exch} \tag{12.9}$$

whereas the Q_{exch} is the heat exchanged rate, E is the internal energy the stress power is present in the fact that through dissipation power is converted into heat. This equation holds for macroscopic continuous domain Ω where by using specific internal energy (per unit mass).

Internal energy is defined

$$E_\Omega = \int_\Omega E\,dm \tag{12.10}$$

For the variation of internal energy with time

$$\frac{dE_\Omega}{dt} = \int_\Omega \frac{dE}{dt}\,dm = \int_\Omega \rho \frac{dE}{dt}\,dv \tag{12.11}$$

whereas ρ is the density, constant elementary masses are denoted by dm, and Ω elementary volume of is denoted by dv.

Based on continuum mechanics theory stress power can be defined as

$$P_{str} = \int_\Omega (\sigma : D)\,dv \tag{12.12}$$

whereas the operator is ":" is contraction, D is the strain rate and σ is the Cauchy stress

Now the equivalent stress power can be expressed in the following terms:

$$P_{str} = \int_\Omega \frac{1}{\det(F)} \cdot \left(P : \frac{dF}{dt} \right) dV \tag{12.13}$$

By using Fourier's law, the exchanged heat rate Q_{ecxh} can be accounted for heat conditions. The following equation is a summary of all possible sources of heat for the domain, including joule heating and heat from chemical reactions

$$Q_{exch} = -\int_{\partial\Omega} (q.n)\,ds - \int_{\partial\Omega} (q_r \cdot n)\,ds + \int_\Omega Q\,dv \tag{12.14}$$

All of these components can now be combined into a heat-balanced equation for terms calculated from Equation 12.8, which can result in

$$\int_{\Omega} \rho \frac{dE}{dt} dv + \int_{\partial\Omega} (q \cdot n) ds + \int_{\partial\Omega} (q_r \cdot n) ds = \int_{\Omega} (\sigma : D) dv + \int_{\Omega} Q dv \quad (12.15)$$

The localization form in the frame will be determined by the following equation

$$\rho \frac{\partial E}{\partial t} + \rho u \cdot \nabla E + \nabla \cdot (q + q_r) = \sigma : D + Q \quad\quad\quad (12.16)$$

12.2.2 Mixing index

The Mixing Index (MI) is a critical factor in analyzing the mixing performance of any mixture. It is the extent to which fluid flow promotes the dispersion of attributing fluids and leads to homogeneity (Sadegh Cheri et al., 2013)

$$\tau^2 = \frac{1}{n} \sum_{i=1}^{i=n} (C_i - C_\infty)^2 \quad\quad\quad (12.17)$$

$$MI = 1 - \sqrt{\frac{\tau^2}{\tau_{max}^2}} \quad\quad\quad (12.18)$$

whereas τ denotes the variation in concentration of each coordinate within the microchannel's cross-section with concentration at infinity. And n denotes the number of coordinate points for which concentration was obtained. C_i denotes the concentration of both species at each coordinate, C_∞ denotes the concentration at infinity. i.e., the concentration at which no mixing was possible, MI is the Mixing Index, which is scaled from 0 to 1. If MI is one, the species are completely mixed; if it is zero, there is no mixing.

12.2.3 Parametric sweep

Parametric sweep is a COMSOL feature that is used in the study. A parametric sweep is conducted in the sequence of stationary studies to identify the solution to a problem that emerges when we need to vary some parameter of significance. In this investigation of the mixing performance of microchannel, we looked at temperature change, species-specific mass flow rates, and water-methanol concentration. The temperature ranges from 393°C to 593°C, concertation ranges from 0 to 2 mol/m³ and the mass flow rate was varied from 2.0×10⁻⁷, 4.0×10⁻⁷, 1.8×10⁻⁶.

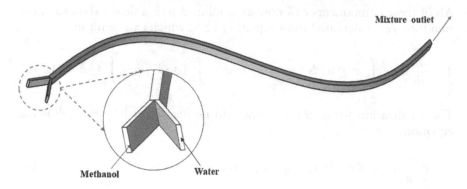

Figure 12.4 Microchannel with V-type inlet port.

12.2.4 Geometric details

Autodesk Fusion 360 is used to create a 3D model of a microchannel. The sine curve $y=2\sin(0.628x)$ was used to design the microchannel, which has a wavelength of 0.1 m and a cross-section of 1 mm×2 mm. The two inlets were sketched with 60° angle between them and of length 5 mm in the XY plane and extruded 2 mm in the Z direction. The built model is shown in Figure 12.4. The built microchannel model has a total volume of 2.1786×10^{-7} m^3 total length of a microchannel of 0.10857 and forms a total of three domains, 17 boundaries, 31 edges, and 81 vertices all formed automatically.

12.3 RESULT AND DISCUSSION

The full-scale simulation of the mixing of water and methanol in a wavy microchannel is performed using COMSOL multiphysics. There is a total of three domains; two for the water and methanol and one for the wavy microchannel. The starting channel is filled with air. Water and methanol are supplied from their respective inlet port with a mass flow rate of 10^{-6} kg/s. The outlet of the microchannel is set to the static pressure of 0 Pa.

12.3.1 Mesh build-up

When accuracy is required in simulation in fluid flow model simulation, the model required a fine resolution to converge. For meshing, a physical control for mesh setup is used in Figure 12.5. The physics-controlled Finer Mesh in the current domain has 424,128 elements. The minimum quality is 0.1091 and the average quality is 0.6734.

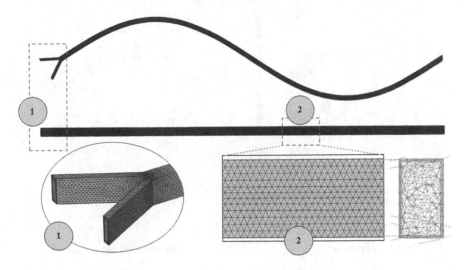

Figure 12.5 Developed mesh on wavy microchannel.

12.3.2 Mixing of two species

The microchannel bottom surface temperature is kept at a constant temperature of 393.15 K. That is the required temperature for the proper evaporation of both species. The heat transfer surface area is 1.0893×10^{-4}. As shown in Figure 12.6, the concentration of both species was found to be higher at the edges and walls. The evaluation of concentrations at edges no doubt gives clear-cut information about the homogeneity of the mixture.

Figure 12.7 shows the change in concentration of water and methanol along the arc length of each edge Figure 12.7a. As shown in Figures 12.7b and c. Along the length of the arc of edges, the concentration of methanol and water reaches an average concentration of 0.5 mol/m^3. The graph shows that the concentration of the methanol initially drops from 1 mole/m^3 to 0.35 m of its arc length, and then gradually rises to 0.5 mole/m^3. The clearer confirmation of the mixture is further evaluated by calculating the concentration difference between water and methanol. As shown in Figure 12.7d. The concentration difference decreases with arc length and reaches zero at the end of the microchannel, confirming the mixture's homogeneity.

The difference in both species' concentrations is depicted in Figure 12.8. As we can see, the mixture is homogeneous and the maximum area is close to zero, indicating that there are no concentration differences in the mixture. However, a substantial concentration is present on two of the edges for a very small area.

The pressure of the mixed fluid falls along the flow direction, and there is little change in velocity. The microchannel design and the V shape inlet port,

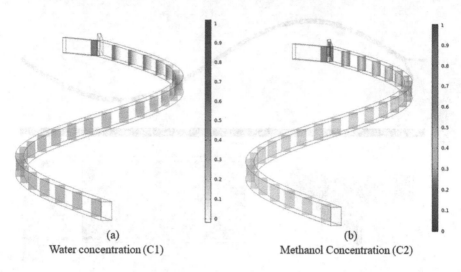

Figure 12.6 Surface contour of incoming fluid streams for variation in concentration.

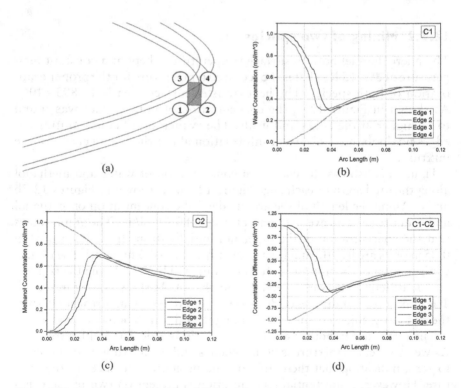

Figure 12.7 (a) Microchannel edges;1,2,3,4, (b) variation of water concentration along Arc length, (c) variation of methanol concentration along Arc length, and (d) difference between the concentration of water and methanol along Arc length.

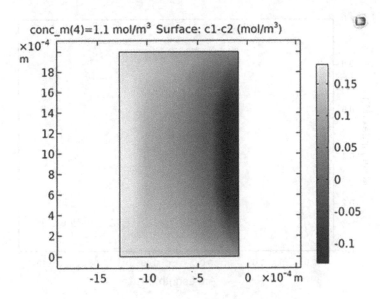

Figure 12.8 Surface contour of concentration difference at the outlet of the microchannel.

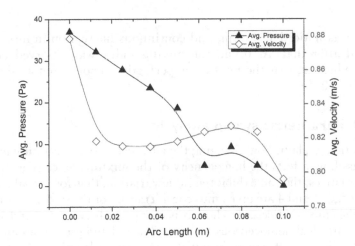

Figure 12.9 Variation of average pressure (Pa) and average velocity (m/s) along Arc length (m).

which maintains velocity to this extent, are responsible for the final velocity outcomes. The sine wave microchannel enables rapid mixing of the entering fluid while reducing velocity to a minimum as shown in Figure 12.9. On the other hand, the pressure along the flow direction reduces.

Along with the flow, the mixture's mixing index increases. Due to the wavey design of the microchannel, very steep growth in the mixing of

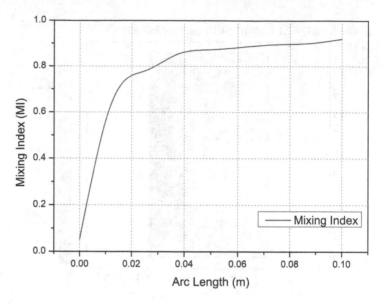

Figure 12.10 Variation of Mixing Index (MI) along Arc length (M).

species was noticed at 0.02 m, and continuous fluctuation in mixing was detected after that. At 0.1 m, the mixing index was observed close to unity, indicating that the mixture is perfectly homogenized as shown in Figure 12.10.

12.3.3 Parametric sweep analysis

This section explains how changing the parameters of temperature and species mass flow affects the homogeneity of the mixture. Since edges 1 and 2 are alternatives, they all exhibit the same variation, thus for the sake of simplicity, edges 1 and 4 are taken into consideration for the study. Figure 12.11 shows the graph that describes how concentrations between 0.9 and 1.1 mol/m³ offer the ideal homogeneous solution. The red and green lines are pointing in the direction of zero at the outlet where the arc length is 0.1, indicating that there is no difference in the mixture, which amply supports the claim of homogeneity.

But at the same time, it is interesting to see from Figure 12.12 that there is a very minute variation in concentration in the observer when the temperature is varying from 393 to 593 K which indicates that this temperature range is suitable for the operation. This Figure also states that low operating temperature is good for achieving homogeneity as the temperature is increasing to 693 K the difference in concentration is increasing.

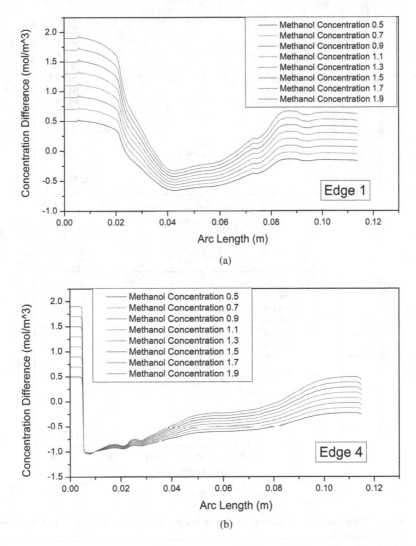

Figure 12.11 The parametric sweep of the concentration (a) along edge 1 and (b) along edge 4.

12.4 CONCLUSION

The mixing of water and methanol in the wavy microchannel is demonstrated in this study. The COMSOL model for the transportation of diluted species is used for the observation of the concentration variation of each species. The laminar flow model was used to examine the velocity and pressure

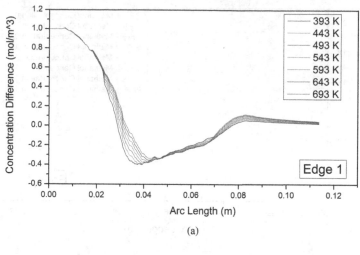

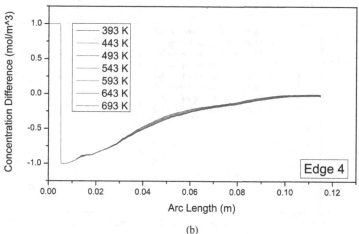

Figure 12.12 The parametric sweep of the temperature (a) along edge 1 and (b) along
edge 4.

in a microchannel, while the heat transfer in the fluid model was employed
to provide an actual scenario of fluid mixing for microreactor applications.

According to the findings, the chosen wavy shape geometry of the microchannel facilitates the proper mixing of water and methanols. At the microchannel's exit, the mixing index of the species is close to unity, indicating
that the mixture is homogeneous. The concentration of both incoming species was found to be highest at the walls in any cross-section. And decreased
toward the center. The V-shaped inlet ports and sine wave microchannel
design are responsible for the smaller drop in velocity and pressure of the
mixture, making this design suitable for microreactor application.

ACKNOWLEDGEMENT

The authors would like to thank the Department of mechanical engineering, Galgotias College of engineering and technology, Greater Noida for providing research infrastructure and support for this study.

REFERENCES

Alam, Afroz, and Kwang Yong Kim. 2012. "Analysis of Mixing in a Curved Microchannel with Rectangular Grooves." *Chemical Engineering Journal* 181–182: 708–16.

Bayareh, Morteza, Mohsen Nazemi Ashani, and Azam Usefian. 2020. "Active and Passive Micromixers: A Comprehensive Review." *Chemical Engineering and Processing - Process Intensification* 147: 107771.

CFD Module User's Guide. 1998. www.comsol.com/blogs.

Chemical Reaction Engineering Module User's Guide. 1998. www.comsol.com/blogs.

Chen, Falin, Min-Hsing Chang, Chih-Yi Kuo, Ching-Yi Hsueh, and We-Mon Yan. 2009. "Analysis of a Plate-Type Microreformer for Methanol Steam Reforming Reaction." *Energy and Fuels* 23(10): 5092–98.

Chen, Yongping, Chengbin Zhang, Rui Wu, and Mingheng Shi. 2011. "Methanol Steam Reforming in Microreactor with Constructal Tree-Shaped Network." *Journal of Power Sources* 196(15): 6366–73.

Edwards, Peter P., V. L. Kuznetsov, and William I. F. David. 2007. "Hydrogen Energy." *Philosophical Transactions of the Royal Society A: Mathematical, Physical and Engineering Sciences* 365(1853): 1043–56.

Hafeez, Sanaa, Elsa Aristodemou, George Manos, Sultan M. Al-Salem, and Achilleas Constantinou. 2020. "Modelling of Packed Bed and Coated Wall Microreactors for Methanol Steam Reforming for Hydrogen Production." *RSC Advances* 10(68): 41680–92.

Heat Transfer Module User's Guide. 1998. www.comsol.com/blogs.

Herdem, Münür Sacit, Mayur Mundhwa, Siamak Farhad, and Feridun Hamdullahpur. 2019. "Catalyst Layer Design and Arrangement to Improve the Performance of a Microchannel Methanol Steam Reformer." *Energy Conversion and Management* 180: 149–61. https://www.sciencedirect.com/science/article/pii/S0196890418312226.

Hsueh, Ching Yi, Hsin sen Chu, and Wei Mon Yan. 2009. "Numerical Study on micro-Reformer Performance and Local Transport Phenomena of the Plate Methanol Steam Micro-Reformer." *Journal of Power Sources* 187(2): 535–43.

Kawamura, Yoshihiro, Naotsugu Ogura, Tadao Yamamoto, and Akira Igarashi. 2006. "A Miniaturized Methanol Reformer with Si-Based Microreactor for a Small PEMFC." *Chemical Engineering Science* 61(4): 1092–1101.

Lee, Hyunyong, Inchul Jung, Gilltae Roh, Youngseung Na, and Hokeun Kang. 2020. "Comparative Analysis of On-Board Methane and Methanol Reforming Systems Combined with HT-PEM Fuel Cell and CO_2 Capture/Liquefaction System for Hydrogen Fueled Ship Application." *Energies* 13(1): 224.

Liu, Jingjing. 2010. "A Risk-Based Life Cycle Assessment of OPAL Petrol and BP Regular Unleaded Petrol." PhD diss., Memorial University of Newfoundland.

Lu, Weiqin, Rongjun Zhang, Sam Toan, Ran Xu, Feiyi Zhou, Zhao Sun, and Zhiqiang Sun. 2022. "Microchannel Structure Design for Hydrogen Supply from Methanol Steam Reforming." *Chemical Engineering Journal* 429: 132286. https://www.sciencedirect.com/science/article/pii/S1385894721038651.

Lund, Henrik. 2022. *Aalborg Universitet Renewable Energy Strategies for Sustainable Development Renewable Energy Strategies for Sustainable Development*. Downloaded from vbn.aau.dk

Meijer, Han E.H., Mrityunjay K. Singh, Tae Gong Kang, Jaap M. J. den Toonder, Patrick D. Anderson. 2009. "Passive and active mixing in microfluidic devices." In *Macromolecular Symposia*, vol. 279, no. 1, pp. 201–209. Weinheim: WILEY-VCH Verlag.

Muradov, N. Z., and T. N. Veziroğlu. 2005. "From Hydrocarbon to Hydrogen-Carbon to Hydrogen Economy." *International Journal of Hydrogen Energy* 30(3): 225–37.

Palo, Daniel R., Robert A. Dagle, and Jamie D. Holladay. 2007. "Methanol Steam Reforming for Hydrogen Production." *Chemical Reviews* 107(10): 3992–4021.

Park, Gu Gon, Sung-Dae Yim, Young-Gi Yoon, Chang-Soo Kim, Dong-Joo Seo, and Koichi Eguchi. 2005. "Hydrogen Production with Integrated Microchannel Fuel Processor Using Methanol for Portable Fuel Cell Systems." *Catalysis Today* 110(1–2): 108–13.

Qian, Miao, Deqing Mei, Zoudongyi Yi, Yanbing Feng, and Zichen Chen. 2017. "Fluid Flow and Heat Transfer Performance in a Micro-Reactor with Non-Uniform Micro-Pin-Fin Arrays for Hydrogen Production at Low Reynolds Number." *International Journal of Hydrogen Energy* 42(1): 553–61.

Ravi, Manoj, Marco Ranocchiari, and Jeroen A. van Bokhoven. 2017. "Die Direkte Katalytische Oxidation von Methan Zu Methanol - Eine Kritische Beurteilung." *Angewandte Chemie* 129(52): 16684–704.

Rohr, Don. 2002. *Natural Gas Processing Technologies for Large Scale Soild Oxide Fuel Cells*. https://www.researchgate.net/publication/315614971.

Sadegh Cheri, Mohammad, Hamid Latifi, Mohammadreza Salehi Moghaddam, and Hamidreza Shahraki. 2013. "Simulation and Experimental Investigation of Planar Micromixers with Short-Mixing-Length." *Chemical Engineering Journal* 234: 247–55. https://www.sciencedirect.com/science/article/pii/S1385894713011200.

Satterfield Cambridge MA (United States). Dept. of Chemical Engineering. n.d. *Heterogeneous Catalysis in Industrial Practice*. 2nd Edition. Massachusetts Inst. of Tech, Oak Ridge.

Schlapbach, Louis, and Andreas Züttel. n.d. *Hydrogen-Storage Materials for Mobile Applications*. http://doc.rero.ch.

Sia, Samuel K., and George M. Whitesides. 2003. "Microfluidic Devices Fabricated in Poly(Dimethylsiloxane) for Biological Studies." *Electrophoresis* 24(21): 3563–76.

Sowndarya, K., S. Monica, M. S. Abhisheka, A. K. Pradikshan, and M. Venkatesan. 2021. "Vane Induced Two Phase Flow Mixing." In *IOP Conference Series: Earth and Environmental Science*, vol. 850, no. 1, p. 012022. IOP Publishing.

Wang, Haoyu, Haiying Shen, Jiechao Gao, Kevin Zheng, and Xiaoying Li. 2021. "Multi-Agent Reinforcement Learning Based Distributed Renewable Energy Matching for Datacenters." In *ACM International Conference Proceeding Series*, Association for Computing Machinery.

Wang, Yancheng, Weidi Zeng, Deqing Mei, and Xiaolong Zhang. 2020. "Numerical Modeling of Microchannel Reactors with Gradient Porous Surfaces for Hydrogen Production Based on Fractal Geometry." *International Journal of Hydrogen Energy* 45(38): 19733–44. https://www.sciencedirect.com/science/article/pii/S0360319920318188.

Zhen, Xiaoyan Li, Yang Wang, Daming Liu, and Zhi Tian. 2020. "Comparative Study on Combustion and Emission Characteristics of Methanol/Hydrogen, Ethanol/Hydrogen and Methane/Hydrogen Blends in High Compression Ratio SI Engine." *Fuel* 267: 117193. https://www.sciencedirect.com/science/article/pii/S0016236120301885.

Chapter 13

Heat transfer and pressure drop penalty study in flattened micro-finned tubes

Ankit R. Singh
Indian Institute of Technology

Anand Kumar Solanki
Gayatri Vidya Parishad College of Engineering

Nitesh Dutt
College of Engineering

CONTENTS

NOMENCLATURE

Roman Symbols

A	Area, m^2
C_p	Specific heat at constant pressure, J/kg/K
D	Tube diameter, m
dH	Hydraulic diameter, m
f	Friction factor
h	Heat transfer coefficient, $W/m^2/K$
I	Inlet turbulence intensity
k	Turbulence kinetic energy, m^2/s^2
kcond	Thermal conductivity, W/m/K
L	Length, m
$\dot{m}$	Mass flow rate, kg/s
Nu	Nusselt number
P	Pressure, N/m^2

DOI: 10.1201/9781003367420-13

p	Wetted Perimeter, m
Pr	Prandtl ratio
q	Heat flow, W
Re	Reynolds number
T	Temperature, K
u, v	Velocities in i and j directions, m/s

Greek Symbols

ρ	Density, kg/m³
μ	Dynamic viscosity, kg/m/s
ε	Turbulence dissipation rate, m²/s³
Δ	Change/Difference
ϵ	Surface roughness, m
μ_t	Eddy viscosity

Subscripts & Superscripts

c/s	Cross-section
i, j	Coordinate directions
in	Inlet
out	Outlet
p	Plain circular tube
s	Surface

Acronyms

HTC   Heat Transfer Coefficient

13.1 INTRODUCTION

Much research studies have been carried out over the years to explore various heat transfer enhancement techniques. Many geometrical modifications, such as extended surfaces, re-entrant cavities, and fins, have been explored. One such geometric modification is a microfin tube with a typical fin height to the inside tube diameter of 0.02–0.04. In addition, microfins with different geometric types, namely rectangle-shaped, triangle-shaped, and trapezium-shaped, are investigated for heat transfer enhancement. The helically coiled tubesare also being explored for heat transfer enhancement. Helically coiled tubes significantly improve heat transfer and minimal pressure losses primarily because of the centrifugal force-induced secondary flow (Reay, 1991). The use of microfins in the helically coiled tubes further improves the thermal-hydraulic characteristics of tubes (Solanki and Kumar, 2018). Enhancing heat transfer is crucial in many other areas (Dutt et al., 2020; Singh et al., 2018, 2020).

The effect of microfins with varying helix angles in the helical and straight tubes is also a crucial area of research in studying thermal-hydraulics characteristics. Yang and Hrnjak (2018) studied the heat transfer and pressure drop with straight micro fins in R410a evaporation. However, it has been further shownthat the enhancement in boiling heat transfer isrestricted to straight fins. Comparedto the smooth tubes, the 0° helical microfins may improve the boiling heat transfer of R32 by up to 60% (Wu et al., 2015).

The microfins induce flow turbulence which enhances the heat transfer. Copetti et al. (2004) reported 2.9 times more heat transfer coefficient for the microfinned tube and 80% enhancement. Further, the microfinned tubes with Reynolds numbers lower than acritical value be have as smooth tubes (Xiao-Wei et al., 2007); but the thermal performance could be twice the smooth tube. A similar observation of enhanced heat transfer using microfins has been reported by many researchers (Dastmalchi et al., 2017; Han and Lee, 2005; Wang and Rose, 2004).

The flat and elliptical tubesproduced from circular tubes have also been explored to study the heat transfer enhancement. Wilson et al. (2003) experimentally studied flow characteristics in the flattened tube with R134a and R410A Refrigerants for a mass flux between 75 and 400 kg/m²/s and quality between 10% and 80%. An 18° helical tube results in the most significant heat transfer coefficients. Nasr et al.(2010) experimented with a horizontal plain tube and flattened tubes for heat transfer and pressure drop characteristics of R-134a refrigerant. They reported that the 5.5 mm internal height tube performed better than other flattened tubes. An experiments were carried out by Razi et al. (2011) to study the heat transfer and pressure drop characteristics of CuO−base oil nanofluids in laminar flow in a flattened tube. The heat transfer cocfficient and pressure drop have increased with nanofluids than base fluid.

Thus, the above literature showsthat the application of microfins will enhance surface heat transfer. Moreover, the surface area to cross-section area ratio is typically higher in the flattened tubes. Hence, flattened tubes with micro fins could be examined as an alternative geometric alteration for heat transfer enhancement. In this chapter, a single-phase steady-state flow in the circular and flattened microfin tubes has been numerically simulated to study the effectiveness of the tubes. The performance parameters such as heat transfer, pressure drop penalty, and friction factor are compared for circular and flattened microfin tubes. Straight trapezoidal micro fins are used in the modeling with an isothermally heated tube wall. Reynold numbersvary between 10,000 and 30,000, creating a turbulent flow.

13.2 NUMERICAL MODELS AND DATA REDUCTION

The governing equations (continuity equation, Navier−Stokes or momentum equation, and energy equation) can be written as Equations(13.1)−(13.3).

The viscous dissipation is not modeled, and the gravity effect is not considered in the formulation.

$$\frac{\partial}{\partial x_i}(\rho u_i) = 0 \tag{13.1}$$

$$\frac{\partial}{\partial x_j}(\rho u_i u_j) = -\frac{\partial P}{\partial x_i} + \frac{\partial}{\partial x_j}\mu\left(\frac{\partial u_i}{\partial x_j} + \frac{\partial u_j}{\partial x_i}\right) + \frac{\partial}{\partial x_j}\left(-\overline{\rho u_i' u_j'}\right) \tag{13.2}$$

$$\frac{\partial}{\partial x_i}(\rho \, u_i T) = \frac{\partial}{\partial x_j}\left[\left(\frac{\mu}{\text{Pr}} + \frac{\mu_t}{\text{Pr}}\right)\frac{\partial T}{\partial x_j}\right] \tag{13.3}$$

There alizable k–ε turbulence model with enhanced wall treatment is applied to resolve turbulent term of the momentum equation. The turbulent Reynolds shear stress and the turbulent kinematic viscosity are given by Equations (13.4) and (13.5), respectively. The turbulence intensity has been predicted by using Equation (13.6).

$$-\overline{\rho u_i' u_j'} = \mu_t\left(\frac{\partial u_i}{\partial x_j} + \frac{\partial u_j}{\partial x_i}\right) \tag{13.4}$$

$$\mu_t = \rho C_\mu \frac{k^2}{\varepsilon} \tag{13.5}$$

$$I = 0.16 \, \text{Re}^{-0.125} \tag{13.6}$$

The flattened tube modelsare generated in the ANSYS design modeler, and the meshings are generated in the ANSYS meshing environment. The SIMPLE algorithm of ANSYS Fluent R20 is used for pressure-velocity coupling. The second-order up wind method is used in discretization of the momentum and energy equations. Inlet temperature of the water is 298.15 K. Inlet velocity is estimated from Reynolds number (Re) by using Equation(13.7). The tube hydraulic diameter (d_H) is estimated by using Equation (13.8). Re varied between 10,000 and 30,000. The tube wall temperatureis 353.15 K. The operating pressure is 101,325 Pa. Outlet gauge pressureis set to 0 Pa. Noslip wall boundary has beenconsidered in the formulation. The convergence criteria are 10^{-6} for energy and 10^{-5} for continuity and momentum equations, respectively.

$$\text{Re} = \frac{\rho u_{in} d_H}{\mu} \tag{13.7}$$

$$d_H = \frac{4A_{cls}}{p_{\text{wetted}}} \tag{13.8}$$

Heat flux (q) at tube surface is estimated by using Equation (13.9). T_{out} and T_{in} are area-weighted temperature at outlet and inlet boundaries, respectively. The convective heat transfer coefficient (h) is estimate dusing Equation (13.10). T_{fluid} is arithmetic mean temperature of inlet and outlet temperatures.

$$q = \dot{m}C_P\left(T_{out} - T_{in}\right) \tag{13.9}$$

$$h = \frac{q}{A_s\left(T_{fluid} - T_{wall}\right)} \tag{13.10}$$

The Nusselt number and Darcy–Weisbach friction factor (f) are estimated using Equations (13.11) and (13.12). Performance Evaluation Factor (PEF), as given in Equation (13.13), is used in this study to compare the effectiveness of flattened micro-finned tubes.

$$Nu = \frac{h d_H}{k} \tag{13.11}$$

$$f = \frac{2(P_{inlet} - P_{outlet})d_H}{\rho u_{in}^2 L} \tag{13.12}$$

$$PEF = \left(\frac{h_{flat}A_{s,\,flat}}{h_{circ}A_{s,\,circ}}\right) \Big/ \left(\frac{f_{flat}}{f_{circ}}\right) \tag{13.13}$$

The circular and flattened tubes with micro fins are schematically represented in Figure 13.1. The appropriate dimensions are also mentioned in the figure. The total circumferential or peripheral length of the flattened cross-section is assumed to be equal to the circular cross-section for a given flattened tube. The ratio of microfin height to circular tube diameter is set to 0.025. The 12 and 16 microfins cases are used to study the effect of variation of circumferential fins. The tube is 1 meter long.

13.3 MESH-SENSITIVE STUDY

A mesh-sensitive study has beencarried out for circular tubes with microfins. The fluid domain has beenseparated and meshed with hexahedral mesh elements. The 12 inflation layers with a first-layer thickness of 1 μm are used to resolve the turbulence close to fluid walls. The meshings of the fluid domain of micro-finned tubes are schematically represented in Figure 13.2.

Validation of numerical techniques with existing correlations (Celen et al. (2013), Dittus and Boelter (1985), and Gnielinski (1976)) for smooth circular tubes without microfins is provided by Singh and Solanki (2022).

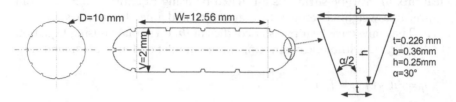

Figure 13.1 Schematics of tubes: (Left) Circular tubes with micro fins, (Central) Flattened tubes with micro fins, (Left) Trapezoidal micro fin.

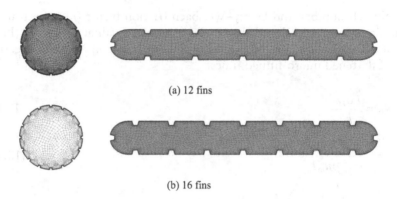

(a) 12 fins

(b) 16 fins

Figure 13.2 The meshings of straight tubes: (Left) Circular tubes with micro fins, (Right) Flattened tubes with micro fins.

Table 13.1 Mesh Sensitive Test for Circular Tube with 12 Microfins at Re = 10,000

Total Mesh Elements	h (W/m².K)	f (–)	Percent Change, h	Percent Change, f
277,250	5,758.016	0.03332	7.64	6.83
865,250	5,423.339	0.0315	1.38	0.99
921,500	5,395.98	0.03136	0.87	0.55
1,131,500	5,372.057	0.03125	0.42	0.19
1,401,250	5,349.389	0.03119	-	-

It has been reported that the numerically predicted heat transfer coefficient (HTC) agrees well within ±20% of the correlation predicted values. Thus, the numerical model used is adequate for predicting fluid flow and heat transfer. Hence, validation of numerical technique has not been explored in this study. However, amesh sensitive test is carried out for the circular tube with micro fins. The mean HTC and friction factor for various meshing elements for a microfinned circular tube with 12 microfins at Re = 10,000 are shown in Table 13.1. The respective percentchanges in HTC and friction factor are calculated from the finer mesh values. It is noted that the percent changes in HTC and friction factor are negligible for 921,500 mesh elements model. Hence, a minimum of 1 million mesh elements are used for the further meshing and analysis.

13.4 RESULTS AND DISCUSSIONS

Variations in the difference between outlet and inlet temperatures with Re are shown in Figure 13.3. Temperature difference seems to decrease with increase in Re for all tube configurations. However, temperature difference is more significant in flattened tube than circular tube, indicating enhanced heat exchange in the flattened tubes. The effect of increasing the microfins beyond 12 fins seems unreasonable as the thermal enhancement is negligible.

The HTC and friction factor for flattened tubes with microfins are depicted in Figures13.4 and 13.5. For comparison, the results for circular tubes with microfins are also presented in these figures. The mass capacity of entry fluid increases with Re, increasing heat transfer from wall to fluid. Consequently, the HTC also increases. It is found that the HTC has enhanced significantly in the flattened tubes compared to circular tubes. However, the effect of increasing microfins beyond 12 fins shows limited enhancement. An enhancement of 141% at Re = 10,000 and 151% at Re = 30,000 is obtained for the flattened tube with 12 microfins. The friction

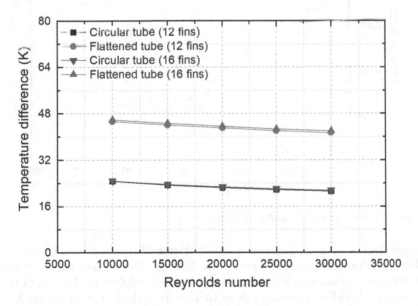

Figure 13.3 Variation in temperature difference with Re.

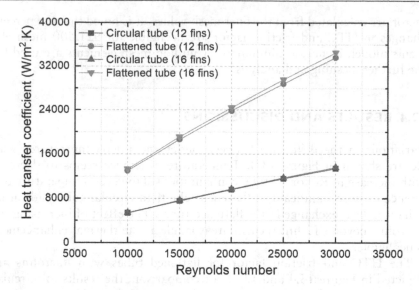

Figure 13.4 Heat transfer coefficient results for all tubes.

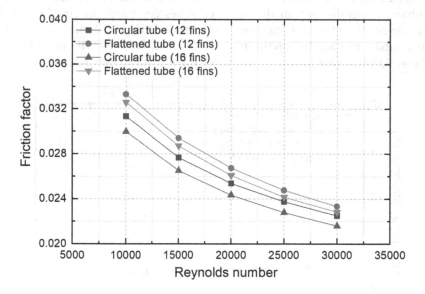

Figure 13.5 Friction factor results for all tubes.

factor decreases with Re. It is more in flattened tubes. Furthermore, it is observed that the increasing microfins decrease the friction factor. A drop of 30% in friction factor is witnessed from Re = 10,000 to Re = 30,000.

Figure 13.6 shows pressure drop variations with Re for all tubes. A significant pressure drop per unit length has developed in the flattened tube,

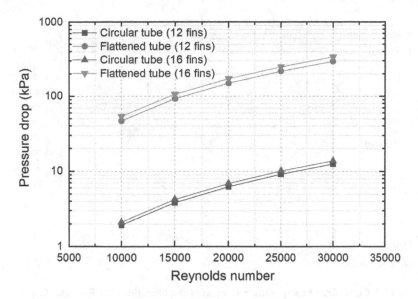

Figure 13.6 Pressure drop per unit tube length results for all tubes.

which further increases with the Re. The fluid pump power is generally given by v.ΔP. Thus, pumping power is directly dependent on the pressure drop per unit tube length. Increased pressure drop results in significant pumping power needed for fluid flow in flattened microfinned tubes.

Contours of temperature, pressure, and velocity along the tube length (Inlet, X = 250mm, X = 500mm, X = 750mm, and Outlet) for flattened tube with 12microfins at Re = 20,000 are shown in Figures 13.7–13.9. The area-weighted average inlet temperature and pressure are 298.15 K and 150, 144.7 Pa, respectively. The values at the outlet are 341.15 K and 0 Pa, respectively. Since the velocity is imposed as an inlet boundary condition, the area-weighted average velocity will remain the same at the inlet and outlet-boundaries. However, the velocity contour shows a significant redistribution of initial uniform-imposed velocity along the length.

The centerline velocity and temperature are shown in Figure 13.10. The figure shows an apparent increase in centerline velocity from 5.86 to 6.97 m/s; Moreover, the increase is obtained within 200 mm length from the inlet. The centerline temperature has also increased from 298.15 K at the inlet to 338.89 K at the outlet. Thus,it can be inferred that the redistribution of fluid field results in significant wall heat transfer to fluid.

The performance evaluation factors (PEF) for flattened tubes at different Re are given in Table 13.2. The PEF measures relative improvement in heat transfer compared to friction factor; the PEF will be equal to one for the circular tube. Thus, PEF more than one shows enhancement in thermal characteristics of modified tubes. The PEF is found to be greater than 2

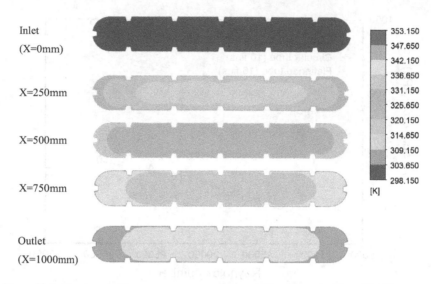

Figure 13.7 Contours of temperature at various axial locations for Re = 20,000.

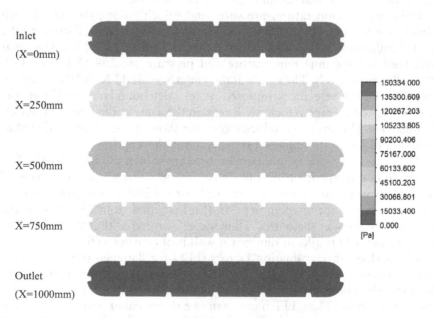

Figure 13.8 Contours of gauge pressure at various axial locations for Re = 20,000.

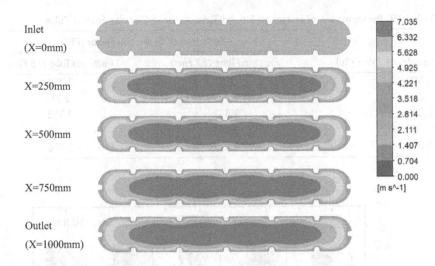

Figure 13.9 Contours of velocity at various axial locations for Re = 20,000.

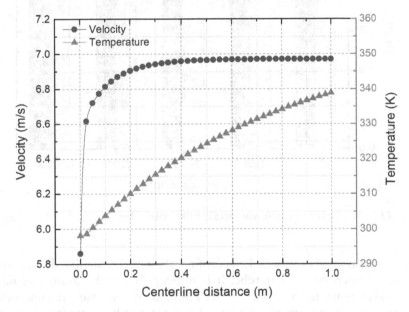

Figure 13.10 Centerline temperature and velocity in the flattened tube for Re = 20,000.

for flattened tubes. Also, PEF increases with Re. An increase of 6.5% is obtained at Re = 30,000. However, the effect of increasing microfins has a limited effect on the further improvement in the performance.

The heat transfer ratio and pressure drop ratio are given in Figures 13.11 and 13.12, respectively. The ratios are computed from circular finned tubes;

Table 13.2 Performance of Flattened Finned Tube Against Circular Finned Tube

	Performance Evaluation Factor (PEF)	
Reynolds Number (Re)	Flattened Tube (12 Fins)	Flattened Tube (16 Fins)
10,000	2.27	2.252
15,000	2.323	2.31
20,000	2.362	2.353
25,000	2.393	2.39
30,000	2.417	2.406

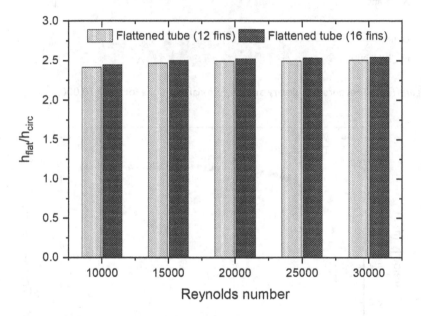

Figure 13.11 Heat transfer coefficient ratio for flattened tubes.

e.g., for a flattened tube with 12 microfins, the ratio is calculated respective to the 12 microfinscircular tube. It is noticed that both parameters have minimal response to an increase in Re. The effect of increasing the microfins has an insignificant contribution to enhancement of heat transfer. A nearly 2.5 times enhancement is estimated for the HTC. In comparison, nearly 25 times increase in pressure drop is noticed for flattened micro-finned tubes. Nearly 5% further increase in the pressure drop has been noticed for the flattened tubes with 16 fins.

Figure 13.13 shows pressure drop penalty variations with Re for flattened tubes and circular tubes with microfins. It is given by $h/\Delta P$; it shows the penalty on heat transfer improvement. Even though the HTC of flattened

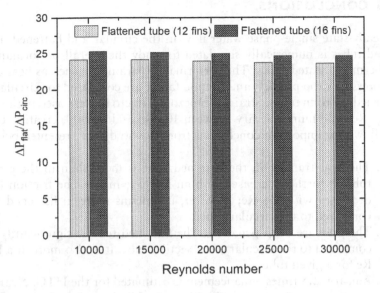

Figure 13.12 Pressure drop ratio for flattened tubes.

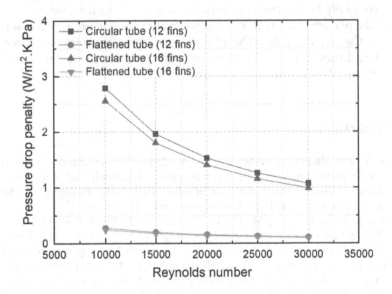

Figure 13.13 Pressure drop penalty for all tubes.

tubes with microfins is comparatively much greater than the circular micro-finned tubes, the pressure drop penalty values for flattened tubes are much lower than the circular tubes. Thus, the pressure drop generated per unit tube length is substantially higher than the enhancement in the heat transfer.

13.5 CONCLUSIONS

A steady-state single-phase water flow in the circular and flattened micro-finned tubes is numerically simulated to study the overall performance of isothermally heated tubes. The performance parameters such as heat transfer, pressure drop penalty, and friction factor are compared for circular and flattened microfin tubes. Straight trapezoidal micro fins are used in the modeling. Reynold numbers vary between 10,000 and 30,000, creating turbulent flow. The important conclusions from this study are presented below:

1. The heat transfer in the flattened tube is more than in the circular tube. It further increases with an increase in Re. The friction factor decreases with the Re; however, it remains more in flattened tubes compared to the circular tubes.
2. The pressure drop penalty in the flattened tube is significantly low-compared to the circular cross-section tube. It is also more at a lower Re for a given tube.
3. A nearly 2.5 times enhancement is estimated for the HTC. Nearly 25 times increase in pressure drop is noticed in the flattened micro-finned tubes. Thus, the pressure drop generated per unit tube length is substantially higher than the enhancement in the heat transfer.
4. The performance factor of the flattened tube has improved compared to the circular tube. The PEF is found to be greater than 2 for flattened tubes. It also increases with the increase in Re for tubes with either 12 or 16 fins.

REFERENCES

Celen, A., Dalkilic, A. S., & Wongwises, S. (2013). Experimental analysis of the single phase pressure drop characteristics of smooth and microfin tubes. *International Communications in Heat and Mass Transfer, 46,* 58–66. https://doi.org/10.1016/j.icheatmasstransfer.2013.05.010

Copetti, J. B., Macagnan, M. H., de Souza, D., & Oliveski, R. D. C. (2004). Experiments with micro-fin tube in single phase. *International Journal of Refrigeration, 27*(8), 876–883. https://doi.org/10.1016/j.ijrefrig.2004.04.015

Dastmalchi, M., Arefmanesh, A., & Sheikhzadeh, G. A. (2017). Numerical investigation of heat transfer and pressure drop of heat transfer oil in smooth and micro-finned tubes. *International Journal of Thermal Sciences, 121,* 294–304. https://doi.org/10.1016/j.ijthermalsci.2017.07.027

Dittus, F. W., & Boelter, L. M. K. (1985). Heat transfer in automobile radiators of the tubular type. *International Communications in Heat and Mass Transfer, 12*(1), 3–22. https://doi.org/10.1016/0735-1933(85)90003-X

Dutt, N., Singh, A. R., & Sahoo, P. K. (2020). CFD analysis of suspended debris during postulated severe core damage accident of PHWR. *Nuclear Engineering and Design, 357*(1), 110390. https://doi.org/10.1016/j.nucengdes.2019.110390

Gnielinski, V. (1976). New equations for heat and mass transfer in turbulent pipe and channel flow. *International Chemical Engineering, 16*, 359–368.

Han, D. H., & Lee, K.-J. (2005). Single-phase heat transfer and flow characteristics of micro-fin tubes. *Applied Thermal Engineering, 25*(11–12), 1657–1669. https://doi.org/10.1016/j.applthermaleng.2004.10.015

Nasr, M., Akhavan-Behabadi, M. A., & Marashi, S. E. (2010). Performance evaluation of flattened tube in boiling heat transfer enhancement and its effect on pressure drop. *International Communications in Heat and Mass Transfer, 37*(4), 430–436. https://doi.org/10.1016/j.icheatmasstransfer.2009.11.011

Razi, P., Akhavan-Behabadi, M. A., & Saeedinia, M. (2011). Pressure drop and thermal characteristics of CuO–base oil nanofluid laminar flow in flattened tubes under constant heat flux. *International Communications in Heat and Mass Transfer, 38*(7), 964–971. https://doi.org/10.1016/j.icheatmasstransfer.2011.04.010

Reay, D. A. (1991). Heat transfer enhancement—A review of techniques and their possible impact on energy efficiency in the U.K. *Heat Recovery Systems and CHP, 11*(1), 1–40. https://doi.org/10.1016/0890-4332(91)90185-7

Singh, A. R., Sahoo, P. K., & Tariq, A. (2018). Investigation of the channel disassembly behaviour of Indian 200MWe PHWR – A numerical approach. *Nuclear Engineering and Design, 339*(1), 137–149. https://doi.org/10.1016/j.nucengdes.2018.09.008

Singh, A. R., & Solanki, A. K. (2022). Numerical study on thermal-hydraulic characteristics of flattened microfin tubes. *Chemical Product and Process Modeling.* https://doi.org/10.1515/cppm-2022-0005

Singh, A. R., Tariq, A., & Majumdar, P. (2020). Experimental study on thermomechanical deformation of PHWR channel at elevated temperature. *Nuclear Engineering and Design, 364*(1), 110634. https://doi.org/10.1016/j.nucengdes.2020.110634

Solanki, A. K., & Kumar, R. (2018). Condensation of R-134a inside micro-fin helical coiled tube-in-shell type heat exchanger. *Experimental Thermal and Fluid Science, 93*, 344–355. https://doi.org/10.1016/j.expthermflusci.2018.01.021

Wang, H. S., & Rose, J. W. (2004). Prediction of effective friction factors for single-phase flow in horizontal microfin tubes. *International Journal of Refrigeration, 27*(8), 904–913. https://doi.org/10.1016/j.ijrefrig.2004.04.013

Wilson, M. J., Newell, T. A., Chato, J. C., & Infante Ferreira, C. A. (2003). Refrigerant charge, pressure drop, and condensation heat transfer in flattened tubes. *International Journal of Refrigeration, 26*(4), 442–451. https://doi.org/10.1016/S0140-7007(02)00157-3

Wu, X., Zhu, Y., & Huang, X. (2015). Influence of 0° helix angle micro fins on flow and heat transfer of R32 evaporating in a horizontal mini multichannel flat tube. *Experimental Thermal and Fluid Science, 68*, 669–680. https://doi.org/10.1016/j.expthermflusci.2015.07.010

Xiao-Wei, L., Ji-An, M., & Zhi-Xin, L. (2007). Experimental study of single-phase pressure drop and heat transfer in a micro-fin tube. *Experimental Thermal and Fluid Science, 32*(1), 641–648. https://doi.org/10.1016/j.expthermflusci.2007.08.005

Yang, C.-M., & Hrnjak, P. (2018). Effect of straight micro fins on heat transfer and pressure drop of R410A during evaporation in round tubes. *International Journal of Heat and Mass Transfer, 117*, 924–939. https://doi.org/10.1016/j.ijheatmasstransfer.2017.10.064

Chapter 14

Heat and mass transfer in convective flow of nanofluid

Alok Kumar Pandey
Graphic Era Deemed to be University

Himanshu Upreti
Graphic Era Hill University

CONTENTS

NOMENCLATURE

C Concentration
C_P Heat capacity (J/kg/K)
D_B Coefficient of Brownian diffusion
D_T Thermophoretic diffusion coefficient
g Acceleration due to gravity (m/s^2)
h_w Coefficient of heat transfer $(\text{W/m}^2\text{K})$
k^* Mean absorption coefficient
k Thermal conductivity (W/m/K^{-1})
Le Lewis number
Nb Brownian motion parameter
Nr Buoyancy parameter
Nt Thermophoresis parameter
Pr Prandtl number
q_r Radiative heat flux
R Radiation parameter
Ra Rayleigh number
S Suction/injection parameter
T Temperature (K)
(u, v) Velocity component in $x-$ and $y-$ axes (m/s)

DOI: 10.1201/9781003367420-14

Greek Symbols

α Thermal diffusivity $\left(m^2 \, / \, s\right)$
β Thermal expansion coefficient $\left(K^{-1}\right)$
γ Chemical reaction parameter
η Similarity variable
θ Dimensionless temperature
λ Convection parameter
υ Kinematic viscosity $\left(m^{-2} \, / \, s\right)$
ρ Density $\left(kg/m^3\right)$
σ^* Stefan Boltzmann constant
ϕ Dimensionless concentration

14.1 INTRODUCTION

A Lie group transformation is the most powerful and systematic technique to generate similarity transform. It is also known as scaling group transformation. It is widely used in non-linear dynamical system, particularly in the domain of deterministic chaos. In this transformation, the group invariant solutions of both initial and boundary value problems are well-defined similarity solutions. This technique is implemented by [1–17] to discuss the various flow problems arising in computational fluid dynamics, aerodynamics, chemical engineering, plasma physics, and many other branches of engineering.

In previous decades, the assessment of heat transfer flow through stretching surface had been considered important due to its various uses in manufacturing and production processes, for example, fiber glass and production of paper, extrusion of polymer and metal, spinning of metal, blowing of glass, aerodynamic plastic sheets extrusion, etc. The cooling of stretching sheets is required to assure the finest worth of material and dedicated temperature control and therefore, information of flow and heat transfer in those systems. Wang [18] introduced the influence of free convection on stretching vertical surface. The author found that with enhancement in Prandtl number, thermal boundary layer width reduces. Gorla and Sidawi [19] illustrated the effect of suction/injection on vertical stretching surface with free convection. They revealed that as augment in injection parameter thickness of boundary layer increases. Khan and Pop [20] have studied nanofluid flow owing to a stretching sheet. The authors analyzed that Nusselt number reduces as boost in Prandtl number. Pal et al. [21] examined the impact of radiation and mixed convection on nanofluid flow along a stretching or shrinking sheet. Several authors have studied flow characteristics of nanofluids due to stretching surface [22–27].

Nanofluids are extensively utilized in several industrial and technological purposes. For instance, cancer therapy, storage of energy, glues, melts of polymers, paints, tarmac, etc. Lin et al. [28] scrutinized the heat transfer analysis of nanofluid in the companionship of linear radiation and power law viscosity. Mahmoodi and Kandelousi [29] designated the influence of linear thermal radiation and viscous dissipation on 2-Dnanofluid flow within a channel. Upreti et al. [30] analyzed the influence of free convection on magnetic Ag-kerosene nanofluid flow via cone. In recent research [31–35], different nanofluids are used to discuss their thermal characteristics.

In the current work, we presented a Lie group approach for the heat and mass transfer analysis for 2-D steady flow of nanofluid along a convectively heated stretching sheet. The flow equations have been tackled through Runge-Kutta-Fehlberg procedure along with shooting scheme. The influence of several objective factors on velocity, temperature and concentration are illustrated by graphs.

14.2 MATHEMATICAL FORMULATION

Consider a steady 2-D convective flow of a nanofluid via a stretching surface. The flow direction toward x–axis which is revealed in Figure 14.1. The surface is in contact with a boiling fluid at temperature T_w and is further exposed to convective heating with h_w as convective heat transfer coefficient. The governing flow equations are:

$$\frac{\partial u}{\partial x_1} + \frac{\partial v}{\partial x_2} = 0 \tag{14.1}$$

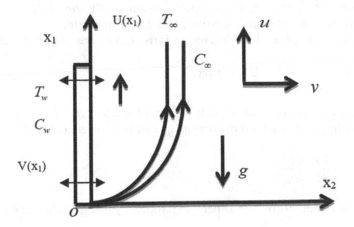

Figure 14.1 Flow model.

$$u\frac{\partial u}{\partial x_1}+v\frac{\partial u}{\partial x_2}-v\left(\frac{\partial^2 u}{\partial x_2^2}\right)$$

$$-\left[(1-C_\infty)\rho_{f\infty}(T-T_\infty)\beta g-(\rho_p-\rho_{f\infty})(C-C_\infty)g\right]=0 \qquad (14.2)$$

$$u\frac{\partial T}{\partial x_1}+v\frac{\partial T}{\partial x_2}-\alpha\left(\frac{\partial^2 T}{\partial x_2^2}\right)$$

$$+\frac{1}{\rho C_p}\frac{\partial q_r}{\partial x_2}-\tau\left[\frac{D_T}{T_\infty}\left(\frac{\partial T}{\partial x_2}\right)^2+D_B\frac{\partial T}{\partial x_2}\frac{\partial C}{\partial x_2}\right]=0 \qquad (14.3)$$

$$u\frac{\partial C}{\partial x_1}+v\frac{\partial C}{\partial x_2}-D_B\left(\frac{\partial^2 C}{\partial x_1^2}\right)-\frac{D_T}{T_\infty}\left(\frac{\partial^2 T}{\partial x_2^2}\right)+k_1(C-C_\infty)=0 \qquad (14.4)$$

with boundary conditions

$$\left.\begin{array}{l}u=U(x_1),\,v=V(x_1),\,\dfrac{\partial T}{\partial x_2}=\dfrac{h_w(T_w-T)}{-k},\,C=C_w \text{ as } x_2=0\\[2mm] u=0,\,T=T_\infty,\,C=C_\infty \text{ at } x_2=\infty\end{array}\right\} \qquad (14.5)$$

In above system, $u,\,v$ represents velocity components, v kinematic viscosity, ρ density, β thermal expansion coefficient, g acceleration due to gravity, T temperature, C concentration, k thermal conductivity, a thermal diffusivity, C_P specific heat, τ ratio of effective heat capacity of the nanoparticle material to the effective heat capacity of the base fluid, D_T thermophoretic diffusion coefficient, D_B Brownian diffusion coefficient, T_∞ ambient temperature, C_∞ ambient concentration, and k_1 reaction rate.

The velocities of suction/blowing and streamwise are introduced as [6,11]:

$$V(x_1)=V_0 x_1^{(m-1)/2},\,U(x_1)=ax_1^m,$$

here V_0 is the suction velocity, if $V_0>0$ and for injection, if $V_0<0$.

According to Rosseland deduction, q_r can be written as [21–27]

$$q_r=-\frac{4\sigma^*}{3k^*}\frac{\partial T^4}{\partial x_2} \qquad (14.6)$$

Now, the temperature T^4 has been expanded by using Taylor series as

$$T^4=T_\infty^4+4T_\infty^3(T-T_\infty)+6T_\infty^2(T-T_\infty)^2+... \qquad (14.7)$$

Neglecting the higher order terms i.e., terms with $(T - T_\infty)^2$ and onwards in Equation (14.7), we acquire

$$T^4 \cong 4T_\infty^3 T - 3T_\infty^4 \qquad (14.8)$$

Solving Equations (14.6) and (14.8) we obtain

$$q_r = -\frac{16T_\infty^3 \sigma^*}{3k^*} \frac{\partial T}{\partial x_2} \qquad (14.9)$$

The stream function $\psi(x_1, x_2)$ and similarity variables are written as:

$$u = \frac{\partial \psi}{\partial x_2} \text{ and } v = -\frac{\partial \psi}{\partial x_1} \qquad (14.10)$$

$$T = T_\infty + \theta(\eta)(T_w - T_\infty) \text{ and } C = C_\infty + \phi(\eta)(C_w - C_\infty) \qquad (14.11)$$

With the help of Equations (14.9)–(14.11), the system of Equations (14.2)–(14.5) reduced to

$$\frac{\partial \psi}{\partial x_2} \frac{\partial^2 \psi}{\partial x_1 \partial x_2} - \frac{\partial \psi}{\partial x_1} \frac{\partial^2 \psi}{\partial x_2^2} - v\left(\frac{\partial^3 \psi}{\partial x_2^3}\right)$$

$$= (1 - \phi_\infty)\rho_{f\infty}\theta\Delta\theta\beta g - (\rho_p - \rho_{f\infty})\phi\Delta\phi g \qquad (14.12)$$

$$\frac{\partial \psi}{\partial x_2} \frac{\partial \theta}{\partial x_1} - \frac{\partial \psi}{\partial x_1} \frac{\partial \theta}{\partial x_2}$$

$$= \left(\alpha + \frac{16T_\infty^3 \sigma^*}{(\rho C_P)_f 3k^*}\right)\frac{\partial^2 \theta}{\partial x_2^2} + \tau\left[\frac{D_T}{T_\infty}\Delta\theta\left(\frac{\partial \theta}{\partial x_2}\right)^2 + D_B\Delta\phi\frac{\partial \phi}{\partial x_2}\frac{\partial \theta}{\partial x_2}\right] \qquad (14.13)$$

$$\frac{\partial \psi}{\partial x_2} \frac{\partial \phi}{\partial x_1} - \frac{\partial \psi}{\partial x_1} \frac{\partial \phi}{\partial x_2} = D_B\frac{\partial^2 \phi}{\partial x_2^2} - k_1\phi + \frac{D_T\Delta\theta}{T_\infty\Delta\phi}\frac{\partial^2 \theta}{\partial x_2^2} \qquad (14.14)$$

with the corresponding boundary conditions

$$\frac{\partial \psi}{\partial x_2} = x_1^m, \frac{\partial \psi}{\partial x_1} = -V_0 x_1^{\frac{(m-1)}{2}}, k\frac{\partial \theta}{\partial x_2} = -h_w(1 - \theta), \phi = 1 \text{ as } x_2 = 0$$

$$\left.\begin{array}{l} \\ \frac{\partial \psi}{\partial x_2} = 0, \theta = 0, \phi = 0 \text{ at } x_2 \to \infty \end{array}\right\} \qquad (14.15)$$

where $T_w - T_\infty = \Delta\theta$ and $C_w - C_\infty = \Delta\phi$.

14.2.1 Transformation of Lie group and similarity solutions

$$\Gamma : x_1^* = e^{\varepsilon\alpha_1} x_1, \; x_2^* = e^{\varepsilon\alpha_2} x_2, \; \psi^* = e^{\varepsilon\alpha_3}\psi, \; u^* = e^{\varepsilon\alpha_4} u,$$

$$v^* = e^{\varepsilon\alpha_5} v, \; \theta^* = e^{\varepsilon\alpha_6}\theta, \; \phi^* = e^{\varepsilon\alpha_7}\phi. \tag{14.16}$$

$$\left(\frac{\partial^2\psi^*}{\partial x_1^*\partial x_2^*} \frac{\partial\psi^*}{\partial x_2^*} - \frac{\partial^2\psi^*}{\partial x_2^{*2}} \frac{\partial\psi^*}{\partial x_1^*} \right) e^{\varepsilon(\alpha+2\alpha_2-2\alpha_3)} = v \frac{\partial^3\psi^*}{\partial x_2^{*3}} e^{\varepsilon(3\alpha_2-\alpha_3)} +$$

$$\left(1 - \phi_\infty \right) \rho_{f\infty} e^{-\varepsilon\alpha_6} g\beta\Delta\theta - e^{\varepsilon\alpha_7} \left(\rho_p - \rho_{f\infty} \right) \phi g\Delta\phi \tag{14.17}$$

$$\left(\frac{\partial\theta^*}{\partial x_1^*} \frac{\partial\psi^*}{\partial x_2^*} - \frac{\partial\theta^*}{\partial x_2^*} \frac{\partial\psi^*}{\partial x_1^*} \right) e^{\varepsilon(\alpha_1+\alpha_2-\alpha_3-\alpha_6)} = (R+\alpha) e^{\varepsilon(2\alpha_2-\alpha_6)} \frac{\partial^2\psi^*}{\partial x_2^{*2}} +$$

$$\tau\left[\Delta\phi D_B \frac{\partial\theta^*}{\partial x_2^*} \frac{\partial\phi^*}{\partial x_2^*} e^{\varepsilon(2\alpha_2-\alpha_6-\alpha_7)} + \Delta\theta \frac{D_T}{T_\infty} \left(\frac{\partial\theta}{\partial x_2^*} \right)^2 e^{\varepsilon(2\alpha_2-\alpha_6)} \right] \tag{14.18}$$

$$\left(\frac{\partial\psi^*}{\partial x_2^*} \frac{\partial\phi^*}{\partial x_1^*} - \frac{\partial\psi^*}{\partial x_1^*} \frac{\partial\phi^*}{\partial x_2^*} \right) e^{\varepsilon(\alpha_1+\alpha_2-\alpha_3-\alpha_7)} = D_B \frac{\partial^2\phi}{\partial x_2^{*2}} e^{\varepsilon(2\alpha_2-\alpha_7)}$$

$$+ \frac{\Delta\theta D_T}{\Delta\phi T_\infty} e^{\varepsilon(2\alpha_2-\alpha_6)} \frac{\partial^2\theta^*}{\partial x_2^{*2}} - \phi k_1 e^{-\varepsilon\alpha_7} \tag{14.19}$$

Now Equations (14.17)–(14.19) are invariant due to Lie group scaling (14.16).

The subsequent expressions are acquired:

$$\left.\begin{array}{l} 7\alpha_1 + 2\alpha_2 - 2\alpha_3 = \alpha_2 - \alpha_3 = 3\alpha_2 - \alpha_3 = -\alpha_6 = -\alpha_7 \\[4pt] \alpha_1 + \alpha_2 - \alpha_3 - \alpha_6 = 2\alpha_2 - \alpha_6 = 2\alpha_2 - \alpha_6 - \alpha_7 = 2\alpha_2 - 2\alpha_6 \\[4pt] \alpha_1 + \alpha_2 - \alpha_3 - \alpha_7 = 2\alpha_2 - \alpha_6 = 2\alpha_2 - \alpha_7 = -\alpha_7 \end{array}\right\} \tag{14.20}$$

From this relation, we acquired $\alpha_2 = \frac{1}{3}\alpha_3 = \frac{1}{4}\alpha_1$, $\alpha_6 = 0 = \alpha_7$. Hence, the boundary conditions are $\alpha_4 = \frac{1}{2}\alpha_1 = m\alpha_1$, $\alpha_5 = -\frac{1}{4}\alpha_1 = \frac{m-1}{2}\alpha_1$, as $m = \frac{1}{2}$ is taken.

Now using above conditions, the boundary conditions Equation (14.15) becomes

$$\frac{\partial \psi^*}{\partial x_2^*} = x_1^{*^{1/2}}, \frac{\partial \psi^*}{\partial x_1^*} = -V_0 x_1^{*^{-1/4}}, -k\frac{\partial \theta^*}{\partial x_2^*} = h_w\left(1-\theta^*\right), \phi^* = 1, \text{ as } x_2^* = 0,$$

$$\frac{\partial \psi^*}{\partial x_2^*} = 0, \theta^* = 0, \phi^* = 0, \text{ at } x_2^* = \infty$$

$$(14.21)$$

And, the group of transformations Equation (14.16) changed into

$$\Gamma : x_1^* = e^{\varepsilon\alpha_1}x_1, x_2^* = e^{\varepsilon\alpha_1/4}x_2,$$

$$\psi^* = e^{3\varepsilon\alpha_1/4}\psi, u^* = e^{\varepsilon\alpha_1/2}u, v^* = e^{-\varepsilon\alpha_1/4}v, \theta^* = \theta, \phi^* = \phi.$$

$$(14.22)$$

With the aid of Taylor's series, we get the relations:

$$x_1^* = x_1 + \varepsilon\alpha_1 x, x_2^* = x_2 + \varepsilon\left(\frac{\alpha_1}{4}\right)x_2, \psi^* = \psi + 3\varepsilon\left(\frac{\alpha_1}{4}\right)x_2, u^* = u + \varepsilon\left(\frac{\alpha_1}{2}\right)u,$$

$$v^* = v + \varepsilon\left(\frac{\alpha_1}{4}\right)v, \theta^* = \theta, \phi^* = \phi.$$

$$(14.23)$$

The assisting equations are:

$$\frac{dx_1}{\alpha_1 x_1} = \frac{dx_2}{\frac{1}{4}\alpha_1 x_2} = \frac{d\psi}{\frac{3}{4}\psi\alpha_1} = \frac{du}{\frac{1}{2}u\alpha_1} = \frac{dv}{-\frac{1}{4}v\alpha_1} = \frac{d\theta}{0} = \frac{d\phi}{0} \qquad (14.24)$$

Now solving above equations, we achieved:

$$\eta = x_1^{*^{-1/4}}x_2^*, \psi^* = f(\eta)x_1^{*^{3/4}}, \theta^* = \theta(\eta), \phi^* = \phi(\eta) \qquad (14.25)$$

Applying similarity transformation (14.25) into the leading Equations (14.17)–(14.19), we acquire the following set of non-linearly ODEs:

$$f''' + Ra(\theta - Nr\phi) + \frac{1}{Pr}(\frac{3}{4}ff'' - \frac{1}{2}f'^2) = 0 \qquad (14.26)$$

$$(1+R)\theta'' + PrNb\theta'\phi' + \frac{3}{4}Prf\theta' + PrNt\theta'^2 = 0 \qquad (14.27)$$

$$\phi'' + \frac{3}{4}Lef\phi' - \gamma\phi + \frac{Nt}{Nb}\theta'' = 0 \qquad (14.28)$$

And supplementary boundary conditions are

$$
\left.\begin{array}{l}
\dfrac{\partial f}{\partial \eta} = 1, f(\eta) = S, \dfrac{\partial \theta}{\partial \eta} = -\lambda\left(1 - \theta(\eta)\right), \phi(\eta) = 1 \ \text{ at } \eta = 0 \\[4mm]
\dfrac{\partial f}{\partial \eta} = 0, \theta(\eta) = 0, \phi(\eta) = 0 \ \text{ as } \ \eta \to \infty
\end{array}\right\}
\tag{14.29}
$$

where, the existing parameters are defined as

$$
\left.\begin{array}{l}
Pr = \dfrac{v}{\alpha}, Ra = \dfrac{(1 - \phi_\infty)\beta g \Delta \theta}{v\alpha}, Le = \dfrac{v}{D_B}, Nr = \dfrac{(\rho_p - \rho_{f\infty})\Delta\phi}{\rho_{f\infty}\beta\Delta\theta(1 - \phi_\infty)}, Nb = \tau D_B \Delta\phi, \\[4mm]
Nt = \tau D_T \dfrac{\Delta\theta}{T_\infty}, R = \dfrac{16\sigma * T_\infty^3}{3k^*k}, \gamma = k_1 \dfrac{U}{v}, S = -\dfrac{4V_0}{3}, \lambda = \dfrac{h_w}{k} x_1^{1/2}, h_w = cx_1^{-1/4}
\end{array}\right\}
$$

The quantities of physical importance such as skin friction coefficient C_f, local Nusselt number Nu and local Sherwood number Sh are expressed as

$$
\left.\begin{array}{l}
C_f = \dfrac{\mu}{\rho_f U}\left(\dfrac{\partial u}{\partial x_2}\right)_{x_2=0} = -\mathrm{Re}_x^{-1/2} f'', \\[4mm]
Nu = -\dfrac{x_1}{T_w - T_\infty}\left[1 + \dfrac{16\sigma * T_\infty^3}{3k^*k}\right]\left(\dfrac{\partial T}{\partial x_2}\right)_{x_2=0} = -\mathrm{Re}_x^{1/2}(1 + R)\theta'(0) \\[4mm]
\text{and } Sh = -\dfrac{x_1}{C_w - C_\infty}\left(\dfrac{\partial C}{\partial x_2}\right)_{x_2=0} = -\mathrm{Re}_x^{1/2}\phi'(0)
\end{array}\right\}
$$

where $\mathrm{Re}_x = \dfrac{Ux_1}{v_f}$ is the local Reynolds number.

14.3 RESULTS AND DISCUSSION

The altered Equations (14.26)–(14.28) with boundary conditions (14.29) were integrated numerically via RKF scheme to attain the missing values of $\left(\dfrac{\partial^2 f}{\partial \eta^2}\right)_{\eta=0}$, $\left(\dfrac{\partial\theta}{\partial\eta}\right)_{\eta=0}$ and $\left(\dfrac{\partial\phi}{\partial\eta}\right)_{\eta=0}$, for some values of the principal parameters. To ensure the validity of the work, the comparison of numerous values of Nusselt number with earlier published results, which were acquired via Kandasamy et al. [6], Wang [18] and Gorla and Sidawi [19]. The comparison elucidates fine agreement as mentioned in Table 14.1.

Table 14.1 Comparison of Values of $-\theta'(0)$ with Earlier Studies When $Nb = Nt = 0$

Pr	Kandasamy et al. [6]	Wang [18]	Gorla and Sidawi [19]	Present Outcomes
0.7	0.454285	0.4539	0.53488	0.53488
2.0	0.911423	0.9113	0.91142	0.91163
3.0	–	–	1.15970	1.15966
7.0	0.895264	1.8954	1.89046	1.89026

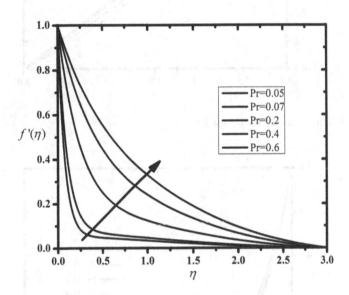

Figure 14.2 Velocity profiles for Pr.

The influence of Prandtl number (Pr) on the velocity, thermal and concentration outlines are illustrated in Figures 14.2–14.4. Figure 14.2 reveals that velocity function is escalated with Prandtl number and each curve are asymptotic which means that given boundary conditions are satisfied. Due to this pattern's momentum boundary layer width increases. Figures 14.3 and 14.4 shows that both thermal and concentration profiles of nanofluid declined with increase in the values of Prandtl number.

The impact of buoyancy parameter (Nr) on the existing physical quantities are depicted in Figures 14.5–14.7. Figure 14.5 shows that velocity curves of nanofluid regularly deaccelerated with increase in the values of Nr. Moreover, the temperature and concentration functions outlines are enhanced with buoyancy parameter in the entire range, which is visible from Figures 14.6 and 14.7, respectively.

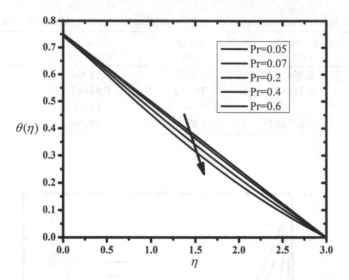

Figure 14.3 Temperature profiles for Pr.

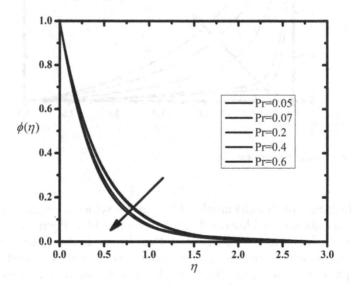

Figure 14.4 Concentration profiles for Pr.

The impact of radiation parameter (R) on flow field, temperature and concentration for $\gamma = 0.5$, $Le = 3$, $Nr = 5$, $\lambda = Nb = Nt = Ra = 1$, $S = 0.8$, are revealed in Figures 14.8–14.10. Figure 14.8 illustrates that on mounting R values, velocity of nanoparticle incessantly boosts corresponding to each value of η in the fixed range. From, Figure 14.9 it is lucid that on improving

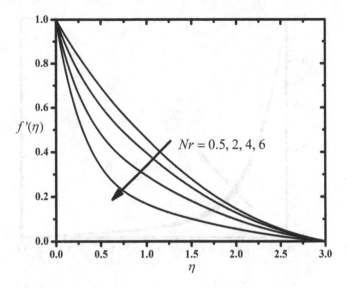

Figure 14.5 Velocity profiles for Nr.

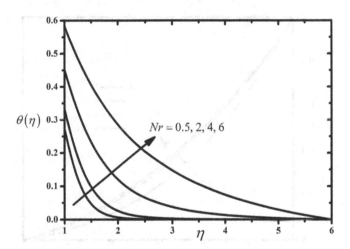

Figure 14.6 Temperature profiles for Nr.

the values of thermal radiation R, temperature profiles gradually increase in a fixed domain. Figure 14.10 reveals that on escalating the values of parameter R, concentration of working fluid lightly diminishes because this basis boundary layer depth of the nanoparticle descends. The influence of chemical reaction parameter (γ) on mass transfer function is depicted in Figure 14.11. The outlines of the figure proofs that concentration function of nanofluid depreciated with increase in γ values.

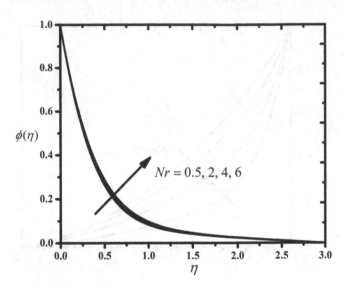

Figure 14.7 Concentration profiles for Nr.

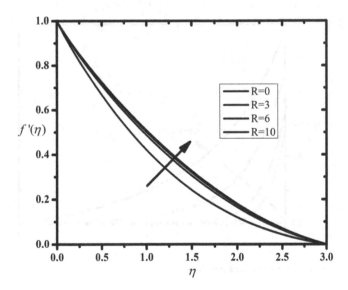

Figure 14.8 Velocity profiles for R.

Figures 14.12–14.14 exhibit the deviation in velocity, temperature and concentration profiles with convection parameter λ, when $Nr = 5$, $\gamma = 0.5$, $Nb = Nt = 1$, $R = 3$, $Ra = 1$, $Le = 3$, $S = 0.8$. Figure 14.12 describes that the consequence of convection parameter λ on fluid velocity profiles. From the outlines it declared that on escalating the values of λ, the velocity of

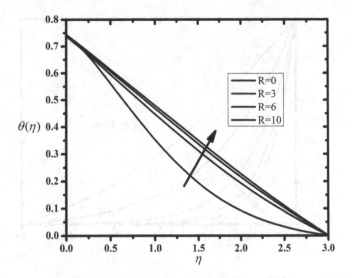

Figure 14.9 Temperature profiles for R.

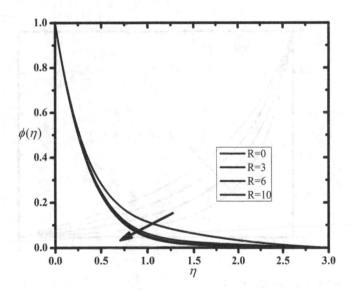

Figure 14.10 Concentration profiles for R.

working fluid frequently rises then the momentum boundary layer also rises. Figure 14.13 exposes that on mounting the values of convection parameter λ, thermal outlines increase due to this reason thermal boundary layer augments and surface temperature rising. Figure 14.14 proves that influence of convection parameteron nanoparticle volume concentration ϕ.

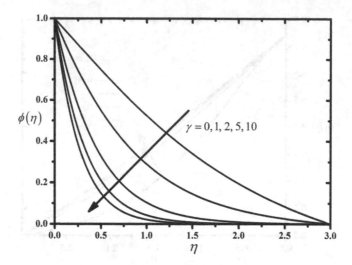

Figure 14.11 Concentration profiles for γ.

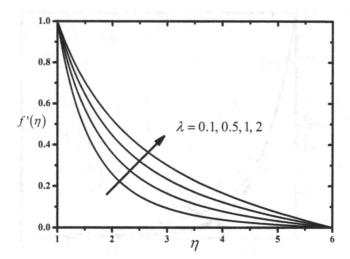

Figure 14.12 Velocity profiles for λ.

It is scrutinized from outlines that on improving the values of λ, the concentration of nanoparticle gradually escalations in the specify domain of similarity variable η. Due to this cause concentration boundary layer width becomes wide in that section. The influence of convection parameter λ, and radiation parameter R, on the of reduced Nusselt number $Re_x^{-1/2} Nu$, skin friction coefficient $Re_x^{-1/2} C_f$, and reduced Sherwood number $Re_x^{-1/2} Sh$, are

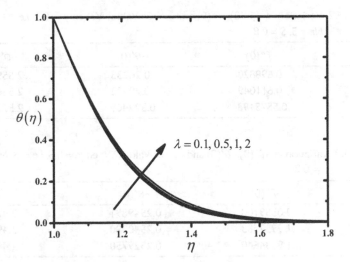

Figure 14.13 Temperature profiles for λ.

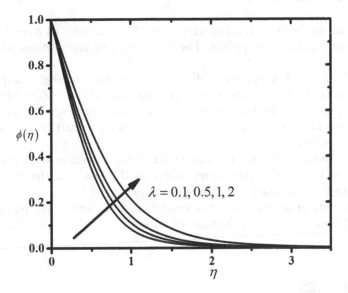

Figure 14.14 Concentration profiles for λ.

shown in Tables 14.2 and 14.3 correspondingly. It is noticeable from these tables heat transfer rate regularly escalates with raising the values of λ, while it depreciates for the large values of R. In the same way the absolute values of skin factor run down, when convection parameter λ, enhances. Moreover, the mass transfer rate is enhanced, on escalating the values of λ and R.

Table 14.2 Variation in $-f''(0), -\theta'(0)$ and $-\phi'(0)$ Values When $\gamma = 0.5$, $Le = R = 3$, $Nr = 5$, $S = 0.8$

λ	$-f''(0)$	$-\theta'(0)$	$-\phi'(0)$
1	0.698820	0.26233	2.55932
2	0.611049	0.29972	2.566061
4	0.5573193	0.322409	2.57049

Table 14.3 Variation in $-f''(0), -\theta'(0)$ and $-\phi'(0)$ Values When $\gamma = 0.5$, $Le = 3$, $Nr = 5$, $S = 0.8$

R	$-f''(0)$	$-\theta'(0)$	$-\phi'(0)$
3	2.0559100	0.2549891	2.476743
6	1.9952783	0.2540380	2.491619
10	1.9634640	0.2529750	2.499989

14.4 CONCLUSIONS

The impact of linear thermal radiation on 2D steady flow of nanofluid over a stretching surface is inspected. The following conclusions were obtained:

- Velocity and temperature distribution profiles of the nanoparticles increased with escalating thermal radiation and convection parameter.
- Volume concentration graph of nanoparticle enhances as increase in convection parameter; however, it is reducing as increase in radiation parameter.
- The reduced Sherwood number $Re_x^{-1/2} Sh$, enhances with growth in thermal radiation parameter whereas, the heat transfer function is reduced as increase in radiation parameter.
- As convection parameter is amplified, heat and mass transfer are enhanced, while skin friction diminishes.

REFERENCES

1. Lie, S. (1874). Veralgemeinerung und neue Verwerthung der Jacobischen Multiplicator-Theorie. Christ. Forh, 255–274.
2. Lie, S. (1960). *Gesammelte Abhandlungen*, vol. 7, Teubner, Leipzig.
3. Lie, S. (1880). Theory of transformation groups I. *Mathematical Annals*, 16(4), 441–528.
4. Yürüsoy, M., Pakdemirli, M., & Noyan, Ö. F. (2001). Lie group analysis of creeping flow of a second grade fluid. *International Journal of Non-Linear Mechanics*, 36(6), 955–960.

5. Kandasamy, R., Ismoen, M., & Saim, H. B. (2010). Lie group analysis for the effects of temperature-dependent fluid viscosity and chemical reaction on MHD free convective heat and mass transfer with variable stream conditions. *Nuclear Engineering and Design*, 240(1), 39–46.

6. Kandasamy, R., Loganathan, P., & Arasu, P. P. (2011). Scaling group transformation for MHD boundary-layer flow of a nanofluid past a vertical stretching surface in the presence of suction/injection. *Nuclear Engineering and Design*, 241(6), 2053–2059.

7. Hamad, M. A. A., & Ferdows, M. (2012). Similarity solution of boundary layer stagnation-point flow towards a heated porous stretching sheet saturated with a nanofluid with heat absorption/generation and suction/blowing: A Lie group analysis. *Communications in Nonlinear Science and Numerical Simulation*, 17(1), 132–140.

8. Das, K. (2013). Lie group analysis for nanofluid flow past a convectively heated stretching surface. *Applied Mathematics and Computation*, 221, 547–557.

9. Kandasamy, R., Muhaimin, I., Khamis, A. B., & bin Roslan, R. (2013). Unsteady Hiemenz flow of Cu-nanofluid over a porous wedge in the presence of thermal stratification due to solar energy radiation: Lie group transformation. *International Journal of Thermal Sciences*, 65, 196–205.

10. Das, K. (2013). Lie group analysis of stagnation-point flow of a nanofluid. *Microfluidics and Nanofluidics*, 15(2), 267–274.

11. Ganesh, N. V., Ganga, B., & Hakeem, A. A. (2014). Lie symmetry group analysis of magnetic field effects on free convective flow of a nanofluid over a semi-infinite stretching sheet. *Journal of the Egyptian Mathematical Society*, 22(2), 304–310.

12. Kundu, P. K., Das, K., & Jana, S. (2014). Nanofluid flow towards a convectively heated stretching surface with heat source/sink: A Lie group analysis. *Afrika Matematika*, 25(2), 363–377.

13. Afify, A. A., & Abd El-Aziz, M. (2017). Lie group analysis of flow and heat transfer of non-Newtonian nanofluid over a stretching surface with convective boundary condition. *Pramana*, 88(2), 31.

14. Rawat, S. K., Pandey, A. K., & Kumar, M. (2018). Effects of chemical reaction and slip in the boundary layer of MHD nanofluid flow through a semi-infinite stretching sheet with thermophoresis and Brownian motion: The Lie group analysis. *Nanoscience and Technology: An International Journal*, 9(1), 47–68.

15. Kandasamy, R., Vignesh, V., Kumar, A., Hasan, S. H., & Isa, N. M. (2018). Thermal radiation energy due to SWCNTs on MHD nanofluid flow in the presence of seawater/water: Lie group transformation. *Ain Shams Engineering Journal*, 9(4), 953–963.

16. Ferdows, M., Nabwey, H. A., Rashad, A. M., Uddin, M. J., & Alzahrani, F. (2020). Boundary layer flow of a nanofluid past a horizontal flat plate in a Darcy porous medium: A Lie group approach. *Proceedings of the Institution of Mechanical Engineers, Part C: Journal of Mechanical Engineering Science*, 234(8), 1545–1553.

17. Tufail, M. N., Saleem, M., & Chaudhry, Q. A. (2021). An analysis of Maxwell fluid through a shrinking sheet with thermal slip effect: A Lie group approach. *Indian Journal of Physics*, 95(4), 725–731.

18. Wang, C. Y. (1989). Free convection on a vertical stretching surface. *ZAMM-Journal of Applied Mathematics and Mechanics/Zeitschrift für Angewandte Mathematik und Mechanik*, 69(11), 418–420.
19. Gorla, R. S. R., & Sidawi, I. (1994). Free convection on a vertical stretching surface with suction and blowing. *Applied Scientific Research*, 52(3), 247–257.
20. Khan, W. A., & Pop, I. (2010). Boundary-layer flow of a nanofluid past a stretching sheet. *International Journal of Heat and Mass Transfer*, 53(11–12), 2477–2483.
21. Pal, D., Mandal, G., & Vajravelu, K. (2014). Flow and heat transfer of nanofluids at a stagnation point flow over a stretching/shrinking surface in a porous medium with thermal radiation. *Applied Mathematics and Computation*, 238, 208–224.
22. Ibrahim, W., & Shankar, B. (2013). MHD boundary layer flow and heat transfer of a nanofluid past a permeable stretching sheet with velocity, thermal and solutal slip boundary conditions. *Computers & Fluids*, 75, 1–10.
23. Haq, R. U., Nadeem, S., Khan, Z. H., & Akbar, N. S. (2015). Thermal radiation and slip effects on MHD stagnation point flow of nanofluid over a stretching sheet. *Physica E: Low-Dimensional Systems and Nanostructures*, 65, 17–23.
24. Pandey, A. K., & Kumar, M. (2017). Natural convection and thermal radiation influence on nanofluid flow over a stretching cylinder in a porous medium with viscous dissipation. *Alexandria Engineering Journal*, 56(1), 55–62.
25. Pandey, A. K., & Kumar, M. (2018). Effects of viscous dissipation and heat generation/absorption on nanofluid flow over an unsteady stretching surface with thermal radiation and thermophoresis. *Nanoscience and Technology: An International Journal*, 9(4), 325–341.
26. Upreti, H., Joshi, N., Pandey, A. K., & Rawat, S. K. (2021). Numerical solution for Sisko nanofluid flow through stretching surface in a Darcy–Forchheimer porous medium with thermal radiation. *Heat Transfer*, 50(7), 6572–6588.
27. Upreti, H., Pandey, A. K., Rawat, S. K., & Kumar, M. (2021). Modified arrhenius and thermal radiation effects on three-dimensional magnetohydrodynamic flow of carbon nanotubes nanofluids over bi-directional stretchable surface. *Journal of Nanofluids*, 10(4), 538–551.
28. Lin, Y., Zheng, L., & Zhang, X. (2014). Radiation effects on Marangoni convection flow and heat transfer in pseudo-plastic non-Newtonian nanofluids with variable thermal conductivity. *International Journal of Heat and Mass Transfer*, 77, 708–716.
29. Mahmoodi, M., & Kandelousi, S. (2015). Effects of thermophoresis and Brownian motion on nanofluid heat transfer and entropy generation. *Journal of Molecular Liquids*, 211, 15–24.
30. Upreti, H., Pandey, A. K., & Kumar, M. (2020). Thermophoresis and suction/injection roles on free convective MHD flow of Ag–kerosene oil nanofluid. *Journal of Computational Design and Engineering*, 7(3), 386–396.
31. Haider, S. M. A., Ali, B., Wang, Q., & Zhao, C. (2021). Stefan blowing impacts on unsteady MHD flow of nanofluid over a stretching sheet with electric field, thermal radiation and activation energy. *Coatings*, 11(9), 1048.

32. Gupta, S., Kumar, D., & Singh, J. (2020). Analytical study for MHD flow of Williamson nanofluid with the effects of variable thickness, nonlinear thermal radiation and improved Fourier's and Fick's Laws. *SN Applied Sciences*, 2(3), 1–12.
33. Rasheed, H. U., Islam, S., Helmi, M. M., Alsallami, S., Khan, Z., & Khan, I. (2021). An analytical study of internal heating and chemical reaction effects on MHD flow of nanofluid with convective conditions. *Crystals*, 11(12), 1523.
34. Lund, L. A., Omar, Z., Khan, I., & Sherif, E. S. M. (2020). Dual solutions and stability analysis of a hybrid nanofluid over a stretching/shrinking sheet executing MHD flow. *Symmetry*, 12(2), 276.
35. Ahmed, A., Khan, M., Irfan, M., & Ahmed, J. (2020). Transient MHD flow of Maxwell nanofluid subject to non-linear thermal radiation and convective heat transport. *Applied Nanoscience*, 10(12), 5361–5373.

Chapter 15

Flow of a second-order fluid due to disk rotation

Reshu Gupta
University of Petroleum and Energy Studies

CONTENTS

NOMENCLATURE

p	Hydrostatic pressure
t_{ij}	Stress tensor
v_i	Velocity vector
e_i	Acceleration vector
$m_i; i = 1,2,3$	Mathematical constants
δ_{ij}	Kronecker's delta
u_r	Radial velocity component
u_θ	Transverse velocity component
u_z	Axial velocity component
ρ	Density of the fluid
$\zeta \left(= \dfrac{r}{d} \right)$	Dimensionless parameter
$\eta \left(= \dfrac{z}{d} \right)$	Dimensionless constant
$R \left(= \dfrac{a\rho d^2}{m_1} \right)$	Reynolds number
$t_1 \left(= \dfrac{m_2}{\rho d^2} \right)$	Elastico-viscous parameter
$t_2 \left(= \dfrac{m_3}{\rho d^2} \right)$	Cross-viscous parameter

DOI: 10.1201/9781003367420-15

α Angular velocity

v An arbitrary constant

$k\left(=\dfrac{\omega_0}{2d\alpha}\right)$ Dimensionless injection parameter

15.1 INTRODUCTION

Recently, a lot of attention paid to the direction of non-Newtonian fluids between rotating discs. As a result, several mathematicians have investigated the same model under a variety of physical and experimental conditions. Batchelor (1951) developed a Von Kármán solution (1921) of the Navier–Stokes equations. Stewartson (1953), Cochran (1934) and Lance and Rogers (1962) discussed the flow characteristics of a rotating disk. Sharma and Singh (1986) have described the forced flow for a second-order fluid between two porous disks. Zangooee et al. (2019) have discussed the flow of MHD nanofluid between two stretchable rotating disks. Usman et al. (2020) analyzed the properties of heat transfer in a rotating rough disk. Ahmed et al. (2019), Hayat et al. (2017, 2019), Ahmed et al. (2019), Hafeez et al. (2020) and Yao and Lian (2019) investigated the flow of different fluids in rotating disks by using numerical and analytical methods. Agarwal and Agarwal (2018) presented the flow of a second-order fluid between two infinite discs. Hojjati and Jafari (2008) studied semi-exact solutions of rotating disks by HPM and ADM. Yao and Lian (2018) studied a new technique of the rotationally symmetric flow of an infinite rotating disk. Turkyilmazoglu (2010a, b) studied the analytic solution of MHD flow over a porous rotating disk.

HPM is a recognized technique to resolve nonlinear DEs and ODEs. This strategy has been proven quite effective in several difficulties in various fields. The standard perturbation and homotopy in topology are combined in this procedure. The equations of flow and heat transfer in many fluid mechanics situations are nonlinear. Many researchers solve these nonlinear equations numerically, and some of them, such as the perturbation technique, are discussed analytically. The Homotopy Perturbation Method is firstly introduced by He (1999) and was further developed and improved by him (2000, 2004, 2006, 2008). The HPM has been shown to solve a large class of nonlinear problems efficiently, accurately, and easily. Ariel (2009) discussed HPM and analytical solutions to the problem of the flow past a rotating disk. Doh et al. (2020), Dinarvand (2010), Rashidi et al. (2012) and Rani et al. (2017) applied various analytical methods in their studies. Sheikholeslami et al. (2012) presented HPM for the three-dimensional problem of condensation fluid on inclined rotating disks. Xinhui et al. (2012) have found the solution for the asymmetric laminar flow and heat transfer of viscous fluid between contracting rotating disks by HAM.

Agarwal (2020, 2021, 2022) and Agarwal and Mishra (2021) applied HPM successfully to solve the ODEs in their models.

The objective of this chapter is an effective approach to HPM for second-order fluid flow due to a rotating disk. The variation of radial, transverse and axial velocity components are calculated for different values of elastico-viscous parameter (t_1), cross-viscous parameter (t_2), Reynolds number (R) & injection parameter (k) and presented graphically. Transverse shearing stress has also been calculated on rotating and stationary disks.

15.2 FLOW ANALYSIS AND MATHEMATICAL FORMULATION

For an incompressible second-order fluid, Coleman and Noll (1974) proposed the constitutive equation, which is as follows:

$$t_{ij} = -p\delta_{ij} + 2m_1 D_{ij} + 2m_2 E_{ij} + 4m_3 C_{ij}, \tag{15.1}$$

Where

$$D_{ij} = \frac{1}{2}(v_{i,j} + v_{j,i}),$$

$$E_{ij} = \frac{1}{2}(e_{i,j} + e_{j,i}) + v_{,j}^m v_{m,j}, \tag{15.2}$$

$$C_{ij} = b_{im} b_j^m.$$

The fluid, which is discussed in Equation (15.1) is restricted between two infinite disks. The disk which is concurring with $z = 0$ plane is rotating with a constant angular velocity α about z –axis while the other disk which is concurring with plane $z = d$ is not rotating. It is assumed that this disk is at rest and permeable also. There is a uniform injection $-v_0$ which is applied perpendicularly to the rotating disk. Assume u_r, u_θ and u_z are velocities in the direction of r, θ and z respectively. Here we choose the cylindrical co-ordinate system, so according to Srivastava (1963) and Wilson and Schreyer (1978), the constitutive Equation (15.1) along with the momentum equation is

$$\frac{\partial u_r}{\partial r} + \frac{u_r}{r} + \frac{\partial u_z}{\partial z} = 0 \tag{15.3}$$

$$\rho\left(u_r \frac{\partial u_r}{\partial r} + u_z \frac{\partial u_r}{\partial z} + \frac{u_\theta^2}{r}\right) = \frac{\partial t_{rr}}{\partial r} + \frac{\partial t_{rz}}{\partial z} + \frac{1}{r}(t_{rr} - t_{\theta\theta}), \tag{15.4}$$

$$\rho\left(u_r\frac{\partial u_\theta}{\partial r}+u_z\frac{\partial u_\theta}{\partial z}+\frac{u_r u_\theta}{r}\right)=\frac{\partial t_{r\theta}}{\partial r}+\frac{\partial t_{\theta z}}{\partial z}+\frac{2t_{r\theta}}{r},\tag{15.5}$$

$$\rho\left(u_r\frac{\partial u_z}{\partial r}+u_z\frac{\partial u_z}{\partial z}\right)=\frac{\partial t_{rz}}{\partial r}+\frac{\partial t_{zz}}{\partial z}+\frac{t_{rz}}{r},\tag{15.6}$$

According to Singh and Kumar (1989), the boundary conditions of the present modal are

$$z=0:u_r=0,u_\theta=r\alpha,u_z=0,$$

$$z=d:u_r=0,u_\theta=0,u_z=-\omega_0.\tag{15.7}$$

Under the similarity transformations as suggested by Von Kármán (1921), the components of velocity are

$$u_r=r\alpha H'(\eta),u_\theta=r\alpha C(\eta),u_z=-2d\alpha H(\eta).$$

$$p=m_1\alpha\left[-p_1(\eta)+R\zeta^2(2t_1+t_2)\left(H''^2+C'^2\right)+v\zeta^2\right].\tag{15.8}$$

After putting the values of u_r,u_θ,u_z in Equations (15.4)–(15.6). The following set of equations will be obtained:

$$R\left(H'^2-2HH''-C^2\right)=H'''-2Rt_1\left(H''^2+2C'^2+HH^{iv}\right)$$

$$-Rt_2\left(H''^2+3C'^2+2H'H'''\right)-2v\tag{15.9}$$

$$2R(H'C-HC')=C''+2Rt_1\left(H''C'-HC'''\right)+2Rt_2\left(H''C'-H'C''\right)\tag{15.10}$$

$$4RHH'=p_1'-2H''+4Rt_1\left(11H'H''+HH'''\right)+28Rt_2H'H''\tag{15.11}$$

By using the similarity transformations from Equation (15.8), the boundary condition can be written as

$$\eta=0:H'=0,C=1,H=0,$$

$$\eta=1:H'=0,C=0,H=k.\tag{15.12}$$

15.3 ANALYSIS OF THE HOMOTOPY PERTURBATION METHOD

The following equation is used to describe the fundamental concepts of this method by He (1999):

$$A(u) - f(r) = 0, r \in \Omega \tag{15.13}$$

along with the following boundary conditions

$$B\left(u, \frac{\partial u}{\partial n}\right) = 0, r \in \delta \tag{15.14}$$

$A(u)$ can be distributed into two parts where $L(u)$ is linear and $N(u)$ is non-linear.

Therefore we can rewrite Equation (15.13) as follows

$$L(u) + N(u) - f(r) = 0, r \in \Omega \tag{15.15}$$

The HPM arrangement is as follows

$$H(v, \varepsilon) = (1 - \varepsilon)\left[L(v) - L(u_0)\right] + \varepsilon\left[A(v) - f(r)\right] = 0 \tag{15.16}$$

where

$$v(r, \varepsilon) : \Omega \times [0,1] \rightarrow R \tag{15.17}$$

ε, which value lies between 0 to 1, is an embedded parameter and u_0 is the initial value that satisfies the boundary condition.

The solution of Equation (15.16) will be in the power of ε which is represented as

$$v = v_0 + \varepsilon v_1 + \varepsilon^2 v_2 + \ldots\ldots\ldots$$

and for the best solution, we will assume ε tends to 1. Therefore

$$u = \lim_{\varepsilon \to 1} v = v_0 + v_1 + v_2 + \ldots\ldots\ldots$$

15.3.1 Implementation of the method

We will solve Equations (15.9) and (15.10) by the homotopy perturbation method.

First, we construct homotopy in Equation (15.9)

$$A(H) = H''' - R\left(H'^2 - 2HH'' - C^2\right) - 2Rt_1\left(H''^2 + 2C'^2 + HH^{iv}\right) \tag{15.18}$$
$$- Rt_2\left(H''^2 + 3C'^2 + 2H'H'''\right) - 2v$$

Now separate linear and nonlinear from Equation (15.18). Therefore, flow in HPM is represented as

$$M(H,\varepsilon) = (1-\varepsilon)\left[L(H) - L(H_0)\right] + \varepsilon\left[A(H) - U(r)\right]$$

$$M(H,\varepsilon) = (1-\varepsilon)\left[H''' - 2v - H_0''' - 2v_0\right] +$$

$$\varepsilon\left[\begin{array}{l} H''' - R\left(H'^2 - 2HH'' - C^2\right) - 2Rt_1\left(H''^2 + 2C'^2 + HH^{iv}\right) \\ -Rt_2\left(H''^2 + 3C'^2 + 2H'H'''\right) - 2v \end{array}\right] \quad (15.19)$$

According to He (2008), the first approximation H_0, v_0 satisfy the boundary condition and $U(r)$ is any analytical function.

Now assume the approximation solution of Equation (15.19) is

$$H(\eta) = \sum_{j=0}^{\infty} H_j(\eta)\varepsilon^j; \; C(\eta) = \sum_{j=0}^{\infty} C_j(\eta)\varepsilon^j; v(\eta) = \sum_{j=0}^{\infty} v_j(\eta)\varepsilon^j. \quad (15.20)$$

Substituting Equation (15.20) into Equation (15.19) and comparing the identical terms of ε, we get the following system of linear differential equations:

Coefficient of $\varepsilon^0 = 0$:

$$H_0''' - 2v_0 = 0 \quad (15.21)$$

Coefficient of $\varepsilon^1 = 0$:

$$H_1''' - 2v_1 - R\left(H_0'^2 - 2H_0H_0'' - C_0^2\right) - 2Rt_1\left(H_0''^2 - C_0'^2 + H_0H_0^{iv}\right)$$

$$-Rt_2\left(H_0''^2 + 3C_0'^2 + 2H_0'H_0'''\right) = 0 \quad (15.22)$$

Coefficient of $\varepsilon^2 = 0$:

$$H_2''' - 2v_2 - R\left[2H_0H_1' - 2(H_0H_1'' + H_1H_0'') - 2C_0C_1\right]$$

$$-2Rt_1\left(2H_0''H_1'' + H_0H_1^{iv} + H_1H_0^{iv} + 4C_0'C_1'\right)$$

$$-Rt_2\left(2H_0''H_1'' + 2H_0'H_1''' + H_1'H_0''' + 6C_0'C_1'\right) = 0 \quad (15.23)$$

Now apply HPM in Equation (15.10)

$$M(C,\varepsilon) = (1-\varepsilon)(C'' - C_0'')$$

$$+\varepsilon\left[\begin{array}{l} C'' - 2R(H'C - HC') + 2Rt_1\left(H''C' - HC'''\right) \\ +2Rt_2\left(H''C' - H'C''\right) \end{array}\right] \quad (15.24)$$

Now using Equation (15.20) in Equation (15.24) and after comparing the like terms ε, we obtain the following system of linear differential equations:
Coefficient of $\varepsilon^0 = 0$:

$$C_0'' = 0 \tag{15.25}$$

Coefficient of $\varepsilon^1 = 0$:

$$\begin{aligned} C_1'' - 2R\left(H_0'C_0 - H_0C_0'\right) + 2Rt_1\left(H_0''C_0' - H_0C_0'''\right) \\ + 2Rt_2\left(H_0''C_0' - H_0'C_0''\right) = 0 \end{aligned} \tag{15.26}$$

Coefficient of $\varepsilon^2 = 0$:

$$\begin{aligned} C_2'' - 2R\left(H_0'C_1 - H_1'C_0 - H_0C_1' - H_1C_0'\right) \\ + 2Rt_1\left(H_0''C_1' - H_1''C_0' - H_0C_1''' - H_1C_0'''\right) \\ + 2Rt_2\left(H_0''C_1' + H_1''C_0' - H_0'C_1'' - H_1'C_0''\right) = 0 \end{aligned} \tag{15.27}$$

Use Equation (15.20) in Equation (15.12), we will get

$$H_n'(0) = 0, \forall n \geq 0; C_0(0) = 1; C_n(0) = 0, \forall n \geq 1; H_n(0) = 0, \forall n \geq 0;$$

$$H_n'(1) = 0, \forall n \geq 0; C_n(1) = 0, \forall n \geq 0; H_0(1) = k; H_n(1) = 0, \forall n \geq 1.$$

Now apply the above boundary condition in Equations (15.21)–(15.23) and Equations (15.25)–(15.27). After solving these equations, we will get the following equations

$$H_0(\eta) = k\left(3\eta^2 - 2\eta^3\right),$$

$$H_1(\eta) = R\left[\frac{k^2}{35}\left(-2\eta^7 + 7\eta^6 - 18\eta^3 + 13\eta^2\right) + \frac{\eta^4}{12} - \frac{\eta^5}{60} - \frac{7\eta^3}{60} + \frac{\eta^2}{20}\right]$$

$$+ \frac{4k^2}{5}R(t_1 + t_2)\left(6\eta^5 - 15\eta^4 + 12\eta^3 - 3\eta^2\right)$$

$$H_2(\eta) = \frac{2k^3R^2}{90850}\left(\begin{array}{l}-252\eta^{11}+1386\eta^{10}-1925\eta^9-1188\eta^7\\+3080\eta^6-1725\eta^3+624\eta^2\end{array}\right)$$

$$+\frac{2kR^2}{151200}\left(\begin{array}{l}100\eta^9-135\eta^8+936\eta^7-1386\eta^6-2268\eta^5\\+5670\eta^4-2932\eta^3+105\eta^2\end{array}\right)+.....$$

$$C_0(\eta) = 1-\eta$$

$$C_1(\eta) = \frac{kR}{10}\left(4\eta^5-15\eta^4+20\eta^3-9\eta\right)+2kR(t_1+t_2)(-2\eta^3+3\eta^2-\eta)$$

$$C_2(\eta) = \frac{k^2R^2}{6300}\left(1060\eta^9-900\eta^8+3060\eta^7-2520\eta^6+2916\eta^5+1560\eta^3+1789\eta\right)$$

$$+\frac{R^2}{6300}\left(20\eta^7-140\eta^6+375\eta^5-420\eta^4+210\eta^3-27\eta\right)+.......$$

$$v_0 = -6k$$

$$v_1 = \frac{3R}{140}\left(-72k^2+7\right)-\frac{2Rt_1}{5}\left(18k^2+5\right)+\frac{3Rt_2}{10}\left(36k^2-5\right)$$

$$v_2 = -\frac{69}{539}k^2R^2-\frac{733}{6300}kR^2-\frac{5664}{525}k^3R^2(t_1+t_2)-\frac{1}{35}kR^2(t_1+t_2)$$

$$-\frac{4344}{525}k^3R^2t_1-\frac{506736}{175}k^3R^2t_1(t_1+t_2)+.....$$

After substituting the above values in Equation (15.20), the values of $H(\eta), C(\eta)$ will be obtained.

The transverse shearing stress at rotor and stator is given by

$$(t_{\theta z})_{\eta=0} = -1-\frac{9}{10}kR-2kR(t_1+t_2)+\frac{1789}{6300}k^2R^2-\frac{3}{700}R^2+\frac{41}{175}k^2R^2(t_1+t_2)$$

$$+\frac{121}{35}k^2R^2t_1+\frac{28}{5}k^2R^2t_1(t_1+t_2)-11k^2R^2t_2-4k^2R^2t_2(t_1+t_2)$$

$$(t_{\theta z})_{\eta=1} = -1-\frac{11}{10}kR-2kR(t_1+t_2)+\frac{1253}{900}k^2R^2+\frac{2}{1575}R^2+\frac{41}{175}k^2R^2(t_1+t_2)$$

$$-\frac{107}{7}k^2R^2t_1-\frac{92}{5}k^2R^2t_1(t_1+t_2)+\frac{447}{35}k^2R^2t_2-4k^2R^2t_2(t_1+t_2)$$

15.4 RESULTS AND DISCUSSIONS

The model which has been presented here includes many parameters, based on these parameters a broad range of numerical outcomes has been derived. A minor sector is presented here concisely based on these results. The effect of different parameters on the radial, transverse and axial velocity components are discussed graphically in Figures 15.1 to 15.12. In this model function $H'(\eta), C(\eta), H(\eta)$ corresponds to the radial velocity component u_r, the transverse velocity component u_θ and the axial velocity component u_z respectively.

The influence of the elastico-viscous parameter (t_1) on the radial velocity component has been shown in Figure 15.1 by taking $t_2 = 10$, $k = 5$, $R = 0.05$. According to this figure, complete gap length is distributed in three parts. The first and third parts are near to rotating disk and stationary disk respectively and the rest part is the second part. The radial velocity decreases in the first and third part and increases in the rest part. It is also evident that radial velocity decreases with a decrease in elastico-viscous parameter near the inner disk but it increases with a decrease in elastico-viscous parameter near the outer stationary disk. It can also be seen that the behavior of velocity is symmetric near both the disks but the directions are contrary.

The variation of transverse velocity component for the different elastico-viscous parameter (t_1) when $t_2 = 10, k = 5, R = 0.05$ is shown in Figure 15.2, which is 1 in the beginning and 0 at the end. It is observed that velocity declined speedily in the beginning and after acceleration, it reaches at 0. It

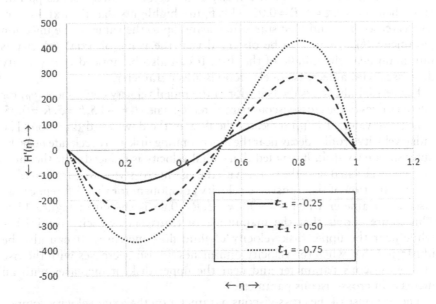

Figure 15.1   An effect of elastico-viscous parameter t_1 on the radial velocity $H'(\eta)$, when $t_2 = 10, R = 0.05, k = 5$.

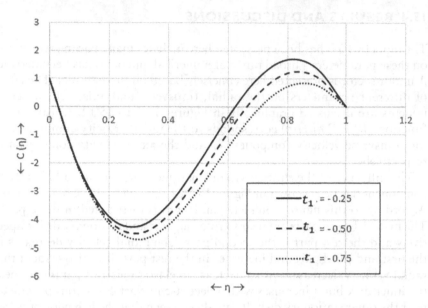

Figure 15.2 An effect of elastico-viscous parameter on t_1 the transverse velocity $C(\eta)$, when $t_2 = 10, R = 0.05, k = 5$.

is also clear from this figure that a decrement in elastico-viscous parameter gives retardation in velocity but velocity enhances near the stationary disk.

Figure 15.3 shows the axial velocity at different elastico-viscous parameter when $t_2 = 10, k = 5, R = 0.05$. This figure highlights that the axial velocity decreases first and then starts increasing up to the value of the injection parameter (k). It can also be observed that the value of axial velocity is minimum near the middle of the disk. It can also be noticed that velocity decreases with a decrease in elastico-viscous parameter.

Figure 15.4 illustrates the behavior of the radial velocity component (u_θ) for altered values of the cross-viscous parameter (t_2) when $t_1 = -0.5, k = 5, R = 0.05$. A negative value of $H'(\eta)$ indicates the flow in the downward direction. The radial velocity profiles decay near the inner rotating disk and accelerate near the outer stationary disk. It is noted that radial velocity is reduced near the inner disk while elevated near the stator with the rise in cross-viscous parameter.

The variation of the transverse velocity component for the different cross-viscous parameter when $t_1 = -0.5, k = 5, R = 0.05$ is shown in Figure 15.5. This figure shows that the magnitude of transverse velocity diminishes while near the upper disk velocity of fluid flow increases. It can also be illustrated that when the velocity diminishes, it also decreases with the rise of cross-viscous parameter and near the upper disk, it increases with an increase in cross-viscous parameter.

The response of the cross-viscous parameter on the axial velocity component (u_z) has been shown in Figure 15.6. This figure reveals that the axial

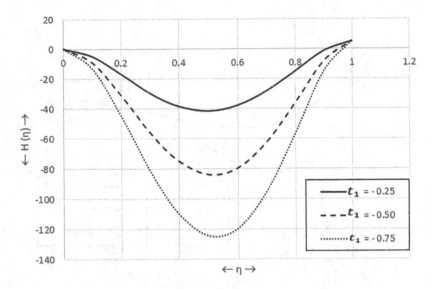

Figure 15.3   An effect of elastico-viscous parameter t_1 on the axial velocity $H(\eta)$, when $t_2 = 10, R = 0.05, k = 5$.

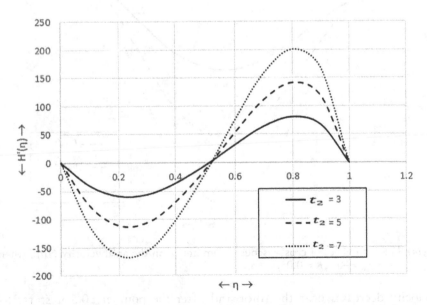

Figure 15.4   An effect of cross-viscous parameter t_2 on the radial velocity $H'(\eta)$, when $t_1 = -0.5, R = 0.05, k = 5$.

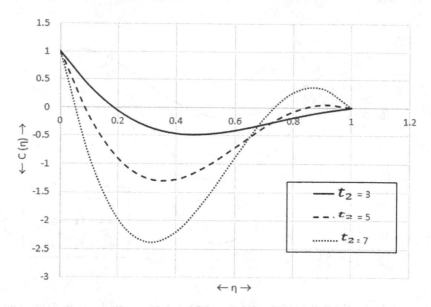

Figure 15.5   An effect of cross-viscous parameter t_2 on the transverse velocity $C(\eta)$, when $t_1 = -0.5, R = 0.05, k = 5$.

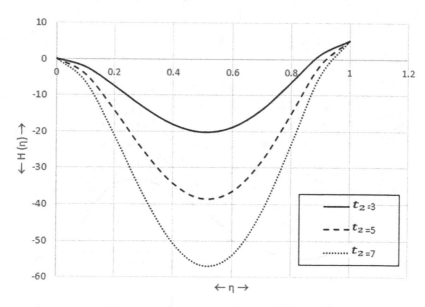

Figure 15.6   An effect of cross-viscous parameter t_2 on the axial velocity $H(\eta)$, when $t_1 = -0.5, R = 0.05, k = 5$.

velocity decreases near the rotor and after the point $\eta \cong 0.5$ it starts to increase. It is also noticed that axial velocity continuously decreases within the boundary layer as an increase in the cross-viscous parameter.

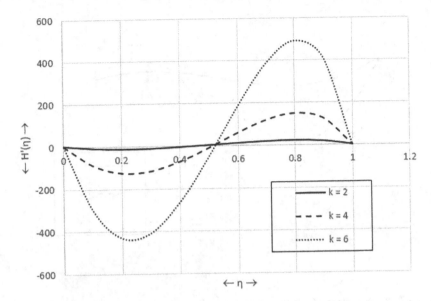

Figure 15.7 An effect of injection parameter k on the radial velocity $H'(\eta)$, when $t_1 = -0.5, t_2 = 10, R = 0.05$.

The influence of the injection parameter (k) on the radial velocity component (u_r) has been shown in Figure 15.7 by taking $t_1 = -0.5, t_2 = 10, R = 0.05$. This figure presents that the radial velocity decreases near the rotor then it increases rapidly but near the stator, it again decreases. It is also observed that up to the point of contact, radial velocity decays with the rise in the injection parameter, and after crossing it velocity increase with an increase in the injection parameter.

Figure 15.8 illustrates the behavior of the radial velocity component (u_θ) for altered values of injection parameter (k) when $t_1 = -0.5, t_2 = 10, R = 0.05$. This figure explains the transverse velocity diminishes near the inner rotating disk, and then it accelerates but retarded near the stationary disk. It can also be seen that transverse velocity decrease as an increase in porosity but near the outer disk, it increases with an increase in porosity.

The effect of the injection parameter on the axial velocity component (u_z) has been shown in Figure 15.9. This figure reveals that the axial velocity decays up to the approximate half of the boundary layer and then starts to increase. It can also be seen that an increase in porosity causes a decrease in axial velocity.

The influence of the Reynolds number (R) on the radial velocity component has been shown in Figure 15.10 by taking $t_1 = -0.5, t_2 = 10, k = 5$. According to this figure, radial velocity decreases in the vicinity of rotating and stationary disks. There is a downward flow up to the middle of the boundary layer $\eta \cong 0.5$ and after that opposite behavior is observed. It can also be depicted that velocity decays with a rise in Reynolds number but after the point of contact it increases with a rise in parameter.

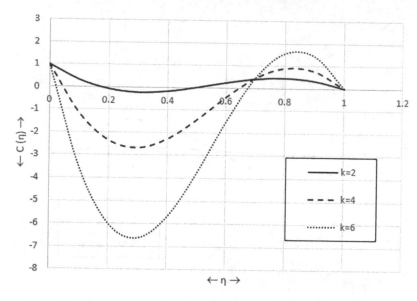

Figure 15.8 An effect of injection parameter k on the transverse velocity $C(\eta)$, when $t_1 = -0.5, t_2 = 10, R = 0.05$.

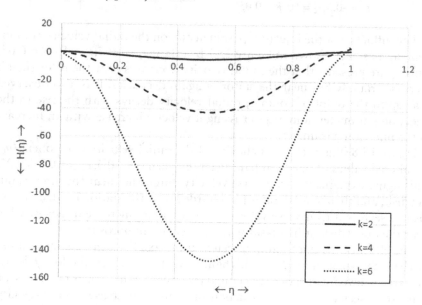

Figure 15.9 An effect of injection parameter k on the axial velocity $H(\eta)$, when $t_1 = -0.5, t_2 = 10, R = 0.05$.

The variation of the transverse velocity component for different Reynolds number when $t_1 = -0.5, t_2 = 10, k = 5$ is shown in Figure 15.11. This figure depicts that at a small Reynolds number $R = 0.01$, velocity decreases throughout the gap length but as increases, the value of the Reynolds number the

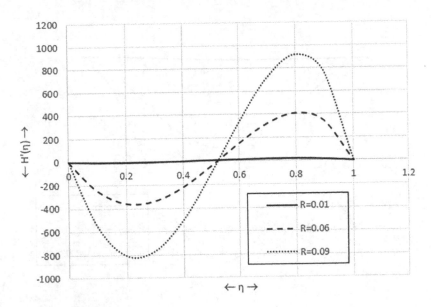

Figure 15.10   An effect of Reynolds number R on the radial velocity $H'(\eta)$, when $t_1 = -0.5, t_2 = 10, k = 5$.

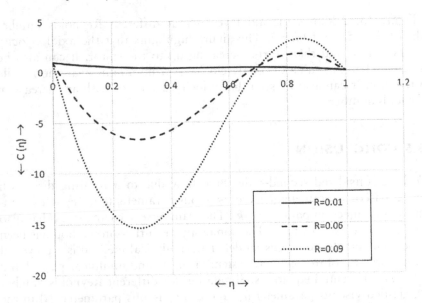

Figure 15.11   An effect of Reynolds number R on the transverse velocity $C(\eta)$, when $t_1 = -0.5, t_2 = 10, k = 5$.

velocity declined speedily near the rotating disk after getting accelerated it reaches at 0. It is also clear from this figure that a rise in Reynolds number gives retardation in velocity but velocity enhances near the stationary disk.

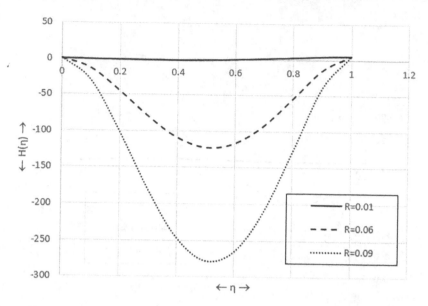

Figure 15.12 An effect of Reynolds number R on the axial velocity $H(\eta)$, when $t_1 = -0.5, t_2 = 10, k = 5$.

Figure 15.12 shows the axial velocity at different Reynolds number when $t_1 = -0.5, t_2 = 10, k = 5$. This figure highlights that the axial velocity decreases first and then starts increasing up to the stator. It can also be observed that the value of axial velocity is a minimum near the middle of the disk. It can also be seen that velocity decreases with an increase in Reynolds number.

15.5 CONCLUSION

We have considered second-order fluid flow due to a rotating disk with elastico-viscous parameter (t_1), cross-viscous parameter (t_2), Reynolds number (R) and injection parameter (k). The transverse shearing stress has also been calculated successfully. The homotopy perturbation method has been successfully employed in this model. This analytical method is a powerful mathematical tool to solve any system of linear and nonlinear partial and ordinary differential equations. Simulation for different Reynolds number (R), elastico-viscous parameter (t_1) and cross-viscous parameter (t_2) in the range of $0 < R < 1, t_1 < 0, t_2 < 0$ are done. The key observations are

- The behavior of radial velocity is the same in all cases. Larger parameter (except t_1) leads to enhancing radial velocity near the stator but diminishes near the rotor but in the case of elastico-viscous parameter t_1 its behavior is reversed.

- In all the cases axial velocity is diminished with the rise in the values of parameter but in the case of elastico-viscous parameter t_1 its behavior is reversed too.
- Near the rotating disk, radial and transverse velocities are minimum while maximum near the stationary disk.
- The axial velocity is minimum at the middle of the boundary layer.

We may end this study by saying that response of the velocity is reasonably correct with the known laws of fluids.

REFERENCES

Agarwal, R. (2020). Analytical study of micropolar fluid flow between two porous disks. *PalArch's Journal of Archaeology of Egypt/Egyptology*, 17(12), 903–924.

Agarwal, R. (2021). Heat and mass transfer in electrically conducting micropolar fluid flow between two stretchable disks. *Materials Today: Proceedings*, 46, 10227–10238.

Agarwal, R. (2022). An analytical study of non-Newtonian visco-inelastic fluid flow between two stretchable rotating disks. *Palestine Journal of Mathematics*, 11, 184–201.

Agarwal R., & Agarwal D. (2018). The flow of a second-order fluid between two infinite discs in the presence of transverse magnetic field. *International Journal of Mathematics Trends and Technology*, 54, 508–518.

Agarwal, R., & Mishra, P. K. (2021). Analytical solution of the MHD forced flow and heat transfer of a non-Newtonian visco-inelastic fluid between two infinite rotating disks. *Materials Today: Proceedings*, 46, 10153–10163.

Ahmed, J., Khan, M., & Ahmad, L. (2019). MHD swirling flow and heat transfer in Maxwell fluid driven by two coaxially rotating disks with variable thermal conductivity. *Chinese Journal of Physics*, 60, 22–34.

Ahmed, J., Khan, M., & Ahmad, L. (2019). Transient thin film flow of nonlinear radiative Maxwell nanofluid over a rotating disk. *Physics Letters A*, 383(12), 1300–1305.

Ariel, P. D. (2009). The homotopy perturbation method and analytical solution of the problem of flow past a rotating disk. *Computers & Mathematics with Applications*, 58(11–12), 2504–2513.

Batchelor, G. K. (1951). Note on a class of solutions of the Navier-Stokes equations representing steady rotationally-symmetric flow. *The Quarterly Journal of Mechanics and Applied Mathematics*, 4(1), 29–41.

Cochran, W. G. (1934). The flow due to a rotating disc. In *Mathematical Proceedings of the Cambridge Philosophical Society*, edited by B. J. Green (Vol. 30, No. 3, pp. 365–375). University of Oxford, Oxford.

Coleman, B. D., & Noll, W. (1974). An approximation theorem for functionals, with applications in continuum mechanics. In *The Foundations of Mechanics and Thermodynamics*, edited by W. Noll (pp. 97–112). Springer, Berlin, Heidelberg.

Dinarvand, S. (2010). On explicit, purely analytic solutions of off-centered stagnation flow towards a rotating disc by means of HAM. *Nonlinear Analysis: Real World Applications*, 11(5), 3389–3398.

Doh, D. H., Muthtamilselvan, M., Swathene, B., & Ramya, E. (2020). Homogeneous and heterogeneous reactions in a nanofluid flow due to a rotating disk of variable thickness using HAM. *Mathematics and Computers in Simulation*, 168, 90–110.

Hafeez, A., Khan, M., & Ahmed, J. (2020). Flow of Oldroyd-B fluid over a rotating disk with Cattaneo–Christov theory for heat and mass fluxes. *Computer Methods and Programs in Biomedicine*, 191, 105374.

Hayat, T., Javed, M., Imtiaz, M., & Alsaedi, A. (2017). Convective flow of Jeffrey nanofluid due to two stretchable rotating disks. *Journal of Molecular Liquids*, 240, 291–302.

Hayat, T., Khan, S. A., Khan, M. I., & Alsaedi, A. (2019). Theoretical investigation of Ree–Eyring nanofluid flow with entropy optimization and Arrhenius activation energy between two rotating disks. *Computer Methods and Programs in Biomedicine*, 177, 57–68.

He, J. H. (1999). Homotopy perturbation technique. *Computer Methods in Applied Mechanics and Engineering*, 178(4): 257–262.

He, J. H. (2000). A coupling method of a homotopy technique and a perturbation technique for non-linear problems. *International Journal of Non-Linear Mechanics*, 35(1), 37–43.

He, J. H. (2004). Comparison of homotopy perturbation method and homotopy analysis method. *Applied Mathematics and Computation*, 156(2), 527–539.

He, J. H. (2006). Homotopy perturbation method for solving boundary value problems. *Physics Letters A*, 350(1–2), 87–88.

He, J. H. (2008). Recent development of the homotopy perturbation method. *Topological Methods in Nonlinear Analysis*, 31(2), 205–209.

Hojjati, M. H., & Jafari, S. (2008). Semi-exact solution of elastic non-uniform thickness and density rotating disks by homotopy perturbation and Adomian's decomposition methods. Part I: Elastic solution. *International Journal of Pressure Vessels and Piping*, 85(12), 871–878.

Lance, G. N., & Rogers, M. H. (1962). The axially symmetric flow of a viscous fluid between two infinite rotating disks. *Proceedings of the Royal Society of London. Series A. Mathematical and Physical Sciences*, 266(1324), 109–121.

Rani, P. J., Kirthiga, M., Molina, A., Laborda, E., & Rajendran, L. (2017). Analytical solution of the convection-diffusion equation for uniformly accessible rotating disk electrodes via the homotopy perturbation method. *Journal of Electroanalytical Chemistry*, 799, 175–180.

Rashidi, M. M., Pour, S. M., Hayat, T., & Obaidat, S. (2012). Analytic approximate solutions for steady flow over a rotating disk in porous medium with heat transfer by homotopy analysis method. *Computers & Fluids*, 54, 1–9.

Sharma, H. G., & Singh, K. R. (1986). Forced flow of a second-order fluid between two porous disks. *Indian Journal of Technology*, 24(6), 285–290.

Sheikholeslami, M., Ashorynejad, H. R., Ganji, D. D., & Yıldırım, A. (2012). Homotopy perturbation method for three-dimensional problem of condensation film on inclined rotating disk. *Scientia Iranica*, 19(3), 437–442.

Singh, B. B., & Kumar, A. (1989). Flow of a second-order fluid due to the rotation of an infinite porous disk near a stationary parallel porous disk. *Indian Journal of Pure and Applied Mathematics, 20*(9), 931–943.

Srivastava, A. C. (1963). Torsional oscillations of an infinite plate in second-order fluids. *Journal of Fluid Mechanics, 17*(2), 171–181.

Stewartson, K. (1953). On the flow between two rotating coaxial disks. In *Mathematical Proceedings of the Cambridge Philosophical Society* edited by B. J. Green (Vol. 49, No. 2, pp. 333–341). University of Oxford,Oxford.

Turkyilmazoglu, M. (2010a). Analytic approximate solutions of rotating disk boundary layer flow subject to a uniform suction or injection. *International Journal of Mechanical Sciences, 52*(12), 1735–1744.

Turkyilmazoglu, M. (2010b). Purely analytic solutions of magnetohydrodynamic swirling boundary layer flow over a porous rotating disk. *Computers & Fluids, 39*(5), 793–799.

Usman, M., Mehmood, A., & Weigand, B. (2020). Heat transfer from a non-isothermal rotating rough disk subjected to forced flow. *International Communications in Heat and Mass Transfer, 110*, 104395.

Von Kármán, T. (1921). Uber laminare und turbulente Reibung. *Zeitschrift für Angewandte Mathematik und Mechanik, 1*, 233–252.

Wilson, L. O., & Schryer, N. L. (1978). Flow between a stationary and a rotating disk with suction. *Journal of Fluid Mechanics, 85*(3), 479–496.

Xinhui, S., Liancun, Z., Xinxin, Z., & Xinyi, S. (2012). Homotopy analysis method for the asymmetric laminar flow and heat transfer of viscous fluid between contracting rotating disks. *Applied Mathematical Modelling, 36*(4), 1806–1820.

Yao, B., & Lian, L. (2018). A new analysis of the rotationally symmetric flow in the presence of an infinite rotating disk. *International Journal of Mechanical Sciences, 136*, 106–111.

Yao, B., & Lian, L. (2019). Series solution of the rotationally symmetric flow in the presence of an infinite rotating disk with uniform suction. *European Journal of Mechanics-B/Fluids, 74*, 159–166.

Zangooee, M. R., Hosseinzadeh, K., & Ganji, D. D. (2019). Hydrothermal analysis of MHD nanofluid (TiO_2-GO) flow between two radiative stretchable rotating disks using AGM. *Case Studies in Thermal Engineering, 14*, 100460.

Chapter 16

Resonance in water pipeline due to transient

Rakesh C. Bhadula
Graphic Era Hill University

Jaipal Singh
DBS (P. G.) College

V.N. Kala
GBPEC

CONTENTS

NOMENCLATURE

A	Cross-sectional area (m^2)
a	Wave speed (m/s)
D	Inner Pipe diameter (m)
f	Darcy–Weisbach friction factor (Dimensional less friction factor)
g	Gravity acceleration (m/s^2)
H^*	Pressure head (m)
Q^*	Discharge (m^3/s)
r	Radius of the pipe (m)
x	Length of the Pipe (m)

16.1 INTRODUCTION

Resonance in water pipeline due to transient has been studied for many years but till today there is no efficient method available to predict the exact location of the transient in pipelines. Gottlieb et al. (1981), investigated transient pressure by a numerical model with the experimental result.

DOI: 10.1201/9781003367420-16

Four different steel and plastic pipe was taken to investigate the behavior of transient in water pipeline network. High transient peaks were recorded immediately upon the collapse of the vapour cavity. Transient peaks were dropped to about 40%, and this level was maintained twice at 1/a second. They showed the presence of peaks that resembled the pressure peaks associated with the implosion of gas bubbles in pumps. Bhadulaet al. (2020, 2021), studied the behavior of chlorine decay in water distribution under transient conditions. It is analyzed that chlorine travel more distance in water pipeline under transient conditions. Goldberg (1983), investigated the solution of a broad cross-section of time-varying wave problems in the distribution system, MOC based methods are likely to continue to see extensive use. They showed the extent of the potential application of the implicit timeline method also deserves more attention. Shimada and Okushima (1984), studied transient behavior by solving the model using the series solution method and the results were compared with the finite difference method. Chudhury and Hussaini (1985), solved the transient problems by MacCormack, Lambda, and Gabutti explicit FD schemes. Brunone et al. (1995) developed a two-dimensional mathematical model, and rapid damping of the pressure peaks was considered after the end of a complete closure of the valve, are closely linked to the shape of the cross-sectional velocity distributions and their Variability in time. Ghidaouiand Kolyshkin (2001), performed linear stability analysis of base flow velocity profiles for laminar as well as turbulent water hammer flows. An analytical solution was obtained for such a base flow velocity profile. Where the transient is generated by the instantaneous reduction in flow rate at the downstream end of a single pipe system. The presence of inflection points in the base flow velocity profile and the large velocity gradient near the pipe wall are the sources of flow instability. Essaidi and Triki (2021), investigated the one-dimensional pressurized pipe flow model to describe the behavior in the elastic and plastic pipeline system. They calculated the pressure with the help of MOC and they analyzed that low-density polyethylene provided a better result than high-density polyethylene. Aliamand Hassan (2014), studied the water hammer phenomenon of two loop networks with different roughness coefficient,different thicknesses as well different diameters of the pipeline network. Wood Don(2005), discussed the transient flow problem with the help of WCM as well as the MOC method, and analyzed that both methods are capable the solving the pipe friction water transient problem. Asli et al. (2012), presented an Eulerian-based computational water hammer model compared with regression of the relationship between the dependent variable and independent variables for water transient wave in water pipeline network. The effect of resonance in the pipeline network was studied where pressure is acting on the pipeline. Resonance behavior is analyzed when the pipe valve is closed and the valve is operated with the sinusoidal with the frequency (Radian per second). In such a case, flow of water was taken as in turbulent nature.

16.2 MATHEMATICAL MODELING

The general equation of motion and equation of continuity for turbulent flow, which is expressed by Ramoset al. (2004), can be written as

$$\frac{\partial H^*}{\partial x} + \frac{1}{gA}\frac{\partial Q^*}{\partial t} = -\frac{nfQ_0^{n-1}}{2gDA^n}Q^* \tag{16.1}$$

$$\frac{\partial H^*}{\partial t} = -\frac{a^2}{gA}\frac{\partial Q^*}{\partial x} \tag{16.2}$$

where H^* is Pressure head and Q^* is known as discharge. Let H_0 and Q_0 be mean or average pressure head and discharge respectively.

Differentiating Equation (16.1) with respect to t and Equation (16.2) with respect to x partially and subtracting to get the differential equation in Q^* and solving the partial differential equation, we get

$$Q^* = real\ part\ of\ \left[q(x).e^{i\omega t}\right];$$

where Q^* has sinusoidal nature with respect to t

where $q = c_1\cosh\beta x + c_2\sinh\beta x$

and

$$H^*(x) = \frac{-\beta a^2}{i\omega gA}(c_1\cosh x + c_2\sinh x) \tag{16.3}$$

where c_1, c_2 are constants and $\beta^2 = \dfrac{-\omega^2 + igA\omega\alpha}{a^2}$.

16.3 RESULT AND DISCUSSION

Pipeline network with the variable cross-sectional area was considered whose boundary is taken as sinusoidal, and radius of the pipe is taken as $r = r_0 + \dfrac{r_0}{5}\sin\left(\dfrac{\pi x}{200}\right)$, and length of the pipeline is taken as $500\,m$ and the wave speed is calculated with help of $a = a_0 + \dfrac{a_0}{10}\cos\left(\dfrac{\pi x}{100}\right)$, where $a_0 = 1{,}000$. Such variable cross-sectional area can be shown with the help of Figure 16.1.

The variation of flow with axial distance in the pipe is shown in Figure 16.2. In this case, it is analyzed that flow shows zig-zag motion along with axial distance this happens due to sudden closer or shut down of the

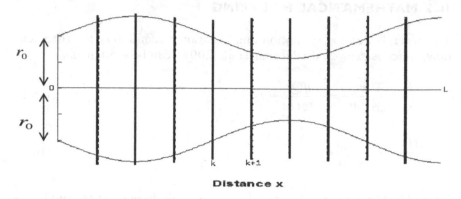

Figure 16.1 Pipe with variable cross-sectional area.

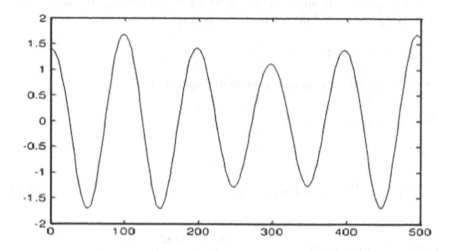

Figure 16.2 Variation of flow with axial distance in the pipe.

pipe valve. Variation of pressure head with axial distance can be analyzed in Figure 16.3. In such case, pressure also shows zig-zag motion along with axial distance, which is caused by sudden closer or opening of the pipe valve. The magnitude of variation of the pressure head is shown in Figure 16.4. The magnitude of the pressure head shows a zig-zag pattern.

Ramos et al. (2004), conducted a laboratory experiment and validated the result with the numerical solution for a polyethylene pipe valve system by taking the length of pipe $L=100$m, and the diameter of the pipe was 44 mm. In the steady state condition discharge of the valve was 1.18l/s while in our model different pipe length is taken for the variable pipe radius, in such case pipe radius is considered as $r = r_0 + \dfrac{r_0}{5}\sin\left(\dfrac{\pi x}{200}\right)$, then the value

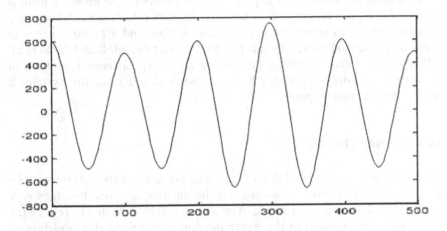

Figure 16.3 Variation of pressure head with axial distance.

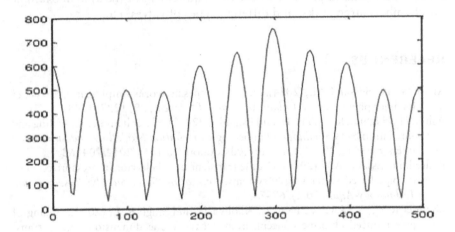

Figure 16.4 Variation magnitude of pressure head.

of the charge was calculated for the different pipe length as well variable radius and it is found the value of the discharge in the steady state condition was similar as the 1.18 l/s for the length of 100 m pipe, which can be seen in the Figure 16.2. A comparison is made in the behavior of pressure heads between our study and Badreddine et al. (2021), model. From Figure 16.3, it is analyzed that the pressure head diminishes to zero, and then it reverses. Forward rotation is the pressure head is observed as the pump rapidly closed and a negative pressure wave is generated in downstream flow while a positive pressure wave is generated in the upstream flow through the section of the pipe. Similar nature of the pressure head is obtained in Figure 16.4. While in Badreddine et al. (2021), a model negative pressure wave was generated in downstream flow and a positive pressure wave was upstream flow, which is the same as our model. Badreddine et al. (2021), studied that in plastic pipeline systems large attenuation effect of pressure head gain and pressure hade loss while in other pipeline systems it is less than the plastic pipes.

16.4 CONCLUSION

Pipeline system is damaged due to transient pressure, water hammer problem cannot be ignored but redesigning the pipeline network becomes very costly as well as time-consuming. Any device can be included to reduce the water hammer problem in the water pipeline network in the modeling of pipeline. However, there is no perfect device for all pipeline networks in all operating conditions, even then we consider various devices or methods to control the transient pressure as per the requirement of the system example air chamber, surge tanks, and different types of valves, etc.

REFERENCES

Aliam G. and Hassan I. M., (2014). Impact of pipes networks simplification on water hammer phenomenon. *Indian Academy of Sciences*, 39, 1227–1244.

Asli K. H., Haghi A. K., Asli H. H. and Eshghi S., (2012). Water hammer modelling and simulation by GIS. *Hindawi Publishing Corporation Modelling and Simulation in Engineering*, 2012(4), 15. http://dx.doi.org/10.1155/2012/704163

Badreddine Essaidi, A.T. (2021). On the transient flow behavior in pressurized plastic pipe-based water supply systems. *Journal of Water Supply: Research and Technology-Aqua*, 70(1), 67–76.

Bhadula R. C., Kala V. N., Pant R., Kholiya D. and Singh S. J., (2021). Ageing of pipe induced chlorine concentration decay: A one dimensional mass transfer modeling. *Materials Today: Proceedings*, 46, 10761–10765. https://doi.org/10.1016/j.matpr.2021.01.670

Bhadula R. C., Mishra A., Kholiya D., Kala V. N. and Garg N., (2020). Chlorine concentration decay under transient condition in water distribution system, *International Journal of Advanced Research in Engineering and Technology*, 11(5), 581–589.

Brunone B., Golia U. M. and Greco M., (1995). Effects of two-dimensionality on Pipe Ansints Modeling. *Journal of Hydraulic Engineering*, 121(12), 906–912.

Chudhury M. H. and Hussaini M. Y., (1985). Second-order accurate explicit finite-difference schemes for water hammer analysis. *Journal of Fluids Engineering*, 107, 523–529.

Essaidi B. and Triki, A., (2021). On the transient flow behavior in pressurized plastic pipe-based water supply systems. *Journal of Water Supply: Research and Technology-Aqua*, 70(1), 67–76. https://doi.org/10.2166/aqua.2020.051

Ghidaoui M. S. and Kolyshkin A. A., (2001). Stability analysis of velocity profiles in water hammer flows. *Journal of Hydraulic Engineering*, 127(6), 499–512.

Goldberg D. E. and Benjamin Wylie E., (1983). Characteristics method using time-line interpolations. *Journal of Hydraulic Engineering*, 109(5), 670–683.

Gottlieb L., Larnæs G and Vasehus J., (1981). Transient Cavitation in pipe line laboratory tests and numerical. In *5th International Symposium on Water Column Separation IAHR, Obermach Germany*, pp. 487–508.

Ramos H., Covas D., Borga A. and Loureiro D., (2004). Surge damping analysis in pipe systems: Modelling and experiments. *Journal of Hydraulic Research*, 42(4), 413–425.

Shimada M. and Okushima S., (1984). New numerical model and technique for water hammer. *Journal of Hydraulic Engineering*, 110(6), 736–748.

Wood Don J., (2005). Water hammer analysis – essential and easy (and efficient). *Journal of Environmental Engg*, 131(8), 1123–1131.

Index